Wine Grape
PRODUCTION GUIDE
for Eastern North America

Editor
Tony K. Wolf, *Virginia Tech*

Plant and Life Sciences Publishing (PALS)
PO Box 4557
Ithaca, New York 14852-4557

NRAES–145
December 2008

ISBN-13: 978-1-933395-12-8

Library of Congress Cataloging-in-Publication Data
Wine grape production guide for eastern North America / editor: Tony K. Wolf.
p. cm. -- (Cooperative Extension NRAES ; 145)
ISBN 978-1-933395-12-8

1. Viticulture--East (U.S.) 2. Viticulture--Canada, Eastern. 3. Vineyards--East (U.S.) 4. Vineyards--Canada, Eastern. 5. Wine and wine making--East (U.S.) 6. Wine and wine making--Canada, Eastern. I. Wolf, Tony Kenneth, date- II. Natural Resource, Agriculture, and Engineering Service. Cooperative Extension. III. Series: NRAES (Series) ; 145.
2.

SB387.76.E27W56 2008
634.80974--dc22

2008046160

Requests to reprint parts of this publication should be sent to PALS. In your request, please state which parts of the publication you would like to reprint and describe how you intend to use the material. Contact PALS if you have any questions.

To order additional copies, contact:

Plant and Life Sciences Publishing (PALS)
PO Box 4557, Ithaca, New York 14852-4557
Phone: (607) 255-7654 • Fax: (607) 254-8770
E-mail: palspublishing@cornell.edu • Web site: http://palspublishing.cals.cornell.edu/

Printed December 2013

Contents

Preface v

Acknowledgments vi

About the Authors vii

1 ▪ Costs and Returns of Vineyard Establishment and Operation 1

Tony K. Wolf, *Virginia Tech*, William M. Boyd, *formerly with Surry Community College*, Keith R. Dickinson, *Penn State Cooperative Extension, Chester County*

2 ▪ Vineyard Site Selection 14

Tony K. Wolf, *Virginia Tech*, John D. Boyer, *Virginia Tech*, Timothy E. Martinson, *Cornell University*, Fritz A. Westover, *Texas AgriLife Extension Service*

3 ▪ Wine Grape and Rootstock Varieties 37

Tony K. Wolf, *Virginia Tech*

4 ▪ Vineyard Design and Establishment 71

Thomas J. Zabadal, *Michigan State University*

5 ▪ Pruning and Training 98

Andrew G. Reynolds, *Brock University (Ontario, Canada)*, Tony K. Wolf, *Virginia Tech*

6 ▪ Grapevine Canopy Management 124

Andrew G. Reynolds, *Brock University (Ontario, Canada)*, Tony K. Wolf, *Virginia Tech*

7 ▪ Crop Yield Estimation and Crop Management 135

Tony K. Wolf, *Virginia Tech*

8 ▪ Nutrient Management 141

Terence R. Bates, *Cornell University*, Tony K. Wolf, *Virginia Tech*

9 ▪ Grapevine Water Relations and Irrigation 169

David S. Ross, *University of Maryland*, Tony K. Wolf, *Virginia Tech*

(continued on next page)

10 ▪ Spray Drift Mitigation 196

Andrew J. Landers, *Cornell University*

11 ▪ Disease Management 216

Wayne F. Wilcox, *Cornell University*, Tony K. Wolf, *Virginia Tech*

12 ▪ Major Insect and Mite Pests of Grapes 241

Douglas G. Pfeiffer, *Virginia Tech*

13 ▪ Vineyard Weed Management 262

Jeffrey F. Derr, *Virginia Tech*

14 ▪ Wildlife Deterrence 272

Tony K. Wolf, *Virginia Tech*

15 ▪ Grape Purchase Contracts and Vineyard Leases 278

Mark L. Chien, *The Pennsylvania State University*

16 ▪ Wine Grape Quality: When Is It Time to Pick? 282

W. Gill Giese, Surry Community College

Appendices 295

Glossary 319

Financial Support 325, 326

Sponsors 327

Peer Reviewers 331

Index 332

About PALS .... inside back cover

PREFACE

The aim of the *Wine Grape Production Guide for Eastern North America* guide is to provide a comprehensive resource for individuals exploring or currently engaged in commercial wine grape production. The ambitious title reflects the commonality of many of the relevant production issues from northern Georgia, through the mid-Atlantic and into New England and southern Canada. The need for a publication of this scope is evident with the interest and expansion of wine industries throughout this large geographic region over the last two decades.

The Guide substantially builds upon an earlier production guide (Wolf and Poling, 1995) in scope, depth and authorship. New chapters on means of reducing spray drift, irrigation technology, weed management, insect pest management and grape ripening processes have been added, while core subjects of the prior publication have been substantially revised and expanded. As overall editor, I take responsibility for the relevancy of the material presented, as well as omission of other subjects that might have been included.

One might expect that the compilation of a 16-chapter manual on wine grape production would exhaustively answer questions related to the subject. In fact, the information presented varies somewhat in technical presentation. The intended audience is an established grower with some experience. To the uninitiated, particularly those without horticulture, soil science, or pest management experience or training, some of the material may be challenging. Conversely, others will seek more in-depth coverage of those areas that are lightly treated here. Research advances and shared industry experience continue to provide new insights and new knowledge, just as new threats —economic, biotic and abiotic—pose new challenges and stimulate new questions. Readers are encouraged to stay abreast of research and extension programs to be competitive in a changing business and cultural environment.

The outlook for wine and wine grape production in eastern North America is bright, but change is inevitable. Locally-produced wines do not have a monopoly on the local wine market. Competitive growers will scrutinize all costs of grape production while continuing to strive for high quality, sustainable crop yields. Vineyard site selection, which has focused on minimizing climatic and biotic threats, will increasingly seek to match specific varieties with specific soils and mesoclimates to increase grape and wine potential. Vineyard cultural management will continue to be refined as we gain more experience and research knowledge of factors that affect grape and wine quality. Understanding the interactions between crop level and environmental stresses will provide bases for adjusting crops to meet seasonal or site-specific constraints. The cost of labor will continue to drive larger vineyards towards increased use of mechanization, while the encroachment of residential development on agriculture will increase the scrutiny of vineyard operations by an increasingly non-agrarian populace. Climate change will undoubtedly affect the distribution of certain diseases and arthropod pests, and it may modify our recommendations about the varieties that should be grown at a particular site. This may not be all negative; sites or regions that were once considered too cold or too short-seasoned for grape production may see increased opportunity.

I personally thank all of the many authors involved with this work; their shared vision of the goal, their patience and tolerance, and their expertise made this book possible. Appreciation is also extended to the numerous colleagues who provided their time and expertise to the peer review of the manuscript. I also thank Marty Sailus and the staff at NRAES for patience and assistance with the publication of this Guide.

— Tony K. Wolf
Editor

ACKNOWLEDGMENTS

A number of people with expertise in wine grape production reviewed the original manuscript and their suggestions resulted in significant improvements. A list of reviewers is included on page 331.

Photos in this book were contributed by the authors and their colleagues. When a photo was contributed by someone other than an author, the name and affiliation of the contributor is indicated on the top right corner of the photo. Photos without an acknowledgement were contributed by the authors.

Portions of chapters 2, 5, 6, 7, 8, and 9 including several photos and illustrations, were adapted from *The Mid-Atlantic Winegrape Grower's Guide*, written by Tony K. Wolf and E. Barclay Poling, and published by North Carolina Cooperative Extension Service, Raleigh, NC, 1995.

Portions of chapter 4 were adapted from *Vineyard Establishment I — Preplant Decisions*, by T. Zabadal and J. Andresen, and *Vineyard Establishment II — Planting and Early Care of Vineyards*, by T. Zabadal, Michigan State University, E. Lansing, MI, 1997.

Appreciation is extended to Patricia Peacock, formerly with Virginia Tech, who provided the original draft of chapter 3, Wine Grape and Rootstock Varieties.

Funding provided by several organizations allowed this book to be published. More information about these organizations can be found on pages 325 through 330.

The book production was managed by Marty Sailus, NRAES Director, and Jeff Popow, NRAES former managing editor.

Institutional support for this project was provided by Virginia Tech and Cornell University.

ABOUT THE AUTHORS

(listed in alphabetical order)

Terence R. Bates is a Viticulture Research Associate in the Department of Horticultural Sciences at the Cornell Lake Erie Research and Extension Laboratory in Portland, NY (Chapter 8).

William M. Boyd is formerly with Surry Community College, Dobson, NC (Chapter 1).

John D. Boyer is an instructor in the Geography Department, Virginia Tech, Blacksburg, VA (Chapter 2).

Mark L. Chien is a Wine Grape Educator with Penn State Cooperative Extension, Lancaster, PA (Chapter 15).

Jeffrey F. Derr is a Professor of Weed Science in the Department of Plant Pathology, Physiology, and Weed Science, Virginia Tech, and is located at the Hampton Roads Agricultural Research and Extension Center, Virginia Beach, VA (Chapter 13).

Keith R. Dickinson is an Extension Educator with Penn State Cooperative Extension of Chester County, West Chester, PA (Chapter 1).

W. Gill Giese is a Viticulture Instructor at Surry Community College, Surry, NC (Chapter 16).

Andrew J. Landers is a Senior Extension Associate and Pesticide Application Technology Specialist in the Department of Entomology and the Department of Biological and Environmental Engineering, Cornell University, at the New York State Agricultural Experiment Station, Geneva, NY (Chapter 10).

Timothy E. Martinson is a Senior Extension Associate, Statewide Viticulture Extension Program, in the Department of Horticultural Sciences, Cornell University, at the New York State Agricultural Experiment Station, Geneva, NY (Chapter 2).

Douglas G. Pfeiffer is Professor of Entomology in the Department of Entomology, Virginia Tech, Blacksburg, VA (Chapter 12).

Andrew G. Reynolds is a Professor of Viticulture with the Cool Climate Oenology and Viticulture Institute, Department of Biological Sciences, Brock University, St. Catharines, Ontario (Chapters 5, 6).

David S. Ross is a Professor and Extension Agricultural Engineer in the Department of Environmental Science and Technology, University of Maryland, College Park, MD (Chapter 9).

Fritz A. Westover is a viticulture extension associate for the Texas Gulf Coast, Texas AgriLife Extension Service, Texas A&M University, Harris County Office, Houston, TX (Chapter 2).

Wayne F. Wilcox is a Professor of Plant Pathology in the Department of Plant Pathology, Cornell University, at the New York State Agricultural Experiment Station, Geneva, NY (Chapter 11).

Tony K. Wolf is a Professor of Viticulture in the Department of Horticulture, Virginia Tech, and is located at the Alson H. Smith Jr. Agricultural Research and Extension Center, Winchester, VA (Introduction and chapters 1, 2, 3, 5, 6, 7, 8, 9, 11, 14).

Thomas J. Zabadal is an Associate Professor and Viticulturist in the Department of Horticulture, Michigan State University, and is located at the Southwest Michigan Research and Extension Center, Benton Harbor, MI (Chapter 4).

—1—

Costs and Returns of Vineyard Establishment and Operation

Authors

Tony K. Wolf, Virginia Tech

William M. Boyd, formerly with Surry Community College

Keith R. Dickinson, Penn State Cooperative Extension, Chester County

This chapter reviews the cost of establishment and operation as well as the expected returns over the economic life of a vineyard. Establishment and operating costs based on 2006 material and labor costs are projected for a 10-acre planting of *Vitis vinifera* grapes such as Chardonnay. The cost estimates were developed within a user-adjustable Microsoft Excel® spreadsheet, which may be obtained from the senior author. Thus, while the 10-acre model vineyard was used to develop the cost and return projections, model parameters, including vine density, training system, projected yields, and vineyard inputs, can be adjusted to review how the economic performance of the vineyard is affected.

Assumptions and Model Components

Readers must understand that a number of assumptions were made in developing the model budget. In fact, no two enterprises will be operated identically. Some prospective growers will choose to purchase used equipment and materials; others may already own some, if not all, of the needed equipment. Some will not require outside financing, while others may choose not to assess their personal labor costs, which, in itself, represents an opportunity cost. These may be legitimate business decisions, but they create an endless number of possible scenarios to describe. Therefore, the budget presented in the following tables was developed using practices and materials that have proved both practical and cost-effective under a wide range of growing conditions. Enterprising growers will find alternative materials or practices to reduce operating costs without compromising vineyard productivity or grape quality. Again, the spreadsheet that accompanies this chapter can be used to adjust various inputs, as defined below, to explore a range of options for improving the economic performance of the vineyard investment.

Land

Land costs vary tremendously throughout the East. This is a result of both development pressure and variance in suitability for cultivation. It is also apparent that many of the smaller vineyards (less than 10 acres) are being established as a means of intensively utilizing small parcels rather than creating economy of scale. Studies in the Finger Lakes region of New York suggest a minimum vineyard size of approximately 40 acres for profitability. However, the majority of start-up vineyards in eastern North America are less than 20 acres in size, and many are less than 10. These small operations reflect the suburbanization occurring in the region. Often, local zoning ordinances produce small "estate" farms that are poor stages for extensive agriculture but that might be used for intensive high-value horticultural crops. In many cases, land is purchased primarily as a residence, not as an agricultural enterprise base. Alternatively, the land may be purchased and the vineyard established with the goal of developing a small winery. For these reasons, and for purposes of simplicity, we have not attempted to factor the purchase of land into the vineyard enterprise; the assumption is that land is already owned. It is possible to add the cost of land to the underlying spreadsheet if desired.

Vineyard Specifics

The model vineyard is established at an excellent site where the hazards of winter cold injury and spring or early fall frosts are minimal (see chapter 2, page 14). Special frost protection measures, winter injury compensation strategies, and costs of hilling and de-hilling graft unions are not included. Experience suggests that even the best-sited vineyard should expect a significant crop reduction 1 year out of 10, on average. The causes for such a loss are variable but might include frost, winter injury, hail, or significant fruit rot or foliar disease caused by disease management failure. Grape growing is farming, and farming is inherently risky. We have not attempted to factor such a loss into the vineyard performance illustrated here, choosing instead to use a conservative crop yield in the economic analysis. A partial to complete crop reduction can be included in the spreadsheet data for one year or more to illustrate how such a loss would impact vineyard cash flow.

The baseline model vineyard is based on cordon-trained Smart-Dyson-trained vines (see chapter 5, page 98) with a row spacing of 10 feet and an in-row vine spacing of 7 feet. A nominal crop (10% of full crop or 0.5 ton/acre) is obtained in the second year, 40% of full crop (2.0 tons/acre) in Year 3, 80% of full crop in Year 4, and full production (5.0 tons/acre) attained in the fifth and subsequent years. Greater yields are certainly possible with Smart-Dyson training, without compromising wine quality; however; we have chosen to use a modest yield run-up and mature, steady-state yield which might reflect average vineyard management throughout a wide geographical area.

The spreadsheet includes an alternative, vertical shoot positioned (VSP) training system (see chapter 5, page 98), which has a yield potential of 4.0 tons/acre with 10-foot rows and about 5.6 tons/acre with 7-foot wide

SIDEBAR

Training System Models Used in Budget Development

Two grapevine training systems are modeled in the underlying workbook. One is a nondivided canopy;the other is a vertically divided canopy called a Smart-Dyson. The systems are illustrated and described in more detail in chapter 5, Pruning and Training. Briefly, the nondivided canopy training system is based on a bi-lateral, cordon-trained vine, with a vertically shoot-positioned (VSP) canopy. A narrow row (7 feet) and the more typical wide row width (10 feet) are compared with the VSP system. Yield per acre, when vines are in full production (Year 5), is estimated at 5.6 tons per acre for the 7-foot row space, and 4.0 tons per acre for the 10-foot row space. The VSP 10-foot (row) × 7-foot (vine) system is commonly used in the mid-Atlantic region; however, this does not imply it is the optimum model for new vineyards. The seven-ft row-width option with the VSP makes more efficient use of vineyard area but requires narrow equipment to navigate the closer row spacing.

Smart-Dyson training uses a vertically divided canopy that originates from a single cordon. A row width of 10 feet and mature vine yields of 6.0 tons per acre are used with the Smart-Dyson model.

Other training systems can be used in the vineyard but have not been specifically fitted to the workbook to detail costs and returns.

rows (40% more canopy per acre) (see sidebar). Training, row spacing, and the in-row vine spacing, as well as the crop yield run-up schedule, can be modified by the user in the underlying spreadsheet. This is particularly useful for looking at yield/return relationships in the sensitivity analyses that are provided. The vineyard receives optimal management, and cultural practices are similar to those recommended throughout this guide. These model vineyard yields are attainable, but they will not be associated with high grape and wine quality if the vineyard is established at an inferior site or is poorly managed.

Labor

The input with the greatest variability amongst vineyards is the time to complete certain tasks. Figures used here for the baseline vineyard description are a synthesis of actual data obtained from growers, similar data used in other states' publications, and our own estimates where other data were lacking. Labor, in the baseline model vineyard, is calculated at two different rates: $17.50/hr for "operator" labor, and $8.00/hr for "unskilled" labor. The operator rate reflects machinery operation and some degree of management skill, such as that used in pest management decisions and personnel supervision. Unskilled labor is for strictly manual tasks that require very little or no training and would not typically involve machinery operation. Both rates include workers' compensation, unemployment benefits, and Social Security and Medicare benefits paid by the employer. Labor rates can also be adjusted in the underlying spreadsheet.

Material Costs, Grape Prices and Debt Service

The model budgets are based on production of premium wine grape varieties, such as Chardonnay or Cabernet Franc. Grapes are priced at the vineyard at $1,400 per ton. Grapes are hand-harvested in the model vineyard at a piece rate of $1.25 per 25-pound picking tray (lug). Rental of a refrigerated transport may be required of the producer but has not been included in production costs. Material costs, including grapevines, trellis materials, and pesticides, are based on figures obtained from vendors in Virginia, New York, and North Carolina in 2006. Material prices are quite variable over time and from region to region and some prices have risen dramatically in 2008 as a consequence of increased energy and certain raw material costs. All material and labor costs can be updated on the underlying spreadsheet to more accurately reflect the conditions at the time this assessment is used.

Machinery

Machinery costs were calculated on the basis of new equipment exclusively purchased for use in the vineyard enterprise (see table 1.1 on page 4). The equipment inventory includes a four-wheel drive 55-horsepower tractor that is equipped with a spray cab and air conditioning, a 50-gallon herbicide sprayer, 300-gallon airblast sprayer for insecticides and fungicides, and other equipment used in both the establishment and annual operation of the vineyard. Included in this list is a new four-wheel drive pickup truck. The purchase price of new equipment is $73,500.

Machinery costs comprise both *fixed costs*, which include annual depreciation, taxes ($43.60 per $1,000 of assessed value) and insurance ($6.00 per $1,000 of value), and annual operating or *variable costs*. Fixed costs are also referred to as ownership costs and occur regardless of whether the equipment is used or not. The fixed costs shown in table 1.1 can be reduced by allocating the equipment usage over a larger vineyard. In fact, operating $73,500 worth of new equipment in a 10-acre vineyard is probably not an economically viable proposition. This point will be illustrated later. Many potential vineyardists will already own some of the necessary equipment or they may purchase used equipment and may also allocate some of the equipment fixed costs to other facets of farm operations. With the exception of the pickup truck (10-year lifespan, all equipment is depreciated over the 25-year expected lifetime of the model vineyard. This might be an overly optimistic service life of a heavily used mower or airblast sprayer, but it should be achievable with a well maintained tractor and infrequently used equipment (for example, a post-driver) that is purchased new at vineyard establishment and used only in the vineyard. The period of ownership is user-adjustable on the "Machinery Cost" sheet of the underlying spreadsheet. This is to reflect unique situations, such as factoring in the cost of a new piece of equipment, say in Year 11 or 12.

Table 1.1 ▪ Value and costs associated with equipment used in 10-acre *V. vinifera* vineyard in the mid-Atlantic region, 2005. Costs are shown as US dollars

Machine	Purchase price	Years owned	Salvage value	Fixed cost/ hr/acre	Variable cost/ hr/acre	Total cost/ hr/ acre
55-hp tractor (4WD, with Cab & AC)	30,000.00	25	3,000.00[b]	44.58	7.11	51.69
Grape Hoe (hip-mounted, hydraulic)[a]	*7,500.00*	*25*				
Utility Vehicle, 4WD (such as "JD Gator")[a]	*7,000.00*	*25*				
4WD, ¾-ton, pickup	20,000.00	10	2,000.00[b]	13.59[c]	9.09[c]	22.68
50-gal herbicide sprayer	2,700.00	25	404.79	107.43	1.86	109.29
100-gal airblast sprayer[a]	*9,000.00*	*25*				
300-gal airblast sprayer	12,500.00	25	1,874.05	30.60	0.21	30.81
5-foot rotary mower	1,600.00	25	281.84	7.32	0.03	7.35
Fertilizer/seed spreader	1,500.00	25	224.89	20.63	0.21	20.85
Post driver	2,000.00	25	352.30	57.68	0.20	57.88
PTO driven auger 12 in.	1,200.00	25	211.38	9.99	0.03	10.02
8-foot flatbed trailer	2,000.00	25	299.85	2.40	0.01	2.41
Total cost of equipment used	73,500.00					

[a] Items in italics are not used in the model vineyard and are not figured in total equipment cost below; they are included to show comparative price or as optional items.

[b] Salvage value shown as 10% of purchase cost. All other equipment salvage values are based on variable depreciation factors.

[c] Costs shown per mile of operation.

Variable costs vary in proportion to the amount of time the equipment is used in the vineyard enterprise. These costs include repairs ($10/hr), fuel ($2.50/gallon for diesel; $3.00/gallon for gasoline), and lubrication (15% of fuel costs) (see table 1.1), all of which can be adjusted by the user. Variable costs used here are derived from standards of the American Society of Agricultural Engineers (2004) and from Edwards (2002), Lazarus (2004) and Wu and Perry (2004).

Cost of Capital

The budget assumes a mix of owner equity and commercial financing to establish and operate the vineyard. Interest on equipment purchase is assumed to be 6%. Commercial financing is also used to subsidize annual operating costs (both fixed and variable costs as well as cumulative debt service) and is financed at 7.5% for one-half of the year in which the loan is taken.

Pest Management Program

Pest management can represent a significant variable cost associated with grape production, especially in reference to disease-susceptible varieties. The disease and insect management program used in the model vineyard is reflective of such a program and would be considered typical of that used in the mid-Atlantic region for Chardonnay in 2004 or 2005. Most of the pesticide sprays are fungicides, with the program designed for a disease-susceptible variety such as Chardonnay. Chardonnay is highly susceptible to powdery mildew and the model spray program is reflective of the need to maintain a tight spray schedule until after veraison. Spray schedules for other, less susceptible varieties could be considerably cheaper. In addition to fungicides, several insecticide applications are budgeted, which may or may not be necessary in a given year. As with other inputs, spray materials and number of sprays per year can be modified on the underlying spreadsheet.

COSTS AND RETURNS, BASELINE MODEL VINEYARD

The establishment and annual operational costs and net returns associated with the basic, model vineyard are detailed in table 1.2 (see pages 6 to 8) and summarized in table 1.3 (see page 9). Again, the model used is a 10-acre, Smart-Dyson-trained vineyard, with vinifera (such as Chardonnay) vines planted 7 feet apart in 10-foot wide rows. Irrigation and electric deer fencing are used, and the vineyard is operated with the equipment inventoried as indicated in table 1.1. Site preparation costs in Year 0 (pre-plant) include minimal land clearing (mowing), soil testing and liming (3 tons/acre), cultivation, and cover crop establishment, all of which are considered fixed costs. Variable costs of vineyard establishment include only the equipment variable costs ($264/acre). Purchase of hand tools, and payment of a nominal consulting fee ($50/acre/year) and loan interest ($55) bring the total costs in the pre-plant phase to $1,527/acre.

Most of the vineyard establishment costs are borne in Year 1, the year the vines are planted. Vines are purchased at $4.00 per unit and total $2,560 for the density (640 vines/acre) used in the vineyard model. The 640 vines is calculated in the underlying spreadsheet and is based upon a certain number of full vineyard rows, not just the row x vine spacing. Combined with planting labor, vine establishment totals $3,291/acre in Year1. Other major expenses in Year 1 are trellis material and construction labor ($2,396/acre) and equipment fixed costs ($3,644/acre). The sum of variable and fixed expenses in Year 1 is $11,628/acre. Interest (7.5%) charged on Year 1 expenses is $436, while the same rate of interest is charged on the carryover expenses from Year 0, or the pre-plant year ($115), for a total of $551 (see table 1.2). Collectively, the Year 1 expenses and interest sum to an annual cash flow of $12,178 and a cumulative (Year 1 + Year 0) cash flow ($12,178 + $1,527) of $13,705, where the brackets denote a negative net return (see table 1.3).

Year 2 is a relatively inexpensive year to operate the vineyard (see table 1.2 on pages 6 to 8). Much of the variable costs are associated with vine training, completion of the trellis, and a modest but important pest management program. Principal fixed costs, such as equipment and other capital amortization, remain. A very modest 1000 lbs of fruit (0.5 ton) per acre is harvested in the second year. This small crop requires both harvest labor ($1.25/lug and lug distribution and retrieval) and a material cost of picking lugs ($5.00/unit). Fifty percent of the lugs are purchased by the vineyard and 50% are loaned by the winery—an arrangement that may or may not apply in all situations. Total expenses in Year 2 are $3,327/acre. Interest on these expenses for one-half of the year is $125/acre, while the interest on accrued cash expenses is $1,028/acre ($1,153 total). Total expenses, including interest, add up to $4,480, which is offset slightly by a crop value of $700/acre (see table 1.3). Thus, the annual cash flow for Year 2 is $3,780 and the cumulative cash flow is $17,485.

Expenses in Year 3 increase due to a more robust pest-management program, increased pruning and canopy management labor, and increased harvest costs (see table 1.2). Interest on one-half of Year 3 expenses, combined with the debt service on preceding years, is $1,455/acre. The Year 3 expenses, before subtracting crop value, are $5,285/acre (see table 1.3). Subtracting the 2.0-ton crop value in Year 3 reduces annual cash flow to $2,485, with the cumulative cash flow increased to $19,970 per acre.

Expenses for Year 4 and subsequent years will be generally similar to Year 3, with a few exceptions. Pest and canopy management costs are considered steady-state at Year 4. Harvest costs increase to $1,039/acre in Year 4 (4.0 tons/acre, $439 in picking costs + $600 in lug purchases). The actual picking costs (no lugs purchased) will further increase to a steady-state rate of $539/acre in Year 5 and thereafter, assuming steady yields of 5.0 tons/acre. Interest payment in Year 4 will approximate $1,670/acre, but this rate will decrease with time as the grower pays off establishment costs. The Year 4 expenses, before subtracting crop value, add up to $6,259/acre (see table 1.2).

The underlying spreadsheet projects costs and returns on out to 25 years without regard for inflation or additional vineyard inputs. In reality, the vineyard will require additional inputs of trellis maintenance, lime, or other soil amendments, vine replacements for various reasons, and possibly other inputs. Labor rates will increase and grape prices may either increase or decrease. What happens beyond a 5- to 10-year horizon becomes quite speculative. We have used what we consider to be a modest

Table 1.2 ▪ Costs to develop 10-acre vineyard

Operation	Labor used	Labor hours	Labor cost	Material cost	Total cost
Site preparation, Year 0					
Lime (3 tons/acre)	Skilled	2.0	35.00	75.00	110.00
Herbicide application	Skilled	0.3	6.00	42.00	48.00
Plow and disc	Skilled	3.0	52.50		52.50
Soil sampling and soil test	Unskilled	1.5	12.00	1.00	13.00
Plant cover crop	Skilled	1.0	17.50	68.40	85.90
Fertilization	Unskilled	2.1	16.80	18.00	35.00
Mowing	Skilled	3.0	52.50		52.50
Equipment (variable costs)					264.00
Equipment (fixed costs)					621.00
Hand tools				62.00	62.00
Deer fence (annualized)					39.00
Equipment shed (annualized)					39.00
Consulting fee					50.00
Interest (7.5%)					55.00
TOTAL, YEAR 0					1,527.00
Year 1					
Vineyard lay-out	Skilled	4.8	84.00	57.00	141.00
Planting	Both	53.3	731.00	2,560.00	3,291.00
Fertilization (hand application)	Unskilled	2.1	16.80	18.00	34.80
Herbicide application	Skilled	0.3	6.00	42.00	48.00
Trellis construction	Both	58.1	708.00	1,688.00	2,396.00
Disease and insect sprays	Skilled	2.6	45.00	161.00	206.00
Canopy management	Unskilled	10.6	84.80	3.00	87.80
Mowing (6 times)	Skilled	3.0	52.50		52.50
Dormant pruning	Unskilled	5.3	42.40		42.40
Equipment (variable costs)					587.00
Equipment (fixed costs)					3,644.00
Safety equipment				19.00	19.00
Weather station				780.00	780.00
Irrigation (annualized)					171.00
Deer fence (annualized)					39.00
Equipment shed (annualized)					39.00
Consulting fee					50.00
Interest (7.5%)					551.00
TOTAL, YEAR 1					12,178.00

Note: All costs in US dollars.

Table 1.2 ▪ Costs to develop 10-acre vineyard *(continued)*

Operation	Labor used	Labor hours	Labor cost	Material cost	Total cost
Year 2					
Herbicide application	Skilled	0.3	6.00	42.00	48.00
Replanting	Both	0.4	4.25	20.0	24.25
Disease and insect sprays	Skilled	3.4	60.00	231.00	291.00
Trellis construction (completion)	Unskilled	10.0	80.00	479.00	559.00
Canopy management	Unskilled	77.7	622.00	3.00	625.00
Fertilization (hand application)	Unskilled	2.1	16.80	18.00	35.00
Mowing	Skilled	3.0	52.50		52.50
Dormant pruning	Unskilled	14.9	119.00	2.00	121.40
Safety equipment				16.00	16.00
Harvest costs (picking and lugs)			55.00	100.00	155.00
Equipment (variable costs)					279.00
Equipment (fixed costs)					822.00
Irrigation (annualized)					171.00
Deer fence (annualized)					39.00
Equipment shed (annualized)					39.00
Consulting fee					50.00
Interest (7.5%)					1,153.00
TOTAL, YEAR 2					4,480.00
Year 3					
Herbicide application	Skilled	0.3	6.00	42.00	48.00
Disease and insect sprays	Skilled	5.6	98.00	440.00	538.00
Canopy management	Unskilled	74.5	597.00	3.00	600.00
Fertilization	Unskilled	2.5	20.00	39.00	59.00
Mowing	Skilled	3.0	52.50		52.50
Dormant pruning	Unskilled	64.0	512.00	2.00	514.00
Safety equipment				16.00	16.00
Harvest costs (picking and lugs)			220.00	300.00	520.00
Equipment (variable costs)					279.00
Equipment (fixed costs)					905.00
Irrigation (annualized)					171.00
Deer fence (annualized)					39.00
Equipment shed (annualized)					39.00
Consulting fee					50.00
Interest (7.5%)					1,455.00
TOTAL, YEAR 3					5,285.00

Note: All costs in US dollars.

Table 1.2 ▪ Costs to develop 10-acre vineyard *(continued)*

Operation	Labor used	Labor hours	Labor cost	Material cost	Total cost
Year 4–25					
Herbicide application	Skilled	0.3	6.00	42.00	48.00
Disease and insect sprays	Skilled	6.0	105.00	516.00	621.00
Canopy management	Unskilled	95.9	768.00	3.00	771.00
Mowing	Skilled	3.0	52.50		52.50
Dormant pruning	Unskilled	57.6	461.00	1.00	462.00
Fertilization	Unskilled	2.5	20.00	39.00	59.00
Safety equipment				16.00	16.00
Harvest costs (picking)[a]			439.00	600.00	1,039.00
Equipment (variable costs)					279.00
Equipment (fixed costs)					942.00
Irrigation (annualized)					171.00
Deer fence (annualized)					39.00
Equipment shed (annualized)					39.00
Consulting fee					50.00
Interest (7.5%)					1,670.00
TOTAL, YEARS 4–25					6,259.00

Note: Summary of specific costs to develop 10-acre vineyard, based on Smart-Dyson vineyard model from pre-plant (Year 0) through fourth year of operation. Operational costs in Years 5 to 25 will be similar to those of year 4. All operations and costs are on a per-acre basis. Labor is either skilled ($17.50/hr) or unskilled ($8.00/hr). Material and labor costs change from year-to-year, but generally increase. The data in this table were current in 2006, but may be substantially greater today.

[a] The cost of harvest labor increases to $539/acre in Year 5 and thereafter, assuming that yields remain at 5.0 tons/acre.

yield expectation (5.0 tons/acre) with the Smart-Dyson training system, so that the long-term returns from the vineyard should error more toward a conservative estimate.

The baseline model vineyard's costs and returns (see table 1.2) are summarized in table 1.3, which includes annual and cumulative interest on operating loans and adjusts the cumulative cash flow to reflect harvest income. The prominent points of the baseline model vineyard, once it reaches a steady-state condition, are that variable costs of operation are approximately $2,800 per acre per year; total fixed costs are an additional $1,200/acre and total costs are roughly $6,000/acre, but this decreases with time due to decreasing interest payments. The *annual* cash flow is about negative $12,000 per acre in Year 1 but becomes positive in Year 4. The cumulative cash flow remains negative until Year 15, when the vineyard returns a "profit" of $982/acre.

Price sensitivity analyses are presented in table 1.4 (page 10) and table 1.5 (page 11) to evaluate the relationships between cost per acre and productivity. At 5.0 tons per acre, the $1,400/ton grapes return $848/ton when all expenses are accounted for (see table 1.4). Table 1.5 allows the user to evaluate a two-way interaction between the value of crop and the productivity (tons/acre). Thus, to generate a net return or profit with the basic vineyard model, one must obtain at least $900/ton at 5 tons/acre, or at least $800/ton at 6 tons/acre to cover all costs (see table 1.5). On the other hand, a positive return (return after variable expenses) can be realized at $1000/ton and 3.0 tons/acre if variable costs only are used for the computations.

Table 1.3 ▪ Summary of establishment and operational costs by year for 10-acre vineyard

COST	Year 0	Year 1	Year 2	Year 3	Year 4	Year > 4
VARIABLE COSTS						
Equipment	264	587	279	279	279	279
Replanting			24			
Weed & cover crop management	100	98	100	100	100	100
Vine fertilization	35	35	35	59	59	59
Dormant pruning		43	121	514	462	462
Canopy management		88	625	600	771	771
Disease & insect management		207	291	538	621	621
Harvest costs			55	220	439	539
Total variable costs	399	1,058	1,531	2,309	2,731	2,831
FIXED COSTS						
Land						
Taxes						
Site preparation	261					
Vineyard lay-out		141				
Planting		3,291				
Trellis construction		2,396	559			
Hand tools	62					
Equipment	621	3,644	822	905	942	942
Irrigation, amortized		171	171	171	171	171
Deer fencing, amortized	39	39	39	39	39	39
Weather station		780				
Safety equipment		19	16	16	16	16
Harvest lugs (50% of need)			100	300	600	
Equipment shed, amortized	39	39	39	39	39	39
Other (consulting)	50	50	50	50	50	50
Total fixed costs	1,073	10,570	1,796	1,521	1,858	1,258
Variable + fixed costs	1,472	11,628	3,327	3,830	4,589	4,089
Total interest	55	551	1,153	1,455	1,670	Decreases
Total costs + interest	1,527	12,178	4,480	5,285	6,259	Decreases[a]
Harvest income			700	2,800	5,600	7,000
Annual cash flow	(1,527)	(12,178)	(3,780)	(2,485)	(659)	Positive returns[b]
Cumulative cash flow	(1,527)	(13,705)	(17,485)	(19,970)	(20,629)	Zeros in year 15

Note: Summary of establishment and operational costs by year for 10-acre, Smart-Dyson-trained, 10 feet × 7 feet planting of grapes valued at $1,400 per ton, which consistently produces 5.0 tons/acre from Year 5 forward. All revenue and costs are shown on a per acre basis. Year 0 is the pre-plant year. Summarized costs may not equal values in columns due to rounding. All costs in US dollars.

[a] Decreases to year 14 whereupon a steady-state value of $4,243 per acre in costs + interest applies.

[b] $1,210 in year 5, increasing to $2,757 per year from year 14 on.

MODELING VINEYARD PERFORMANCE

It will be apparent to the reader that the establishment of a vineyard is a capital-intensive enterprise and the steep capital investment requires a very long-term approach to realizing profits. We chose to use a potentially high-yielding training system (Smart-Dyson divided canopy) with modest yield expectations of 5.0 tons/acre. Greater yields have consistently been obtained without compromising wine quality or significantly delaying harvest in research conducted in Virginia. The 10-foot row space is dictated by the 7-foot high canopy, which would require a minimum of 8 to 9 feet between rows to avoid violating a 1:1 minimum ratio between row width and canopy height (see Pruning and Training, page 98). One could argue that the "standard" VSP would be cheaper to install and manage, and would have similar returns per acre if the rows were established 7 feet apart. If yield/acre is therefore increased about 40%, to 5.6 tons/acre, the performance of VSP is comparable to that of Smart-Dyson (see figure 1.1 on page 12), if Smart-Dyson is constrained to 5.0 tons/acre. If cropped at 6.0 tons/acre, which is achievable with good management, the Smart-Dyson training has superior returns.

Table 1.4 ▪ Costs associated with model vineyard

	Total expenses	Variable expenses only
Cost per acre	$4,242	$2,831
Cost per ton	$848	$566

Net returns (loss) per ton of grapes, as a function of crop value and factored against either total expenses or variable expenses only. For example, if you are paid $1400/ton of grapes, you would net $552 per ton if you factored in all expenses and were producing 5.0 tons/acre.

Price/ton ($)	Total expenses	Variable expenses only
500	(348)	(66)
600	(248)	34
700	(148)	134
800	(48)	234
900	52	334
1000	152	434
1100	252	534
1200	352	634
1300	452	734
1400	**552**	834
1500	652	934
1600	752	1,034
1700	852	1,134
1800	952	1,234
1900	1052	1,334
2000	1152	1,434

Note: Cost of production and cost per ton when factored against total expenses or variable expenses only: baseline model vineyard (10-acre, Smart-Dyson training, Chardonnay spaced 10 feet × 7 feet) at steady-state (> year 4) production of 5.0 tons/acre. Expenses include harvest costs and interest payments. All costs in US dollars.

Significant improvements in vineyard economic performance can be achieved by lowering the capital investment at the outset. The easiest means of doing this is to reduce equipment costs—purchase used equipment rather than paying full price. There are obvious maintenance issues with this route, but reducing the cost of equipment by 50% results in the elimination of 3 years in the amortization of the baseline vineyard (see figure 1.2 on page 13). Other options for reducing equipment costs to the vineyard operation include finding alternate uses for equipment, such as performing custom services to neighboring farms, hiring custom applicators for specific jobs, and leasing certain equipment. A further improvement of vineyard efficiency results by increasing the number of acres from 10 to 20 (see figure 1.2).

The authors have made a conscientious effort to accurately reflect the costs and returns that one might expect to incur in the development and operation of a small, intensively managed wine grape vineyard in the mid-Atlantic region of North America. We remind the reader that actual operations may significantly differ in design, capital inputs, and stability of yields and material pricing. The analysis presented here, as well as on the underlying spreadsheet, is intended to serve as an aid to development of a business plan for a commercial vineyard, not as a plan in and of itself.

Table 1.5 ▪ Sensitivity analysis of break-even costs for Smart-Dyson-trained Chardonnay vines spaced 10 feet × 7 feet (baseline model vineyard)

A. Yield and price needed to recover *all costs*

Yield (tons/ acre)	U.S.Dollars paid per ton											
	$700	$800	$900	$1,000	$1,100	$1,200	$1,300	$1,400	$1,500	$1,600	$1,700	$1,800
1	(3,542)	(3,442)	(3,342)	(3,242)	(3,142)	(3,042)	(2,942)	(2,842)	(2,742)	(2,642)	(2,542)	(2,442)
2	(2,842)	(2,642)	(2,442)	(2,242)	(2,042)	(1,842)	(1,642)	(1,442)	(1,242)	(1,042)	(842)	(642)
3	(2,142)	(1,842)	(1,542)	(1,242)	(942)	(642)	(342)	(42)	258	558	858	1,158
4	(1,442)	(1,042)	(642)	(242)	158	558	958	1,358	1,758	2,158	2,558	2,958
5	(742)	(242)	258	758	1,258	1,758	2,258	2,758	3,258	3,758	4,258	4,758
6	(42)	558	1,158	1,758	2,358	2,958	3,558	4,158	4,758	5,358	5,958	6,558
7	658	1,358	2,058	2,758	3,458	4,158	4,858	5,558	6,258	6,958	7,658	8,358
8	1,358	2,158	2,958	3,758	4,558	5,358	6,158	6,958	7,758	8,558	9,358	10,158

B. Yield and price needed to recover *variable costs only*

Yield (tons/ acre)	U.S.Dollars paid per ton											
	$700	$800	$900	$1,000	$1,100	$1,200	$1,300	$1,400	$1,500	$1,600	$1,700	$1,800
1	(2,131)	(2,031)	(1,931)	(1,831)	(1,731)	(1,631)	(1,531)	(1,431)	(1,331)	(1,231)	(1,131)	(1,031)
2	(1,431)	(1,231)	(1,031)	(831)	(631)	(431)	(231)	(31)	169	369	569	769
3	(731)	(431)	(131)	169	469	769	1,069	1,369	1,669	1,969	2,269	2,569
4	(31)	369	769	1,169	1,569	1,969	2,369	2,769	3,169	3,569	3,969	4,369
5	669	1,169	1,669	2,169	2,669	3,169	3,669	4,169	4,669	5,169	5,669	6,169
6	1,369	1,969	2,569	3,169	3,769	4,369	4,969	5,569	6,169	6,769	7,369	7,969
7	2,069	2,769	3,469	4,169	4,869	5,569	6,269	6,969	7,669	8,369	9,069	9,769
8	2,769	3,569	4,369	5,169	5,969	6,769	7,569	8,369	9,169	9,969	10,769	11,569

Note: Values in table matrix are the net returns or (losses) per acre as a function of both yield per acre and price paid per ton of grapes. The upper table **(A)** is based on both variable and fixed costs of production. The lower **(B)** table is based only on the variable costs of production used in the baseline vineyard budget. Areas shaded green lie outside the range of yield and crop value typically found for high quality Chardonnay in eastern North America.

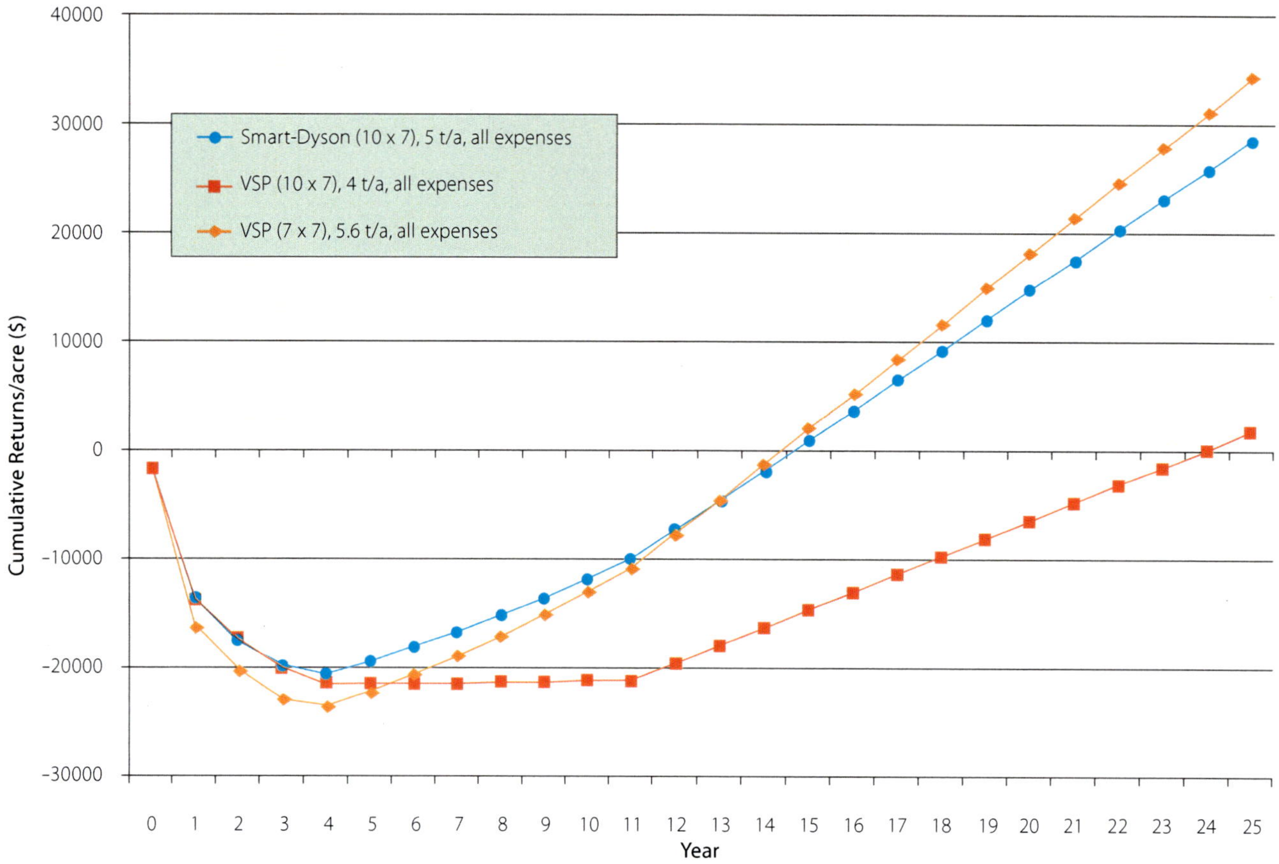

Figure 1.1 ▪ Illustration of cumulative returns (US Dollars) associated with three different vineyard design scenarios. Example 1 (blue line, circles) is the baseline model, a 10-acre vineyard with Smart-Dyson-trained Chardonnay vines spaced 7 feet apart in 10-foot wide rows. Vines are cropped at 5.0 tons/acre at maturity and the crop is valued at $1400/ton. In this example, all expenses are accounted for. Example 2 (red line, squares) is a VSP-trained version of example 1, producing only 4.0 tons/acre at maturity. Example 3 (gray line, diamonds) is similar to example 2; however, row width has been narrowed to 7 feet for greater land use efficiency, which increases yield to 5.6 tons/acre. Caution must be used in extrapolating vineyard performance data beyond 5 or 6 years due to uncertainties about cost and return inflation, climatic factors, market demand, etc.

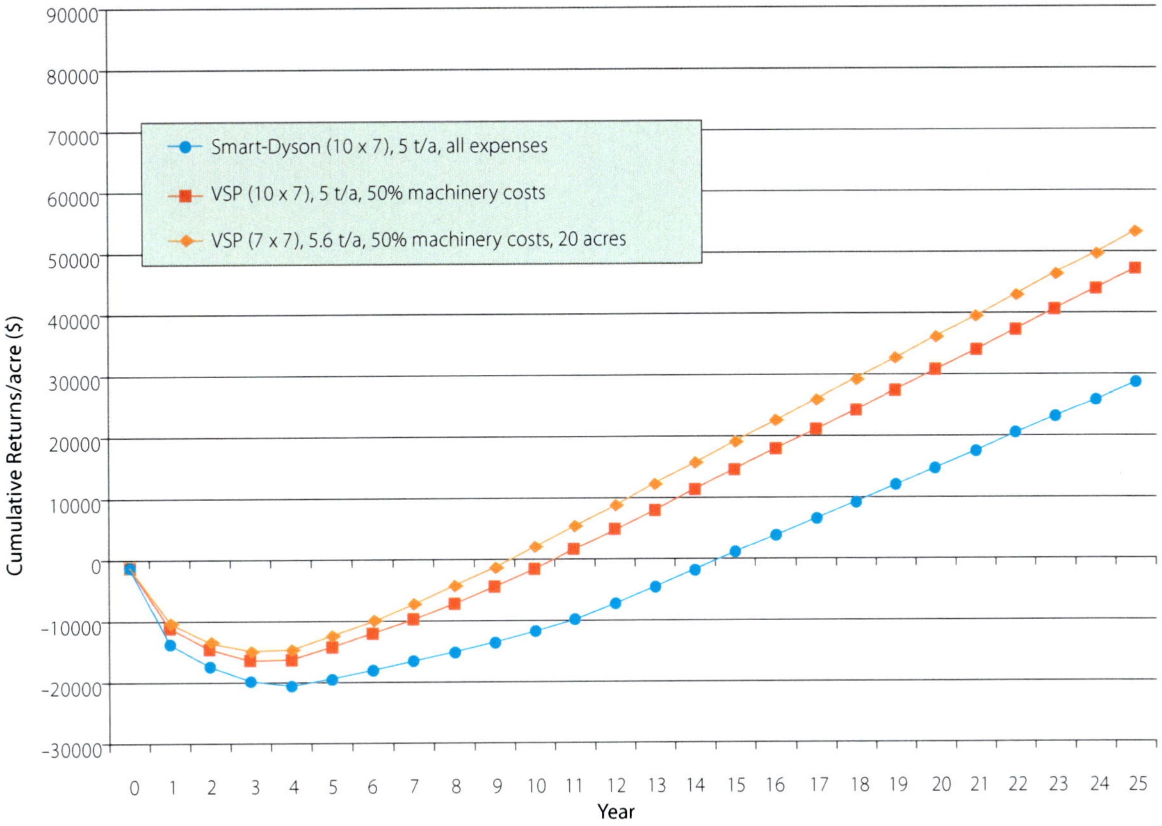

Figure 1.2 ▪ Illustration of cumulative returns (US Dollars) associated with three different vineyard design scenarios. Here, the baseline Smart-Dyson vineyard is established using (1) full equipment price (blue line, circles) and compared with (2) the same vineyard where only 50% of equipment price is factored in (red line, squares), and (3), where scenario #2 is applied to a 20-acre operation in which row width was narrowed to 7 feet (gray line, diamonds), rather than the baseline 10-acre vineyard. Caution must be used in extrapolating vineyard performance data beyond 5 or 6 years due to uncertainties about material and labor costs, crop production and market demand for grapes.

References

American Society of Agricultural Engineers (ASAE). 2002. ASAE standards: Standards, engineering practices and data adopted by the American Society of Agricultural Engineers, 50th edition American Society of Agricultural Engineers, St. Joseph, MI.

Capps, E. R., T. K. Wolf, and B.J. Walker. 1998. The economics of wine grape production in Virginia. Virginia Cooperative Extension Publication 463-008. Virginia Tech, Blacksburg, VA 24061.

Edwards, William. 2002. "Estimating Farm Machinery Costs", Iowa State University Extension Publication Number A3-29, April 2002.

Lazarus, William. 2004. *Farm Machinery Cost Estimates for 2004*, University of Minnesota Extension Service, Publication number FO-06696.

White, G. B. 2005. Cost of establishment and production of vinifera grapes in the Finger Lakes Region of New York—2004. Department of Applied Economics and Management, Cornell University, Ithaca, NY 14853. E.B. 2005-06.

Wolf, Tony K. and E. Barclay Poling. 1995. *The Mid-Atlantic Winegrape Grower's Guide*. North Carolina State University, Raleigh, NC. 126p.

Wu, J., and G.M. Perry. 2004. Estimating farm depreciation: Which functional form is best? *American Journal of Agricultural Economics* 96:483-491.

—2—

Vineyard Site Selection

Authors
Tony K. Wolf, Virginia Tech
John D. Boyer, Virginia Tech
Timothy E. Martinson, Cornell University
Fritz A. Westover, Texas AgriLife Extension Service

Grapevines can be exposed to environmental stresses and biological pests that result in injury or destruction to the vine or that lead to a reduction in overall crop quality and yield. Damaging winter temperatures, spring and fall frosts, droughts or excessive rainfall, destructive hurricanes, and higher than optimal summer temperatures are all events that can occur in some regions of eastern North America. Grapevines also can be severely injured by certain atmospheric pollutants, if grown near the origin of those pollutants, and most vineyards are chronically threatened by diseases, insects, and vertebrate animals such as deer and birds. Despite these challenges, grapes are grown successfully in many areas of eastern North America. Prudent vineyard site selection, however, is required to reduce the threat of physical and biotic challenges to grape production.

The goal of this chapter is to describe the principal physical and biological features that affect grape production and that should be evaluated in the site selection process. In practice, most readers will realize that vineyard site selection involves compromises; few sites are ideally suited to grape production in all respects. Furthermore, those who wish to establish a winery also should recognize that the best vineyard sites might not necessarily be the most accessible to winery customers.

The text is divided into three sections: (a) a discussion of the broad "macroclimate" of the region; (b) a discussion of local climate, or "mesoclimate," and soil factors that have a bearing on site selection; and (c) a description of pests and other threats that should be considered when choosing a vineyard site. This information is primarily intended for wine grape producers, but the basic concepts are also applicable to table grape production.

This chapter is based on a bulletin of the same title developed for Virginia. Thus, the principles apply most closely to the Mid-Atlantic region of North America. Additional information pertinent to the Finger Lakes region of New York provides specific site-selection principles used in that environment. While much of the data is specific to Virginia, similar data are available for other eastern states and Canadian provinces. This chapter will show how those data can be used to assess the merits and deficiencies of a proposed vineyard site. Readers should consult their own state extension specialists or commercial services to determine if more specific guidelines are available for vineyard site selection in their location.

CLIMATE

The term "climate" refers to the average course of the weather at a given location over a period of years. Climate is determined by assessing temperature, precipitation, wind speed, and other meteorological conditions. "Weather" is the state of the atmosphere at a given moment with respect to those same meteorological conditions. "Macroclimate" refers to the prevailing climate of a large geographic region. Many of the grape-producing areas of eastern North America are subject to a macroclimate that is primarily continental. Exceptions occur along the Eastern Seaboard of the United States, including most of Long Island and the Delmarva Peninsula. Continental climates have temperature and precipitation patterns that, in many cases, are modified by large land masses (continents). Air temperatures of continental climates can fluctuate rapidly on a day-to-day basis because land does not readily affect or buffer air temperatures. Maritime climates, on the other hand, are macroclimates that are directly influenced by their proximity to large bodies of water, which act as tremendous heat sinks. Water absorbs heat during summer and slowly cools in the fall. This stored heat will tend to warm the air near the water. This warming may extend the growing season and may raise mid-winter air temperatures enough to prevent vine damage from low temperature events. The depth, surface area, and salinity of these bodies of water will, to a great extent, determine how much heat the water can absorb and release before freezing. As air temperatures rise in the spring, large bodies of water will warm more slowly than the surrounding land. Relatively warm air is cooled as it moves over cold water. The cooled air retards plant development on the leeward side of the water and reduces the risk of spring frost damage.

Temperature moderation associated with water bodies has influenced grape production in specific regions. In Virginia and Maryland, the Tidewater and Eastern Shore counties are subject to a maritime climate due to their proximity to the Atlantic Ocean and the Chesapeake Bay. Macroclimate in New York is strongly influenced by Lake Erie (western New York), Lake Ontario (Finger Lakes), and by the Atlantic Ocean (Long Island and the lower portion of the Hudson River). In the Finger Lakes region, the larger lakes (Canandaigua, Keuka, Seneca, and Cayuga Lakes) have maximum depths ranging from 100 feet to more than 600 feet; these lakes rarely freeze over completely. Though much smaller than the Great Lakes, the Finger Lakes provide local moderation of winter lows within a narrow band of land that surrounds them.

The mesoclimate, or local climate, is more specific than the macroclimate. Horizontal distances of as little as 500 feet, such as those between opposing aspects or hillsides in hill and valley terrain, may affect mesoclimates. The vineyard mesoclimate is influenced by topography, the compass orientation (aspect), the degree of inclination (slope), barriers to air movement and, to a lesser extent, soil type, soil moisture, and the nature of ground cover.

The term "microclimate" is used to describe the specific environment within, and immediately exterior to, grapevine canopies. Grapevine canopies consist of shoots and their leaves and fruit present during the growing season. The microclimate within vine canopies can differ significantly from the climate immediately outside the canopy with respect to air temperature, wind speed, humidity, and, particularly, the quantity and quality of sunlight (see chapter 6, page 124).

MACROSCALE SITE SELECTION

Many readers will have a narrowly defined interest in vineyard site selection. Landowners, for example, may simply wish to determine whether their land is suitable for grape production. Others might ask, "Where should I establish a vineyard, and why?" The answers to both questions have some commonalties and start with a review of the region's macroclimate, paying particular attention to the length of the growing season and the occurrence of temperature extremes.

Length of Growing Season

The length of the growing season will determine whether grapes will ripen. Decreasing fall temperatures reduce the capacity of a vine to synthesize sugar and ripen grapes. Ultimately, frost will kill leaves that have not naturally senesced, thus preventing further sugar accumulation in fruit and perennial portions of the vine. Vineyard sites must, therefore, have sufficient intensity and duration of heat to ensure crop ripeness. Depending on variety, as few as 150 frost-free days, or more than 180 days, may be required to fully ripen the crop. A minimum of 180 frost-free days is recommended for very late-ripening varieties, although very early-maturing varieties, such as some Muscats and Viognier, may ripen with a season as short as 155 days. Additional frost-free days after harvest are desirable, though, to permit further gains in carbohydrate accumulation in roots, trunks, and other perennial organs. The average length of the growing season, conservatively defined as the time from the last average spring occurrence of 32°F to the first fall occurrence of 32°F, is shown in figure 2.1 for Virginia and North Carolina and figure 2.2 for the state of New York.

Growing Degree Days and Other Climate Indices

Viticulturists have sought a convenient means of characterizing climate for predicting the grape ripening suitability of a region. The underlying basis of this goal is that grape species and varieties vary in their adaptation to seasonal duration, heat summation, precipitation, and humidity. Some tolerate extremes of temperature more than others do, and some are more resistant to rain-induced splitting and rots (see chapter 3, page 37). Climatic indices include various methods of summing heat, use of the mean temperature of the warmest month, and exploration of other factors such as latitude and elevation, relative humidity, or aridity. An in-depth review of these indices illustrates that some are more or less suited to continental versus maritime climates, latitudinal extremes of commercial grape growing, or arid versus more humid regions. While the goal of developing and using climatic indices may be to better match a site to varieties for which it is best suited, the science does not always drive the decision as to what growers ultimately decide to plant.

Heat summation, typically expressed as the number of growing degree days (GDD) in a season, has been used to help characterize a region's heat accumulation during that season, which, broadly stated, is considered to be the period from April through October. The basis for this heat summation is the general response of grapevines to temperature. Typically, growth processes are

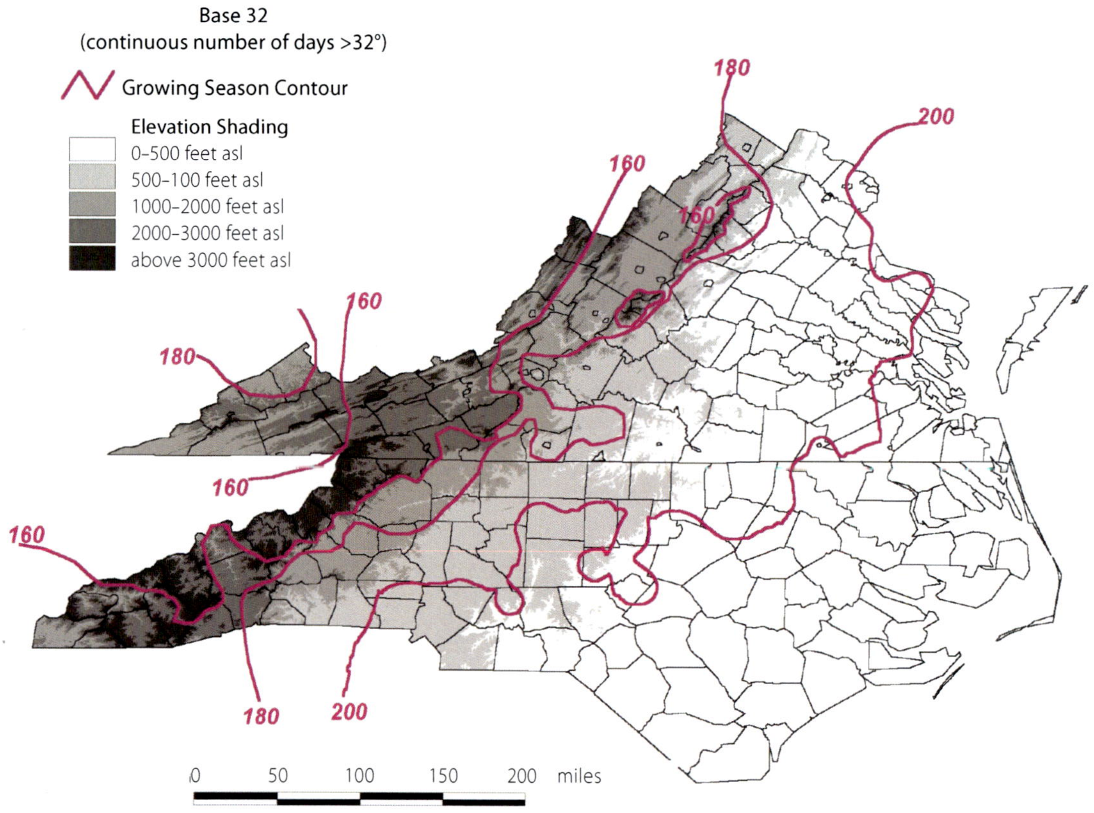

Figure 2.1 ▪ Median growing season length isotherms and elevation shading for Virginia and North Carolina. The isotherm values are the median number of frost-free days from spring to fall. A temperature of 32°F is used to define frost-free days although grapevines can, under certain atmospheric conditions, withstand slightly cooler temperatures without injury.

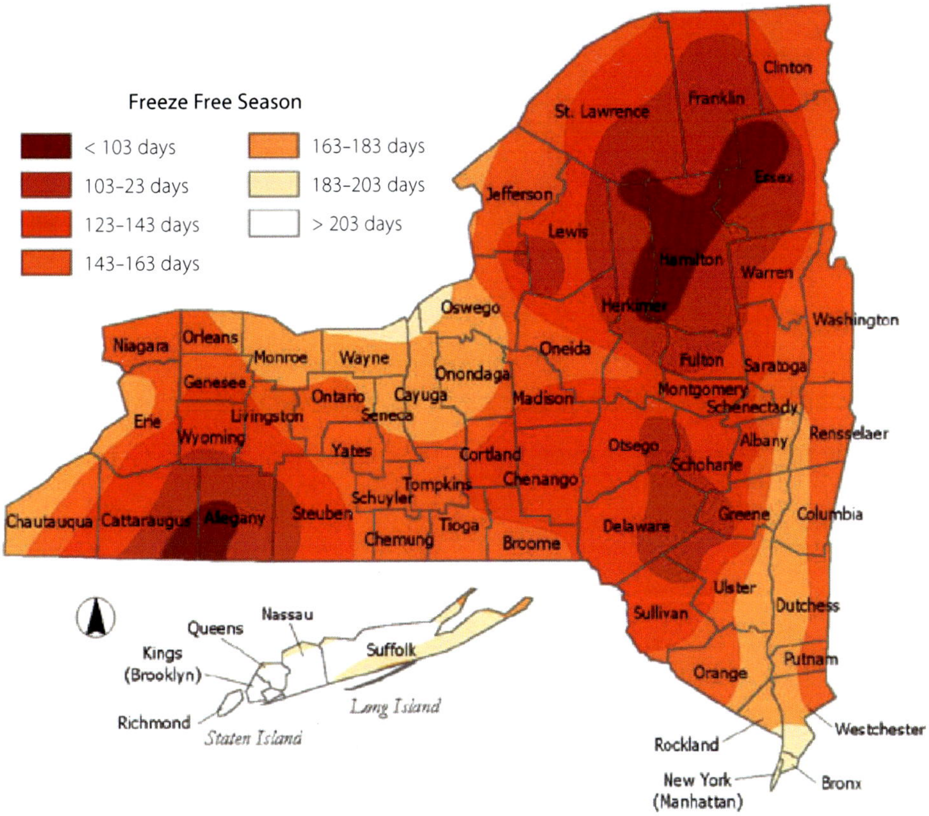

Figure 2.2 ▪ Average number of frost-free days in New York. *Source:* HTTP://WWW.GARDENING.CORNELL.EDU/FRUIT/HOMEFRUIT.HTML

responsive to temperatures greater than 50°F, and are generally unresponsive to cooler temperatures. Although this system was developed to characterize the range of thermal conditions in California's viticultural regions, it has had global, if not precise, applicability. To calculate GDD, based on 50°F, average the minimum and maximum temperatures for a given day to derive a mean daily temperature. Subtract 50 from that mean with the remainder being either 0, if the mean was equal to or less than 50°, or a positive integer if the mean was 51° or greater. A mean daily temperature of 65°F, for example, would produce 15 heat units. Calculate this each day for the period from April 1 through October 31, with the positive integers of each day summed. An approximation of the same data can be obtained more rapidly by averaging the minimum and maximum monthly temperatures for the same period, subtracting 50, and multiplying by the number of days in each month. This was the method originally used by Amerine and Winkler who developed the index in California. From 1,800 to 2,500 GDD are required to ripen American varieties, some French hybrids, and cool-climate vinifera varieties. More heat (2,500 to 3,500 GDD) is required to adequately ripen many of the other varieties and some French hybrids. While crude, the GDD system of temperature summation has been used for over 50 years to classify regions as to their grape growing potential and to help match varieties with the region's thermal regime.

A more refined technique of temperature summation was described by Gladstones (1992) and is based on "biologically effective degree days" (BEDD), which are calculated in degrees Celsius.The computation of BEDD parallels the GDD summation to an extent. The vine growth response to increasing temperatures above 10° C (50°F), however, is not linear; the response to increasing temperature flattens above 19°C (66°F) and eventually declines at temperatures in excess of 30°C (86°F). In simple terms, this means that adding more heat above 19°C is not going to make the grapevine grow faster. Gladstones therefore calculated biologically effective degree days as the temperature summation when the daily mean temperature is between 10°C and 19°C (excluding temperatures below 10°C and above 19°C). The BEDD values are further adjusted for latitude, which affects day length (duration of photosynthesis), and the diurnal temperature range. The BEDD basis of site characterization has not been widely used in eastern North America, although it might have utility here. Further precision might be gained in the future by examining the occurrence of high temperature extremes in the ripening phase of berry development and by incorporating the use of models that sum heat on an hourly, as opposed to a daily or monthly basis. These techniques are facilitated by low-cost, battery-operated data loggers. Collection of seasonal temperature data after the vineyard is established can be useful for predicting grapevine phenology, pest occurrence and other biological events, but one must still rely principally on published data from nearby sites and on local vineyard experiences during one's initial vineyard site evaluation.

Frost

Depending on the time of year, grapevines may be injured by fall frosts, spring frosts, or winter cold, all of which constitute "low-temperature" injury. We will define "frost" injury as the condition in which buds are broken and leaves are present and "winter" injury as that occurring when buds are still closed and vines lack obvious shoot growth. Frost injury and winter injury often have similar causes associated with topography and meteorological conditions. We will discuss these factors in greater detail.

The risk of spring frost damage is increased when unseasonably warm weather promotes early budbreak and shoot growth, which are then followed by more seasonable low temperatures. Grape shoots are very susceptible to freeze injury if temperatures dip below 32°F (see figure 2.3). Under very dry air conditions, the injury might not occur until temperatures reach 25°F or 26°F, but shoots rarely survive lower air temperatures. Spring frosts generally do not kill vines because secondary buds will subsequently break and their shoots will provide sufficient foliage to support the vine; however, secondary buds typically produce shoots with a very low fruiting potential. Thus, the principal impact of spring frost is the significant loss of crop for that season. At the end of the season, fall frosts essentially arrest further sugar accumulation, as previously mentioned. The principal impact of early fall frosts therefore is unripe fruit. Beyond simply ripening the crop, it is also desirable to extend the presence of healthy grapevine foliage post-harvest in order to increase carbohydrate storage in perennial portions of the vine. Those reserves will support the early season development of the vine in the following year.

A goal of site selection is to identify areas or regions that have a relatively low likelihood of late spring or early fall frosts. One method of evaluating a region's risk of frost is based on the range between a site's average mean and average minimum temperatures for a given month. That range, defined for spring months as a "spring frost index" (SFI) by Gladstones (2000), is a measure of the region's continentality, or tendency to produce large fluctuations in temperature over short periods of time—the greater the range, the greater the frost hazard. Budbreak is influenced primarily by air temperature—the warmer the temperature, the earlier the budbreak. Warm weather is a problem only if it is interspersed with sub-freezing temperatures; this condition would tend to lower the average minimum temperature for a month, thereby increasing the SFI value. The average mean temperature for a month is based on the daily mean tempera-

Figure 2.3 ▪ Frost injury to young shoot.

ture (daily high plus daily low, divided by 2), averaged for all days of that month. The average minimum is simply the minimum daily temperature averaged for all days of the month. The average mean and average minimum monthly temperatures are reported for many U.S. weather stations by the National Oceanic and Atmospheric Administration (www.NOAA.gov). To illustrate, average April mean and minimum temperatures, as well as corresponding SFIs, are shown for eight Virginia locations in table 2.1. Although there is no precisely defined threshold for stating what frost risk is tolerable, SFI values less than 11.0 have relatively low frost risk, whereas those of 12.0 or greater are at higher risk. A comparison of two locations in Virginia, for example, illustrates differences in SFI that occur within a relatively small area. Mount Weather, at 1,720 feet above sea level (asl) on the Blue Ridge Mountain, has a low frost risk because it lies within an optimal "thermal elevation belt," as described in following text. Dulles Airport, 30 miles east of Mount Weather, lies in a large cold-air basin; the increased risk of frost near Dulles Airport is reflected in its relatively large SFI. The SFI may not be precise, but the values shown in table 2.1 are consistent with grower experiences with spring frost occurrence in Virginia. Although the SFI applies to spring frost, the same technique can be used to gauge a site's risk of fall frost, such as that occurring in the month of October. Generally, one could expect a strong correlation between spring and fall frost risk at a particular site.

Frequency of Extreme Low Temperatures

Grapevines can be injured or killed by cold winter temperatures. Winter injury historically has been the primary environmental limitation to *Vitis vinifera* varieties in many locations in eastern North America. Injury may include death of overwintering buds, injury to the vascular tissues of canes, cordons and trunks, or even complete vine kill (see figure 2.4). The frequency of damaging low

Figure 2.4 ▪ Riesling vineyard that experienced temperatures from -17°F to -20°F in February, 1996.

Photo A is a dormant bud cut cross-sectionally to reveal a cold injured primary bud; Photo B is a healthy cane *(left)* compared to a cold-injured cane *(right)*. Discolored tissue beneath the bark on the injured cane illustrates the injury to the phloem and vascular cambium.

Table 2.1 ▪ Illustration of Spring Frost Index (SFI) for April at eight Virginia locations

Station	Elevation (feet above sea level)	Average daily mean (°F) for April	Average daily minimum (°F) for April	Spring Frost Index (SFI)	Relative frost risk
Painter	30	56.2	45.9	10.3	Low
Norfolk NAS	33	58.7	49.5	9.2	Low
Mount Weather	1720	49.3	40.2	9.1	Low
Dulles Airport	290	53.1	40.2	12.9	High
Winchester	680	51.7	38.5	13.2	High
Abingdon	1920	52.9	39.3	13.6	High
Farmville	450	56.9	42.5	14.4	High
Chatham	640	53.9	38.7	15.2	High

winter temperatures will determine if grape production is possible and, if so, what the species and variety limitations are to such production. Due to the dynamics of vine cold hardiness, it is difficult to say precisely what temperature will cause cold injury to a specific variety on a specific date. Above the threshold of injurious cold for a given variety, the cooler a region's minimum daily temperatures during fall, the greater the extent of cold acclimation achieved by mid-winter. "Cold acclimation" refers to the metabolic and visible changes that occur with cold-adapted plants as they transition from a non-hardy condition to a dormant cold-hardy condition. One visible manifestation of acclimation is the maturation of green, succulent shoots into brown, woody canes. To illustrate air temperature effects on acclimation, varieties grown in New York State will tolerate mid-winter temperatures that are up to 8°F lower than those tolerated by the same varieties grown in Virginia. Thus, the definition of a critical, injurious temperature in one region will not necessarily be applicable to warmer or cooler regions. With that caveat, experience in Virginia, coupled with numerous controlled freezing tests over the past 15 winters, have led to the use of a critical temperature of −8°F as a guide for predicting the onset of significant cold injury in *V. vinifera* varieties grown in Virginia. The −8°F threshold in Virginia is used as an example here and is not absolute. It will vary by region, as previously stated, by grape species, grape variety, time of year, and as a result of the air temperatures immediately preceding the cold episode. The absolute cold hardiness achieved by a given variety will also be affected by the health and physiological status of the vines. Stressed vines, for example, will not achieve full potential winter cold hardiness. With respect to grape species, the general relationship of winter cold hardiness among commercial species is,

American spp. > interspecific hybrids > *V. vinifera*

While this implies that all American species (for example, *V. labruscana*, 'Concord') are more cold-hardy than all hybrids, the relative differences may be modest. Similarly, there is a range of hardiness among varieties of *V. vinifera*, with some of the more cold hardy comparable in hardiness to some hybrids. It is also important to understand that the −8°F figure does not imply that injury will be absent at warmer temperatures. Finally, the occurrence of −8°F events is counted by each cold episode, not necessarily the number of days that experienced -8°F. For example, a cold front that resulted in two consecutive nights at −8° or lower would be counted as one occurrence.

The frequency of −8°F events, by decade, over a 30-year period is depicted in figure 2.5 for Virginia. It is important to understand that figure 2.5 is a generalized map that may likely mask local or mesoclimate effects. In practice, a region that experiences −8°F three or more times per decade would have an increased need for careful mesoscale site selection if cold-tender varieties were to be grown. If a proposed vineyard site in Virginia experienced −8°F three or more times per decade, it would not be considered suitable for commercial production of cold-tender grape varieties unless special protection (trunk hilling or burying canes) were provided. A site that experienced −8°F two times in a 10-year period would be considered only fair, while a site that experienced −8°F only once per decade would be good.

Again, as one moves north from the mid-Atlantic or explores more cold-hardy grape species and varieties, the critical threshold temperature for these decisions gets lower. For example, "excellent" vineyard sites in New York's principal grape regions would attain −10°F no more than once in a 10-year period, while "acceptable" sites might reach −10°F no more than four times per decade, with a long-term minimum temperature of −15°F occurring no more than once per decade. Conversely, moving south from the mid-Atlantic should be expected to result in a higher critical threshold than the −8°F used in this example.

The temperature moderating influence of the Atlantic Ocean and other large bodies of water such as Lake Erie, Lake Ontario, the Finger Lakes, and Long Island Sound, reduces the frequency of extreme low temperature events in those areas of New York State. As previously stated, varieties grown in New York will acclimate to cooler winter temperatures than those same varieties grown further south; however, as illustrated in figure 2.6, 10-year cold isotherms of −14°F in the Finger Lakes region indicate the need to focus on local mesoclimate data when selecting sites for winter cold injury avoidance. South of the lake plains of western New York, and north and east of the Finger Lakes, higher elevations shorten the growing season, thus limiting the production of *V. vinifera* varieties.

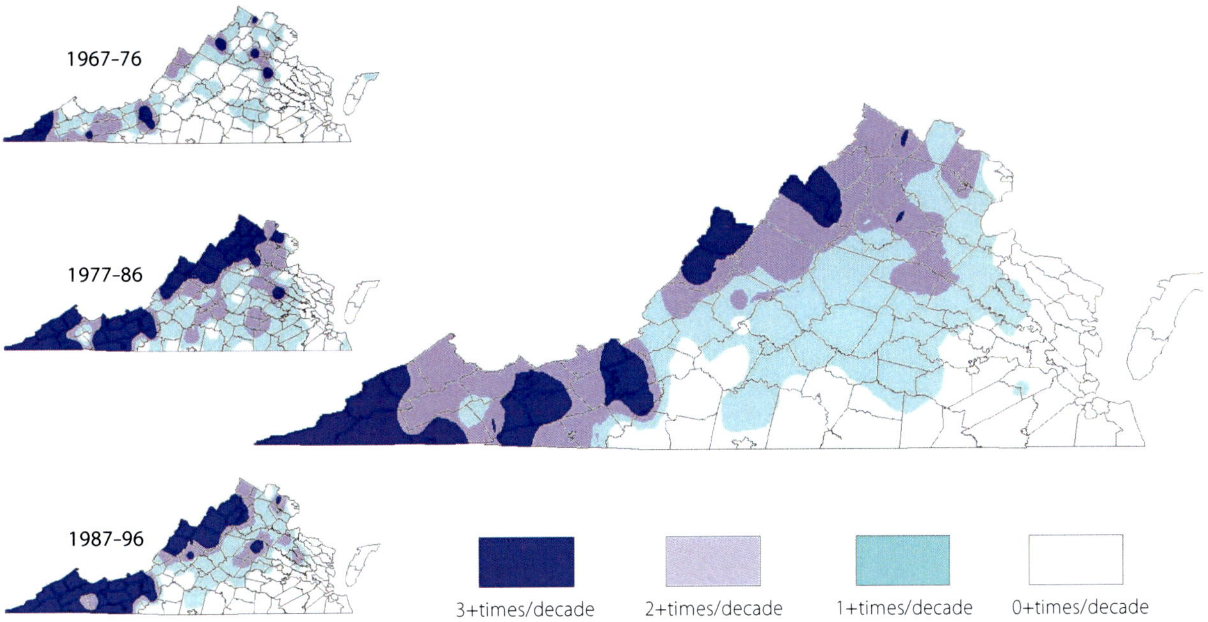

Figure 2.5 ▪ Average number of -8°F episodes per decade in the period 1967 to 1996. The large central image is an average of the three smaller 10-year records.

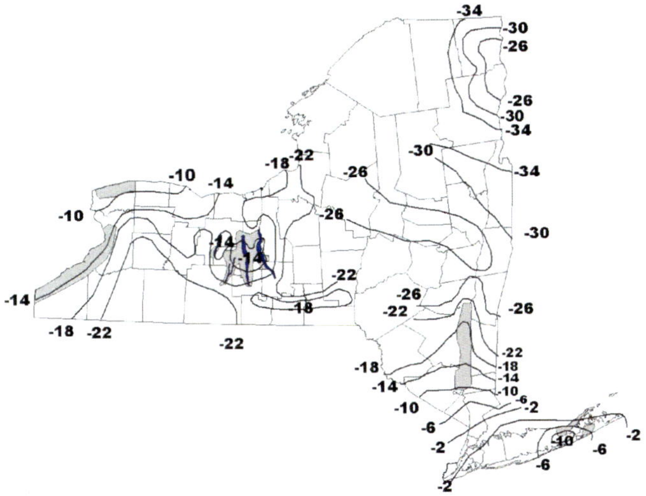

Figure 2.6 ▪ Ten-year extreme low temperature (°F) isotherms in New York.

Note: Redrawn from Dethier et al. 1966, cited at HTTP://WWW.NYSAES.CORNELL.EDU/HORT/FACULTY/POOL/GRAPEPAGESINDEX.HTML

Temperatures from July through October

Air temperatures greater than 86°F can reduce the vine's ability to photosynthetically convert carbon dioxide into sugars and other carbohydrates. Nighttime temperatures greater than about 64°F tend to increase the vine's respiration of this potential energy. In fact, respiration can consume up to 60% of the energy generated by photosynthesis. Many Virginia sites, to use a regional example, have average July maximum temperatures in excess of 86°F and some, particularly those in the more humid regions of the Tidewater, have average July minimum temperatures in excess of 64°F (see figure 2.7). In practical terms, we would not expect vines grown in such regions to be as physiologically productive as vines grown at cooler day and night temperatures.

Optimal primary fruit chemistry (sugar/acid concentrations and pH) and, in red-fruited varieties, color development, are promoted by daytime temperatures of 68°F to 77°F and nighttime temperatures of 59°F to 68°F. Gladstones (1992) suggests an optimal mean daily temperature of 64°F to 70°F in the final month of ripening (August through October, depending on location and variety). Aside from moving to a more northerly latitude, the most effective means of locating sites with lower daily temperatures is to increase elevation. This trend is apparent with the Virginia data of table 2.2 (page 22) that show a decrease in average July temperatures (maxima, minima, and means) with increase in station elevation. But note that both the recorded low temperature and the length of the frost-free period also decrease with increased elevation. Thus, there is a tradeoff between finding sites with cooler summer temperatures and selecting sites in which vines will not

Table 2.2 ▪ Selected climatological indices for various Virginia locations

Station name	County	Elevation (feet, asl)[a]	Recorded low (°F)	Record low date	July daily average max (°F)	July daily average mean (°F)	July daily average min (°F)	Days >90°F (number)	GDD[b]	FFP[c] (days)
Suffolk Lake Kilby	Suffolk	22	-5	21-Jan-85	88.1	78.5	68.8	30	4491	222
Painter 2 W	Accomack	30	-1	2-Feb-71	87.0	78.3	69.5	22	4257	209
Williamsburg 2 N	York	70	-7	21-Jan-85	89.0	78.1	67.1	36	4358	202
Fredericksburg	Spotsylvania	90	-11	22-Jan-84	89.6	77.5	65.4	40	3839	177
Lawrenceville 3 E	Brunswick	137	-10	21-Jan-85	88.3	76.5	64.6	36	3901	179
Warsaw 2 NW	Richmond	140	-6	21-Jan-85	88.1	77.6	67.1	35	4153	195
Ashland	Hanover	220	-11	5-Feb-96	87.1	76.6	66.0	26	3887	191
Louisa	Louisa	420	-21	5-Feb-96	86.6	74.4	62.1	27	3359	172
Farmville 2 N	Cumberland	450	-8	21-Jan-85	89.0	77.2	65.4	39	4036	187
Warrenton 3 SE	Fauquier	500	-11	16-Jan-72	84.2	75.0	65.8	24	3502	193
Piedmont Research Stn	Orange	520	-11	5-Feb-96	86.2	76.1	65.9	24	3636	198
Winchester 7 SE	Frederick	680	-18	19-Jan-94	87.1	74.6	62.0	24	3249	167
Charlottesville 2 W	Albemarle	870	-10	19-Jan-94	88.0	76.9	65.8	27	4035	211
Appomattox	Appomattox	910	-14	19-Jan-94	87	75.9	64.7	25	3666	193
Bedford	Bedford	975	-10	21-Jan-85	86.2	75.7	65.1	17	3657	203
Philpott Dam 2	Henry	1123	-10	21-Jan-85	86.8	76.4	65.9	26	3804	190
Lexington	Rockbridge	1125	-12	21-Jan-85	87.1	74.7	62.3	22	3355	168
Roanoke ROA (Airport)	Roanoke	1149	-11	21-Jan-85	87.5	76.2	64.9	28	3870	192
Rocky Mount	Franklin	1315	-11	21-Jan-85	86.5	75.5	64.4	23	3600	180
Stuart	Patrick	1375	-13	21-Jan-85	85.4	74.7	63.9	17	3601	191
Luray 5 E	Page	1400	-14	5-Feb-96	87.9	74.1	60.2	23	3468	157
Mount Weather	Loudoun	1720	-15	18-Jan-82	79.9	72.1	64.2	4	2846	188
Abingdon 3 S	Washington	1920	-21	21-Jan-85	85.0	72.8	60.5	13	3174	163
Wytheville 1 S	Wythe	2450	-20	21-Jan-85	83.4	70.8	58.1	8	2670	146
Wise 3 E	Wise	2548	-24	21-Jan-85	82.1	71.4	60.6	2	3044	171
Floyd 2 NE	Floyd	2624	-19	21-Jan-85	80.9	69.5	58.1	3	2421	141
Big Meadows	Madison	3539	-29	20-Jan-94	75.4	66.1	56.7	0	1888	143

Source: NOAA, Climatography of the United States No. 81: *1971–2000 (http://www.ncdc.noaa.gov/oa/climate/climatedata.html#CLIMATOLOGY).*

Note: The data are restricted to Virginia here to illustrate differences within a state. Similar data are readily available for other states.

[a] asl – above sea level.

[b] Growing Degree Days, based on 50°F.

[c] Frost-free Period, or growing season length, based on temperatures above 32°F.

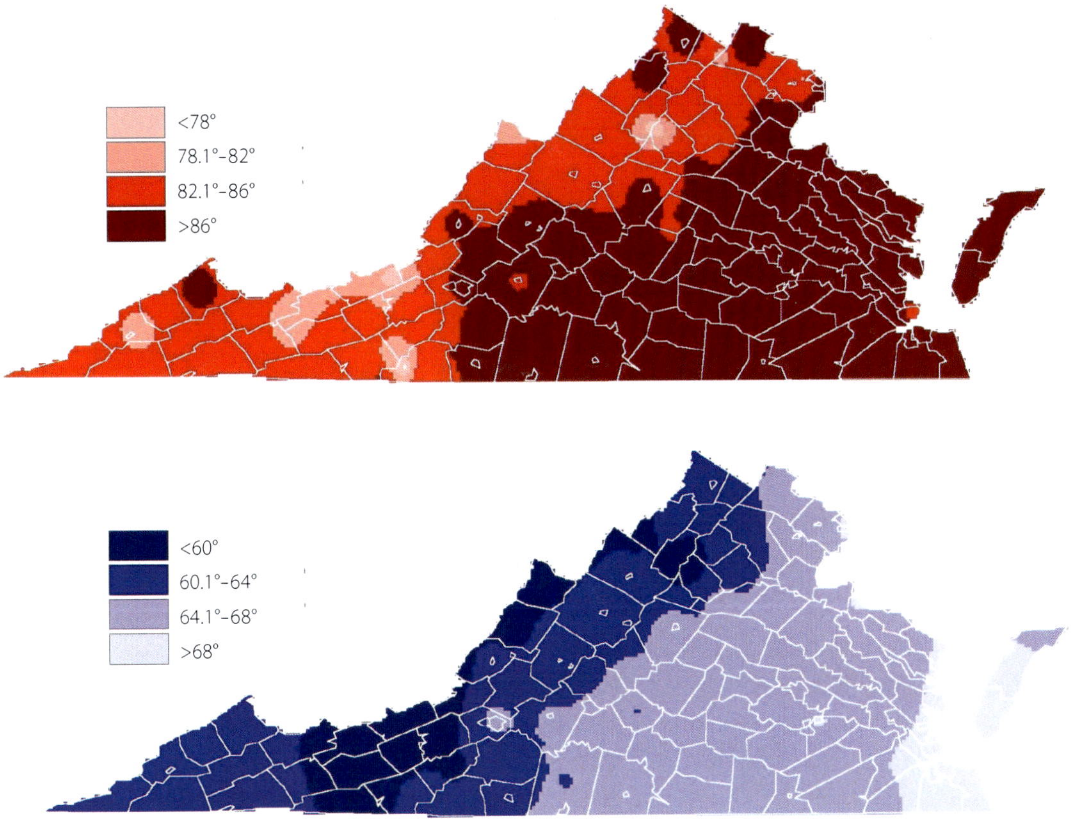

Figure 2.7 ▪ Average daily maximum *(upper)* and minimum *(lower)* July temperatures throughout Virginia.

experience winter injury. The compromise is reached by situating vineyards at increased elevation (see following discussion on Mesoscale Site Selection: Elevation and Topography) that also afford excellent cold air movement out of the vineyard.

Precipitation

The amount of water that grapevines require varies with the age of the vine, the amount of fruit produced, the presence of competitive weeds, and humidity (see chapter 9, page 169). Mild drought stress prior to veraison (onset of fruit ripening) can be of benefit in slowing vegetative growth and advancing grape ripening. More severe drought stress, however, leads to reduced carbohydrate (including grape sugars) production, poor fruit and wine quality, reduced vine vigor, and diminished yields. Excessive precipitation after veraison can significantly increase the splitting and subsequent rotting of grapes. When grapes are ripening, excess rainfall will lead to decreased sugar, flavor, and aroma in the grapes and will ultimately reduce wine quality. The vineyardist has several measures to counter the negative effects of excess precipitation. The most effective measure is to choose to grow varieties that are relatively insensitive to the ill effects of late-season rains (see chapter 3, page 37). Varieties that are prone to rain-induced splitting and rotting are not recommended in those areas that receive abundant late-summer and fall precipitation. Proposed vineyard sites should also be evaluated for surface and internal soil water drainage to facilitate removal of excess moisture from the root zone (see discussion of soils to follow). Finally, precipitation patterns can be evaluated on the macroscale to determine the threat of excessive precipitation in the ripening months of August through October.

MESOSCALE SITE SELECTION: ELEVATION AND TOPOGRAPHY

State maps are helpful in defining the regional basis of vineyard site suitability, but these maps are too general to pinpoint the best site to locate a vineyard. The topography, including the absolute and relative elevations of a particular site, will greatly affect the suitability of a proposed site. The local or mesoscale conditions of a site, including the soils of the site, are the next parameters to evaluate.

Elevation has a profound influence on the minimum and maximum temperatures in a vineyard, particularly in hilly and mountainous terrain. Because frosts and freezing temperatures can so dramatically reduce vineyard profitability, elevation is one of the most—perhaps the most—important features of vineyard site suitability. The physics of topographic effects on air temperature are well documented and the horticultural significance generally well appreciated. Under radiational cooling conditions, with calm winds and clear skies, the earth loses heat to space and cools the adjacent layer of air. If the vineyard is on a slope, the cold, relatively dense air moves downhill (see figure 2.8). This movement can be pronounced in mountainous areas and may even produce local winds. The sinking cold air displaces warmer air to higher elevations that can produce thermal inversions and thermal belts. Above these relatively warm zones, air temperature again decreases at an average rate of 3.6°F per 1,000 foot increase in elevation. The sinking cold air collects in low-lying areas and creates frost pockets. Vineyards in low-lying frost pockets are much more prone to spring and fall frost damage, and winter cold injury, than are vineyards that have been established at higher elevations.

The higher elevations, particularly in the mountainous regions, also afford cooler daytime temperatures during the summer and fruit maturation period in the fall. But the cooler temperatures that are found at higher elevations during fruit maturation can easily become liabilities in winter if the vineyard is situated too high. Air temperature continues to cool with increased altitude above thermal inversions. Those temperatures can be lethal to grapevines, especially with advective freeze events. Advective freezes are characterized by windy conditions in which little or no temperature stratification occurs, except the general pattern of decreasing temperature with increasing altitude. Thus, in most mountain/valley complexes, there is an optimal elevation, sometimes referred to as a "thermal belt," that is most suited for vineyard development. The advantages of increased elevation diminish above this zone, and sites below the zone are subject to increased risk of radiational frosts.

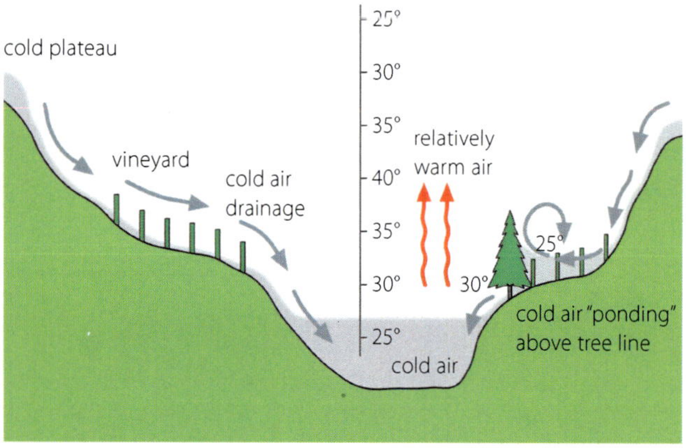

Figure 2.8 ▪ Illustration of site topography effects on air temperatures (°F) during a radiational cooling event.

The relative elevation considers the proposed site's elevation relative to surrounding terrain, and must be considered in tandem with absolute elevation. Good relative elevation cannot overcome the risks associated with inferior absolute elevation, but poor relative elevation can significantly reduce the quality of an otherwise good absolute elevation site. The latter situation has occurred with small valleys that are "perched" in mountainous areas. Even though these valleys may fall within the "best" absolute elevation range predicted for a local area, they are still "ponds" for cold air drainage and are thus subject to increased frequency of frost and winter injury.

To summarize, the candidate site's absolute and relative elevations are both important considerations in the site selection process. Optimal, absolute elevation zones exist in mountain/valley complexes, above and below which vineyards should not be established (see sidebar). Superior relative elevations must also be sought within these zones. For sites that do not have upper limits to optimal elevation, for example those in the mid-Atlantic region which do not exceed 1,500 feet above sea level, one should attempt to find vineyard sites that are in the upper 20% of the elevation range. An exception to this rule would be the desirability of lower elevations around large bodies of water where the stored heat of the water can be used to reduce the risk of low temperature injury to adjacent vineyards. Compromises must occasionally be made with vineyard site selection criteria; however, the elevation of the candidate site must not be compromised.

SIDEBAR

Where is the "Thermal Belt" Found in Hilly Terrain?

Thermal zones or "belts" are commonplace in mountain/valley topography, even in hilly terrain, and become apparent during radiational cooling events. Thomas Jefferson wrote about these topographic features in the vicinity of his home and gardens at Monticello (850 feet above sea level) in the Southwest Mountains east of the Blue Ridge in Virginia. Horticulturists have long recognized the discrete presence of thermal belts and have used them to advantage in sheltering tender plantings from radiational frost injury at lower elevations as well as advective freezes that are more common at higher elevations. Identifying the upper and lower boundaries of thermal belts can be approached in several ways. One approach involves using local experience to determine where tree fruit was or is currently grown, and is based on the premise that historically profitable tree fruit production sites are often found within thermal belts defined by local topography. Other approaches involve data collection from the area of interest. A simple illustration of the presence of a thermal belt on a hillside can be obtained on a cold, calm morning by walking from the bottom to the top of a prospective hillside planting with a handheld thermometer and a Global Positioning System receiver to record elevation. Record the air temperature 4 feet above ground at 50-foot elevation intervals as you walk up the hillside. A similar exercise can be performed in a vehicle that is equipped with an outside air temperature display as you drive up a mountain road. A more time-consuming approach involves establishing recording thermometers or thermographs from high to low elevation intervals on a slope or hillside of interest. Depending upon the total relief, this might require five or more recording stations spaced at 50- to 100-foot intervals. Low-cost dataloggers are available but must be properly sheltered and installed at points with an unobstructed view of the sky to obtain accurate air temperature readings. Recording daily air temperature data over several months, particularly in spring or fall, can be revealing as to relative elevation effects on temperature profiles on a mountainside or hillside.

Local experience and research were used to help define the upper and lower limits of thermal belts in a small mountain/valley region of Virginia as part of a project to develop a Geographical Information System for vineyard site evaluation. The lower limit of the thermal belt was estimated to lie at approximately one-quarter to one-third the elevation range in a local area (for example, a county). The elevation in the study area generally ranged from about 425 feet to 2,225 feet above sea level (asl); a range of 1800 feet. Our definition of the "best" thermal zone started at 875 feet asl (0.25 x 1,800)+425, and extended up to about 1500 feet asl. Experience suggested that above 1500 feet, the threat of advective freezes outweighed the elevation benefits to radiational freeze avoidance. The upper limit of the thermal belt was increased to approximately 2,200 feet asl in the southern portion of Virginia due to higher elevations. The definition of "best" vineyard elevation for Virginia counties that did not exceed 1,500 feet asl was an elevation in the upper 20% range of absolute elevation (for example, ≥ 520 feet asl in a county that ranged from 200 to 600 feet asl).

Slope

The slope is the inclination or declination that land varies from the horizontal, usually expressed as a percentage; a five-foot fall over a 100-foot horizontal distance would be a 5% slope. Flat land has a 0% slope, while a 100-foot rise over a 100-foot run has a slope of 100%. Slope can be accurately measured with an inexpensive handheld inclinometer. A slight to moderate slope is desirable because it accelerates drainage of cold air from the vineyard. As mentioned earlier, cold air is denser than warm air and, much like a fluid, will tend to flow downhill. Generally, the steeper the slope, the faster cold air will move downhill, assuming there are no barriers to air movement (see figure 2.8). The slope of the land is also important for surface and, to some extent, internal soil water drainage. As discussed later, surface and internal soil drainage are extremely important, and a slope is conducive to that movement. Slopes steeper than about 15%, however, are not recommended because equipment is hazardous to operate on steep slopes either because of risk of roll-over or the downhill drift of towed equipment into the vineyard row. Terracing slopes is possible but adds significantly to vineyard establishment and management costs. Soil cultivation and terraces on steep slopes may alter the integrity of soil horizons and also may increase the risk of soil erosion. Advice on soil erosion control measures can be obtained from local USDA Farm Services offices.

Aspect

The "aspect" of a slope refers to the prevailing compass direction that the slope faces (for example, east or southeast). Aspect will affect the angle at which sunlight hits the vineyard and, thus, the total heat gain of a vineyard. Aspect is probably most critical in the cooler regions of the East, particularly in the more northerly latitudes (New York, New England, and Ontario) where south- and southwest-facing aspects would receive more sunlight and heat than the corresponding northerly slopes. The extended daylight on southerly slopes could be of benefit with late-ripening varieties. Even in warm climate grape regions (much of Maryland, Virginia, and more southerly states), vineyards should be exposed to direct sunlight for at least a portion of the day, and eastern exposures are optimal. The early morning exposure advances the start of temperature- and light-dependent photosynthesis and results in more rapid drying (as from dew or rain) of foliage and fruit, potentially reducing disease problems. Eastern slopes also tend to be more sheltered from the hot afternoon sun, resulting in possible benefit to the retention of volatile aromatic compounds in fruit. On the other hand, for late-maturing varieties, such as Cabernet Sauvignon or Norton, there may be some advantage with southern and western exposures to promote fruit ripening in the waning heat and daylight of autumn.

Vineyards with southern and western aspects can warm earlier in the spring, and the vines may undergo budbreak earlier than vineyards with northern slopes. The earlier budbreak on southern and western aspects likely results from earlier warming of soils and air temperature.

In locations that do not have a danger of spring frost, early budbreak may be desirable because it translates into earlier bloom and harvest of the fruit. In frost-prone areas, early budbreak can increase the potential for frost damage in the spring. For example, growers have reported that the buds break and commence growth up to seven days earlier for a variety planted on a southern aspect than for the same variety planted on a nearby northern or an eastern aspect.

Aspect also has a slight, but measurable, effect on winter temperatures. In a long-term Georgia study, minimum temperatures on northerly slopes were 1.0°F to 2.5°F cooler than corresponding elevations of southerly slopes during freezes with temperature inversions. In the same study, the frost-free growing season was, on average, about two weeks longer on the slope with the southern aspect than on the corresponding slope with the northern aspect.

To summarize, there are both pros and cons to most vineyard aspects, as illustrated in table 2.3. While aspect may be quite important in areas that are marginally warm enough to ripen the intended variety, other factors such as elevation, current land use, and soil characteristics will typically rank as more important in the site selection process.

Land Use

Although current land use is not a direct indicator of vineyard site suitability, it is of prime importance to the feasibility and cost of establishing a vineyard. Land use ranges from the unusable urban areas and bodies of water to prime farmland. While not prime, and potentially quite rocky, some forest land can be cleared for vineyard use. Forests, however, often are forested because they are too steep, too rocky, or otherwise unsuited to cultivation.

Vineyard Suitability GIS Maps

To assist with site selection, some state agencies have generated maps that classify areas for vineyard suitability. Depending on resolution, these maps illustrate areas or regions that may have greater or lesser potential for commercial grape production. The maps are based on a Geographical Information System (GIS) in which the

Table 2.3 ▪ Relative effects of compass direction (aspect) of vineyard site on vine phenology and physical parameters

Parameter	Aspect			
	North	South	East	West
Time of budbreak	Retarded	Advanced	Retarded	Advanced
Daily maximum vine temperature	Less	Greater	Less	Greater
Speed of foliage drying in morning	-	-	Advanced	Retarded
Radiant heating of fruit	Less	Greater	Less	Greater
Radiant heating of vines in winter	Less	Greater	Less	Greater
Minimum winter air temperatures	Lower	Higher	-	-
Length of growing season	Shorter	Longer	-	-

individual themes of elevation, land use, slope, and aspect are combined into a single graphic representation that is scored for overall suitability. An example of such a GIS map is illustrated in figure 2.9 for Nelson County, Virginia. Check with state, university, or provincial sources to determine if viticultural suitability maps exist for your proposed vineyard property.

MESOSCALE SITE SELECTION: SOILS

Soil affects grapevine productivity and wine quality, but soil, like climate, comprises many components. Soil can be described in terms of depth, parent rock origin, organic matter content, texture, chemical properties, and hydrology and in terms of microbial and other invertebrate fauna density and diversity. All of these variables may ultimately affect vine growth and wine quality, but precise relationships are not well characterized for all such variables. Furthermore, the confounding influences of vineyard management, climate, varieties and clones, and fertilizer and irrigation practices, as well as variation in fruit harvest and winery practices, may easily obscure the more subtle, unique soil contributions to wine quality. For these reasons, and given our relatively brief experience with wine production, the ideal vineyard soil for eastern regions is imperfectly defined. Nevertheless, some properties are decidedly more important than others by virtue of their known influence on vine performance or because some are more easily improved than others (see

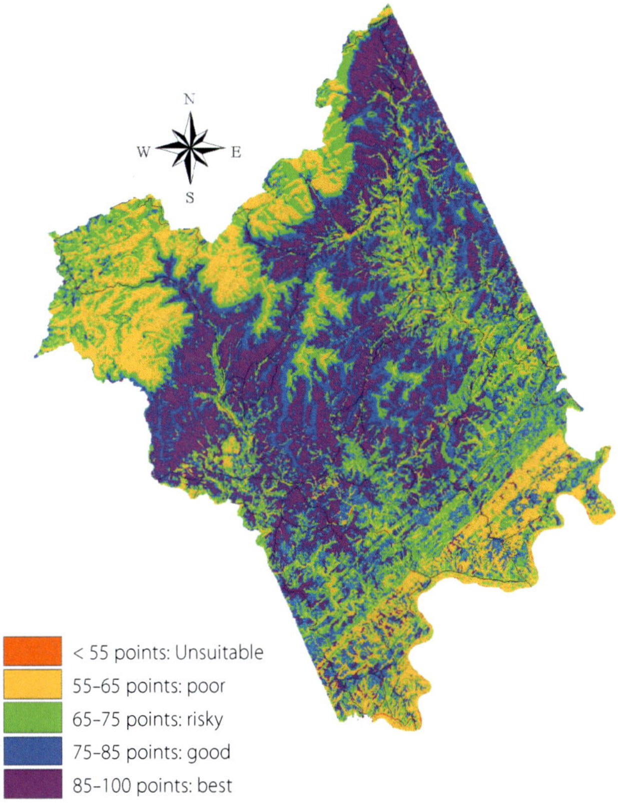

Figure 2.9 ▪ Geographical Information System (GIS) approach to mapping of vineyard suitability.

This "composite rating" is a synthesis of elevation, slope, land use, and other features which, collectively, help define the suitability for grape production within Nelson County, Virginia.

Table 2.4 ▪ Soil features and their importance in vineyard site selection

Soil feature[a]	Importance in site selection[b]	Desirable value	Undesirable value	Ability to modify[c]
Internal water drainage	*****	> 2 in. / hour	< 2 in. / hour	+ (tile drainage is possible but expensive)
Water holding capacity	****	< 0.10 in./ in. of soil (?)	> 0.15 in./ in. of soil (?)	++ (can be increased)
Fertility	****	Low to moderate	Highly fertile	+++ (can be increased)
Effective rooting depth	***	> 3 ft.	< 1 ft. in the absence of irrigation	-- (deep ripping may increase rooting depth)
Moist bulk density	***	< 1.5 g/cm³	≥ 1.5 g/cm³	-- (can be decreased)
Texture (relative proportion of sand, silt and clay)	***	Loam, sandy loam, sandy clay loam	High proportion of silt (>50% silt)	---
Soil pH	***	6.0 - 6.8	< 5.0	+++ (can be adjusted)
Organic matter	**	1.0 - 3.0%	> 5.0%	+++ (can be increased)
Soil organisms	**	Variable	?	+++ (can be increased)
Parent material	*	Granite, sandstone	See text	---
Surface composition	*	Uncertain	See text	---

[a] Features are listed in order of decreasing importance to grape productivity and fruit and wine quality.

[b] Relative importance with multiple asterisks indicating greater importance.

[c] Relative ease of adjustment, where +++ denotes readily adjusted and - indicates increasingly difficult or impossible (---) to practically adjust. (?) indicates a proposed or otherwise uncertain value.

table 2.4). Each of the criteria listed in table 2.4 should be evaluated in the site selection process. Note, however, that like above-ground features, few soils will be ideally suited for all criteria. Soils cannot be evaluated independently of the other vineyard site considerations discussed in this chapter, and some compromises in soil quality may be necessary to ensure that the vineyard site selection process does not become exclusive. The criteria of table 2.4 are discussed below in what we believe is a descending order of importance.

Sources of Soils Data

The principal published source of detailed soils data in the United States is the survey compiled by the Soil Conservation Service, US Department of Agriculture (USDA). This information is published by county and can usually be found at local Cooperative Extension offices, through the offices of the USDA Natural Resources Conservation Service (NRCS), or online at http://soils.usda.gov/. The soils data provide detailed maps to identify and describe each soil series and subclassification found in the county. Ideally, site specific information can be obtained by physically digging pits with a backhoe on the candidate property and evaluating the soil profile for color, depth, texture, bulk density, and degree of existing plant rooting. Interpretation of soil appearance and properties requires specific skills and knowledge. Growers who are interested in soil effects on ultimate wine quality are strongly encouraged to consult a professional soil scientist in this investigative process. Individuals and companies are often advertised in trade publications and might be listed through state or local Cooperative Extension and NRCS offices. Purchase of the candidate property can be made contingent upon an acceptable soil report.

Soil Moisture

The best vineyard soils are those that permit deep and spreading root growth and that provide a moderate supply of water, released incrementally over time. Soils to be avoided include those that are compacted and that severely restrict rooting, soils that are chronically or seasonally water-logged, and soils that are extremely droughty (in the absence of irrigation). Excess moisture leads to excessive vegetative growth, increased fruit acidity, increased risk of winter injury, and diluted fruit and wine flavors. Fruit rots are increased by berry splitting and by the more favorable disease conditions that exist within the dense canopies of overly vigorous vines. At the other extreme, drought stress can lead to insufficient vine growth, reduced yields, impaired fruit ripening, and sun burning of fruit. In the absence of supplemental irrigation, deep, well drained soils afford a reservoir of moisture that can be exploited by vines over an extended dry period. Considering the erratic nature of summer rainfall patterns in eastern North America, the grower should anticipate a surplus of summer rainfall but be prepared to supplement that precipitation with irrigation during droughts. Finding a soil that accommodates our irregular water supply entails identifying deep, well drained soils. In other humid regions (for example, Bordeaux), that goal has been achieved by establishing vineyards on alluvial deposits. Even clayey soils can produce world-class wines if these soils are well drained, as demonstrated in the Pomerol appellation of Bordeaux.

The topography of a site will have a bearing on the underlying soil hydrology. With undulating topography, the superior vineyard sites will typically be situated on convex land patterns—features that tend to shed surface water rather than collect it. Concave land forms—swales, ravines, or gullies—are usually areas of water import, and soils in these zones may be deeper due to erosion from higher ground. Thus, locating vineyards on convex land patterns is an excellent means of restricting water availability to the vines.

Soil moisture is affected by the rate of internal water drainage (permeability) and by the retention, or water-holding capacity of the soil. "Permeability" is a qualitative measure of water movement (for example, "rapid," "moderately rapid," and so on) and is typically estimated by NRCS from soil texture or other means. Hydraulic conductivity is a more specific measure of a soil's ability to transmit water under standard conditions and units. Hydraulic conductivity is assessed under saturated soil conditions and is called "Saturated Hydraulic Conductivity" or K_{sat}. K_{sat} and permeability values may be very similar for a given soil; however, K_{sat} values are considered more precise.

Permeability is perhaps the most important consideration in a candidate vineyard soil. "Drainage" refers to the speed with which free soil moisture drains from the soil profile. Deep soils with good porosity or pore spaces drain better than dense, compacted soils do, or those with textural discontinuities that limit water movement. Permeability rates are determined in part by soil texture. Sands, for example, typically drain faster than clayey or silty soils do; however, sandy soils have very low water-holding capacities and may lead to drought stress sooner than soils with higher proportions of clay or silt. Thus, we seek a compromise between soils that drain well and those that have reasonable water holding capacity. Texturally, loams, loamy sands, and sandy loams would generally fit this description (see discussion of soil texture).

A visual indication of the quality of internal soil drainage is soil color, a sometimes subtle feature that is best observed by digging pits and examining the face of the pit. Well drained soils allow deep infiltration of oxygen and the soils will appear uniformly brown, yellow, orange, or otherwise "bright" to some depth (4 feet or more). Poorly drained soils may reveal mottled shades of gray or blue and may have off-odors. Clues to the quality of soil drainage may also be obtained by examining the native vegetation on the proposed site. Well drained soils support well established grass sod, most agronomic crops, and mixed forest, whereas perennially wet soils often support sedges, willow, or sycamore, and may have poorly developed sod or sod that is spongy during wet weather.

Soil survey data present permeability data in inches of water movement per hour through a moist soil. A speed of two inches or more per hour is a good candidate soil (table 2.4). One useful rule of thumb is to reject soils that are not suitable for a conventional gravity-fed septic field.

Available water capacity (Plant Available Water, PAW) is defined as the difference between the amount of soil water at field capacity and the amount at the wilting point. Field capacity is the amount of moisture held by a soil, as by capillary action, after excess moisture has drained out of the soil profile by gravity. Soil moisture at or below the wilting point is held with such high tension that most plants cannot extract it from the soil. Available water capacity is commonly expressed as "inches per inch of soil" and is a function of soil texture and organic matter content. Sands and sandy soils hold relatively little moisture, whereas clays and silts contain a large reservoir of PAW. Typical PAW values range from 0.10 to 0.20 inches of water per inch of soil profile. If one accepts the premise that "surplus" moisture is a greater problem in their location than is moisture "deficit," one would seek a soil that has a PAW on the lower end of this range. Greater water-holding capacity would be desirable, however, in the absence of irrigation. Soils that have high proportions of silt (greater than 40%) and clay (greater than 35%) often impede water drainage; examples include those created by glacial lake deposits on the Niagara Peninsula and central and eastern sub-districts of the Lake Erie coast. Sites that have historically been used for agriculture may have localized drainage problems that are related to increased soil compaction from passage of farm equipment. Drain tiles are often installed in a site or section of a site where PAW is higher than desired due to these factors. Drain tiles restore soil aeration by aiding in removal of excess or "perched" water and decrease the length of time in which soil moisture reaches field capacity. Earlier spring drainage and after-storm "seasonal" drainage achieved by tiling decreases the amount of time a grower must wait to re-enter with farm equipment. To determine the feasibility and efficacy of drain tile installation, drainage design should be undertaken by a professional drainage contractor or engineer.

Soil Depth

Soil depth is important for providing a buffer against drought. A deep soil (for example, more than 3 feet) offers a greater volume of potential soil moisture than does a shallow soil (for example, less than 12 inches). Grapevines can be grown on shallow soils; however, vines on such soils will be the first to suffer drought stress if supplemental irrigation is not available. Deeper soils also allow grapevines to develop a large perennial root structure, which in turn, fosters a large productive above-ground framework.

Bulk Density

Bulk density is the mass, or weight, of dry soil per unit of bulk volume. It is expressed in g/cm^3 or kg/m^3. In practical terms, bulk density is a measure of the compactness of soil. Naturally or mechanically compacted soils interfere with internal water drainage and may restrict root growth. Bulk density values of about 1.6 g/cm^3 or more are restrictive to root growth of most plant species, including grape. Suitable values would be less than 1.5 g/cm^3 (see table 2.4, page 28). Bulk density data are typically included in the soil surveys; however, the procedures for determining bulk density are not complicated and can be found in agronomy handbooks.

Soil Fertility

Fertility should be evaluated in the site selection process, but this criterion is less important (see table 2.4) because fertility can be modified. On a relative scale, low to moderately fertile soils are superior to highly fertile soils for producing high quality grapes and wine. Evaluation consists of the collection of soil samples, analytical testing by a state or commercial service, and interpretation of results. Soil testing reveals the availability of plant essential nutrients, soil pH, cation exchange capacity (CEC), and the percentage of organic matter. Good vineyard soils provide an adequate and balanced supply of macro- and micro-nutrients have moderate CEC values, and have a pH within the optimal range for the intended species of grape (see chapter 8, page 141). Most of these features can be modified by soil amendments. Exceptions are CEC and the rare situation in which one or more nutrients may be at supra-optimal levels. For example, soils with high organic matter content (greater than 5%), may release excessive amounts of nitrogen that could cause excessive vegetative growth of vines.

Organic Matter

Organic matter contributes porosity, structure, nutrients, and moisture-holding capacity to soil and also aids in supporting a diverse microbial and invertebrate animal (for example, earthworm) ecology. Organic matter provides a pool of slowly available nitrogen and other nutrients to support vine growth. Mineral soils—those which by definition contain less than 20% organic matter—typically range from less than 1% up to 5% organic matter in eastern vineyards. Organic matter values greater than 5% may be counter-productive in that excessive nitrogen released by organic matter decomposition may lead to supra-optimal vine growth. Organic matter values of 3% to 5% may also lead to surplus growth, but if all other properties are acceptable, the vineyardist might simply choose to accommodate the expected greater growth by decreasing vine density or by using more elaborate training systems. Soils that have been exploited by deficit farming or that are inherently low in organic matter can be amended profitably with compost, green manures, or other forms of organic matter and, therefore, should not be rejected as vineyard soils.

Soil Texture

Texture refers to the relative proportions of sand, silt, and clay—the essential triad of soil textural classification (see figure 2.10). While there are interesting ideas on how texture affects wine quality, the direct effects are poorly defined. Indirect effects on soil hydrology are probably more important than more subtle direct effects. For example, sands and soils with a qualifying sandy prefix (for example, sandy loam) will have better drainage characteristics and a lower PAW capacity than clays or silty loams.

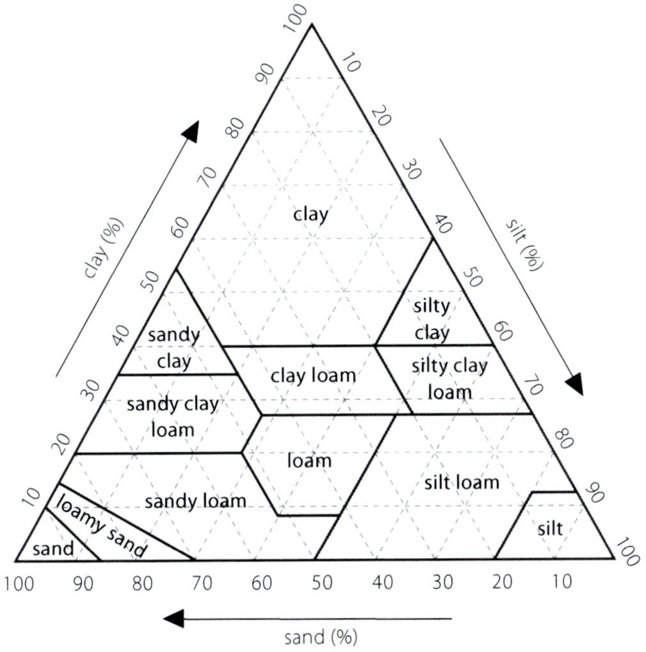

Figure 2.10 ▪ Soil triangle used to describe soil texture.
Soil that consists of 50% sand, and 25% clay would be described as "sandy clay loam."

Soil Biology

"Healthy" soils contain a rich diversity of plant and animal species, most of which are inconspicuous but some of which (for example, earthworms) are easily observed. While some soil animals and fungi can cause disease in grapevines, the vast majority of soil fauna and flora is essential to nutrient recycling and mineralization of organic matter. The maintenance of this living aspect of soil is essential to the maintenance of a healthy vineyard. Unfortunately, many of our farming practices (for example, tillage, use of pesticides, crop monoculture, and soil compaction by machinery) tend to reduce diversity. Commercial labs can evaluate soil microbial diversity and provide an interpretation and action plan to increase diversity if the soil bioassay is low for a particular soil organism functional group (for example, fungi versus bacteria). Unfortunately, the interpretation of soil biological properties is an emerging science and is not sufficiently advanced to make cultural recommendations for potential and existing vineyards.

Soil Origin

The principal residual soils of eastern North America are derived from granite, limestone, sandstone, and shale. In many cases, the soil will consist of some mix of these four types. The parent rock from which soil is derived may indirectly affect grape production through availability of nutrients or by affecting soil hydrology. Soils derived from granite, for example, are considered superior to soils derived from greenstone (that is, Catoctin) by virtue of having less silt, a coarser structure, and, therefore, a lower PAW capacity. Limestone soils, if deep (4 feet or more), can lead to very vigorous vine growth. On slopes (classified as "C" or "D" series in soil survey maps), however, the limestone soils may be very suitable for vineyards. All other factors equal, the soil parent material has no measurable direct effect on grape and wine quality in the eastern region (table 2.4).

Surface Characteristics

In some grape regions, the surface composition of soil directly impinges on grape and wine quality. A stony surface can be important for absorbing and redirecting heat to the vines during the cooler nights of autumn. This feature might be worth considering in cooler sites, but the materials will hinder the operation of rotary hoes and other under-trellis cultivation equipment. Soils that have not been recently cultivated, such as pasture and wooded areas, will impose greater resistance to plowing and site preparation than will recently cultivated land. Some time may be necessary to remove large rocks and roots from a new vineyard site and to establish a cover crop prior to planting grapevines.

Nematodes and Other Soil-Borne Pests

Sites that are in woods or that were previously planted to other fruit crops, including grapes, should be evaluated for the presence of nematodes and other soil-borne grape pests and pathogens. Nematodes are small, wormlike parasites and several genera, notably *Xiphinema* spp., can transmit destructive viruses to grapevines. Soil sampling for nematodes and sample submission instructions can be obtained through a local cooperative extension office. A passive control option involves planting and maintaining non-host plants, such as perennial grass, to the site for up to several years before grapevines are planted to depress the nematode populations. A more active approach involves planting a series of green manure crops, including a brassica (rapeseed, *Brassica napus* 'Dwarf Essex'), that releases a chemical (glucosinolate) that is toxic to nematodes when the crop is incorporated into the soil (see chapter 4, page 71). Soil fumigation or treatment with nematicides is a third alternative, but one that carries potential risks, both to the user and to the overall soil biology. Fumigation is, therefore, not recommended.

Other potential soil-borne pathogens include the oak root fungus (*Armillaria mellea*), which has the potential to infect grapevines in areas where affected oaks, peaches, or other hardwoods previously grew. Deep ripping of the soil followed by scrupulous removal of affected roots from these sites is essential for reducing inoculum levels. Armillaria root rot is not well documented in eastern North America at this time.

POTENTIAL VINEYARD PESTS AND OTHER THREATS

The focus of this chapter has been on the physical and climatological requirements of vineyards. Beyond those factors, consideration must also be given to biological and abiotic threats that can affect grapes. These include diseases, certain insects, nematodes and vertebrate pests, and specific atmospheric pollutants. Vineyard operators must also be sensitive to neighbors who might be concerned about commercial vineyard operations, especially pesticide spraying.

Pierce's Disease

Pierce's disease (PD) is a destructive bacterial disease that affects all bunch grapes in the warmer regions of the southeastern United States, reaching as far north as southern Virginia and westward through parts of Texas and into California. Pierce's disease is caused by a bacterium, *Xylella fastidiosa*, which is transmitted from vine to vine and from alternative host plants to vines principally by leafhoppers, which are small, winged insects. In disease-susceptible vines, the water conducting tissues of the grapevine are blocked either by the bacteria or by defensive gums produced by the vine. This blockage leads to the characteristic disease symptoms and, in advanced cases, death of the vine (see chapter 11, page 216).

Cold winter temperatures limit the occurrence of PD. Areas that have an average minimum January temperature of 30°F or less, are thought to be less at risk of PD than areas with higher winter temperatures. If we look at the historical 30-year January average minimum isotherm (see figure 2.11), the documented cases of PD in the Virginia/North Carolina zone fall to the south (warmer) side of this line, while areas to the north and west of the isotherm have remained apparently free of the disease. Note that the 30°F isotherm of figure 2.11 is based on a 30-year record. If the 30°F isotherm is redrawn with data from only the 1997 through 2001 winters, the high-risk zone moves significantly farther north—a reflection of the warmer winters during that period.

If the 30°F isotherm is relevant to the advance of PD, commercial grape production based on non-Muscadine bunch grapes would be risky in much of southeastern Virginia and eastern North Carolina.

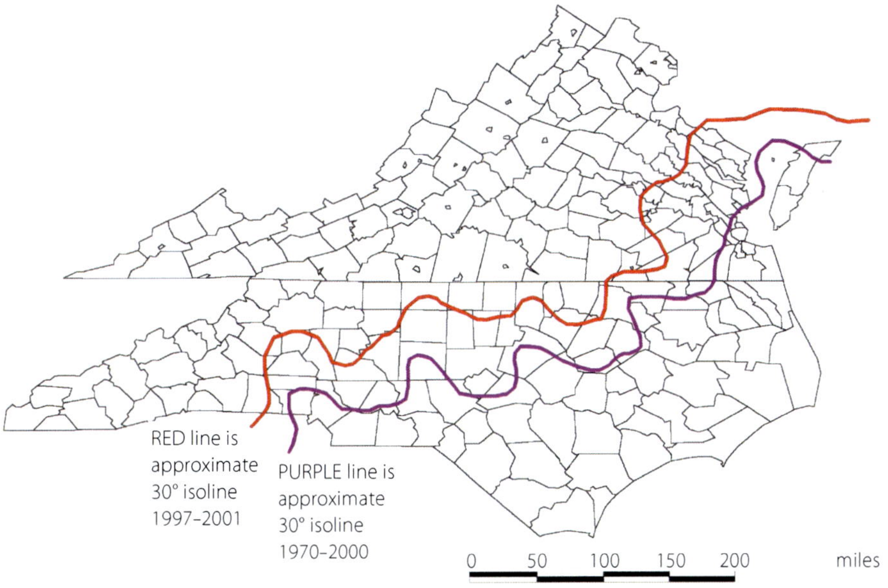

Figure 2.11 ▪ **Illustration of Pierce's disease risk in Virginia and North Carolina.**

Average January minimum of 30°F isotherm based on 30-year (1970–2000) record (purple line) and on a more recent 5-year period (1997–2001, red line). Risk of Pierce's disease is greater for areas south (below) of the 30° isolines.

North American Grapevine Yellows

North American Grapevine Yellows (NAGY) is another destructive disease of grapes whose incidence varies within the region. Affected vines typically die within two or three years of symptom onset, and some Chardonnay vineyards have experienced annual losses of 5% or more of the original planting, with attrition rates approaching 30% over a 5- to 10-year period (see chapter 11, page 216). NAGY has been observed in several eastern states, but the most severely affected vineyards have been in Virginia. The disease is caused by phytoplasmas, single-celled organisms similar to bacteria but lacking rigid cell walls. Leafhoppers and possibly some related insects are thought to transmit phytoplasmas from wild grapevines or other hosts into the vineyard, and possibly from vine-to-vine within the vineyard. To date, wild grapevines (*Vitis cordifolia* and *V. riparia*) have been the most consistent alternative hosts detected, and an abundance of these species appears to increase the frequency with which cultivated vines are diseased. Risk assessment and action for avoiding NAGY are not well defined at this point. The occurrence of the disease appears to be greatest in Chardonnay vineyards of Virginia and, to a lesser extent, Maryland, southeast Pennsylvania, and the Finger Lakes of New York. The disease has been observed throughout the established growing regions of Virginia, but is most frequent in the western piedmont and Shenandoah Valley. Consult grape experts or established growers in your area to determine if NAGY might pose an economic threat to your proposed vineyard.

Deer, Raccoons, and Opossums

Vineyard site selection should also consider the proposed site's proximity to other threats and pests that may adversely affect the vineyard. Among the vertebrate pests, whitetail deer, raccoons, opossums, and various species of birds are of primary concern (see chapter 14, page 272). Deer browse the shoots and ripening fruit. The injury to shoots is particularly troublesome in young vineyards where the shoots are being trained as trunks. The loss of fruit can also be significant, particularly where the resident deer population is high. Deer depredation is greatest in remote vineyards located near woods or other cover; however, deer damage can occur in any vineyard site. Electric fencing, odor-active repellents, and permit shooting all offer a measure of hope for avoiding deer depredation (see chapter 14, page 272). Experience, however, suggests that a 10- to 12-foot high, woven wire fencing is the only effective means of eliminating deer depredation under high-pressure situations. Fencing, if properly constructed, offers the added benefit of excluding small animals, such as raccoons and opossums, which can also cause significant fruit loss. The site selection process should consider the potential for deer damage based on local experts and should also consider how fencing and fence setbacks from adjacent cover will affect usable vineyard space.

Birds

Vineyard sites that are located near roosting areas (trees and overhead power lines, for example) and those that are visited by flocking migratory birds can suffer significant crop loss. Vineyards located near wild turkey habitat can also experience significant fruit loss to these protected animals. Certain viticultural areas in the East are at higher risk for bird damage if control measures are not implemented. Almost all vineyards on Long Island apply bird netting prior to harvest. Some growers have had significant success in netting only the two sides of vertically shoot-positioned canopies. This minimizes problems with shoots growing through the netting and allows sprayer and mower operations to continue after netting. Audible and visible scare devices are commercially available and include recorded distress call broadcasters, propane cannon, plastic, reflective ribbon, and various balloons and other props that mimic predators. Each of these devices appears to offer some measure of protection, but none is entirely effective on its own. Bird netting is expensive and cumbersome to apply and remove but does offer near-complete protection (see chapter 14, page 272).

Black Walnuts and Butternuts

The roots of black walnut (*Juglans nigra*) and butternut (*Juglans cinerea*) trees produce an allelopathic compound called juglone that inhibits the growth of certain plants including grapevines. Grapevines can be killed when they absorb juglone. Vines affected by juglone will have weak growth and wilting, pale or yellowed leaves. Their occurrence in the vineyard is superimposed on the radiating root system of the offending trees (see figure 2.12). Research has not clearly shown how juglone causes the inhibition of growth and/or plant death, but the alteration or inhibition of oxygen uptake and photosynthesis is suspected. Exercise caution when designing your vineyard around these trees. We recommend removing trees near the proposed vineyard by a margin of twice the height of the tree. For example, if the trees are 60 feet high, keep the vines at least 120 feet from the base of those trees. If you intend to remove the tree, wait several years before planting vines in the area of the root system. The roots will continue to release juglone until they completely decompose.

a. Black walnut tree and affected vines *(arrows)*. Vines in foreground were planted where black walnut trees had grown previously.

b. Close-up of juglone-affected vine.

Figure 2.12 ▪ Symptoms of juglone toxicity in Chardonnay vines.

Neighboring Properties and Pesticide Drift

Consideration must also be given to your immediate neighbors in the site selection process. Equipment such as air-blast sprayers and bird-scare cannons are noisy and can cause a disturbance. Neighbors also are concerned about pesticide drift from vineyards onto their property. One of the unpleasant features of our industry growth is the increased incidence of friction that occurs when non-farming neighbors come into conflict with vineyards over issues of spray drift. To the uninformed, the material being sprayed is suspect at least, and may be perceived as acutely toxic to someone who is acting more on emotion than on factual knowledge. Citizen concerns most typically relate to pesticides, but other "hazards" include soil erosion and use of fertilizers. Even the coming and going of vineyard labor has been cited as a "nuisance" associated with living next to vineyards. The general public is not fond of pesticides and individual concerns may be founded in fear or bias rather than fact. It is advisable to discuss the prospective vineyard enterprise with neighbors and to be forthright about vineyard production practices, especially the need for protective sprays.

Drift or movement of pesticides off-target to sensitive areas is a prescription for heightened antagonism against vineyards and must be avoided for both legal and public relations reasons (see chapter 10, page 196). Particularly sensitive areas are houses, playgrounds, animal confinement areas, ponds, riparian areas or wetlands, and roads and parking areas. Avoid planting right up to neighboring property. A safe setback is probably on the order of 50 to 300 feet, but the distance requirement will be a function of the average spray particle size, the wind speed during spraying, and other particulars of the spray conditions. Augmenting the buffer with a vegetative wind screen is an effective means of further reducing drift and also provides some visual-screening of vineyard operations. Fast-growing evergreens such as Leyland Cypress (*Cupressus* x *leylandii*) constitute one choice for this purpose; however, check with local extension agents or horticulturists to be certain that the selected buffer trees do not have peculiar pest or stress problems of their own. Confirm the terminal growth height of the trees/shrubs used as vegetative barriers and plan enough of a setback with vineyard rows so that vines are not competing for moisture and sunlight. Consider buffers, such as streams and ponds, between the vineyard and surface water, or where topographic site features might concentrate runoff from the vineyard and discharge it into streams or ponds. Some pesticides are highly toxic to aquatic animals, even if they have low mammalian toxicity. Grassed buffer strips or other vegetation will minimize the likelihood of pesticides being washed out of the vineyard through surface runoff.

Figure 2.13 ▪ 2,4-D herbicide injury to grapevines.

Conversely, certain herbicides may severely damage the vineyard if they drift in from neighboring property. In particular, 2,4-D, dicamba, and other phenoxy-type herbicides are often used for broadleaf weed control in pastures and no-till corn planting. In its more volatile form, 2,4-D can drift some distance and cause severe damage to the vineyard (see figure 2.13). Although long-range drift is possible, the damage observed in most vineyards has typically resulted from application to fields immediately upwind of the vineyard. Significant vineyard injury has also occurred due to use of these materials along highways, railroad right-of-ways, and golf courses located near the affected vineyard. It is important to know the herbicide use patterns of your immediate neighbors and to alert them to your vineyard enterprise. Applicators are ultimately responsible for ensuring that off-target drift does not occur, but this has not prevented significant damage from occurring in some vineyards.

Atmospheric Pollutants

Grapevines are sensitive to hydrogen fluoride (HF) (see figure 2.14), a colorless, odorless gas that is released by some industrial processes, notably the heating of fluoride-containing soils in brick manufacture. Vineyards should not be located within 2 miles of brick kilns, ceramic producing facilities, and metal smelting facili-

Figure 2.14 ▪ Hydrogen fluoride injury on Syrah leaves.

ties, unless these facilities are equipped with HF abatement devices. Other atmospheric pollutants that may affect grapevines include sulfur dioxide and ozone; the latter produces a dark interveinal stippling on the upper surface of leaves termed oxidant stipple. Both gases are produced directly or indirectly by fossil fuel consumption, notably transportation and electricity generation. Unlike HF, these pollutants are typically produced by non-point sources, and, thus, site selection is not effective as a means of avoiding adverse effects.

CONCLUSIONS

Eastern North America exhibits a varied climate and wide diversity of topography, wildlife and biological pests, and several anthropogenic pollutants, all of which present a challenging environment for commercial grape production. The aim of this chapter is to outline the nature of vine and vineyard threats so that the vineyardist can minimize, if not totally eliminate, risk. The quest for an "ideal" vineyard site may take years and will ultimately involve compromises. Certain features, such as elevation, soil drainage (internal and surface), and length of growing season must never be compromised. Others, such as the vineyard's aspect, disease pressure, or some soil features, can be accepted as less than ideal, because the choice of variety, soil amendments, and other inputs can be incorporated to modify or ameliorate these features. As stated early in this chapter, few, if any, sites will be ideal in all respects.

References

Gladstones, J. 1992. *Viticulture and Environment.* Winetitles, Adelaide 310 p.

Gladstones, J. 2000. Past and future climatic indices for viticulture. *Proceedings of the 5th International Symposium for Cool Climate Viticulture and Oenology*, Melbourne, Australia.

NOAA, 2002. National Oceanic and Atmospheric Administration. *Climatography of the United States No. 81*, Monthly station normals of temperature, precipitation, and heating and cooling degree days, 1971–2000.

Further Reading

Happ, E. 2000. Site and varietal choices for full flavour outcomes in a warm continent. *The Australian and New Zealand Wine Industry Journal*, 15:54–62.

Wolf, Tony K. and John D. Boyer. 2001. Site selection and other vine management principles and practices to minimize the threat of cold injury, pp. 49–59 In: Rantz, J. (ed.) Proceedings of the American Society for Enology and Viticulture. 50th Ann. Meeting, Seattle 19–23 June 2000, ASEV, Davis, CA.

Zabadal, T. J., I. E. Dami, M. C. Goffinet, T.E. Martinson, M.L. Chien. 2007. Winter injury to grapevines and methods of protection. *Michigan State University Extension Bulletin E2930*. 106 p.

—3—

Wine Grape and Rootstock Varieties

Authors
Tony K. Wolf, Virginia Tech

This chapter describes many of the commonly grown wine grapes and their rootstocks in eastern North America. As a preface to the discussion, a natural question would be why some varieties are included and others are not? Certainly the list of grape varieties is vastly more extensive than that provided here. The varieties that are listed are, or have been, grown under a diverse range of conditions found in eastern North America. Some may not have strong commercial importance today for planting but are still widely recognized in state and regional publications as benchmarks for disease susceptibility or for historical reasons. Examples include Chancellor and Rougeon. Others may have limited or even negative commercial experience because of cold-tenderness, fruit-rot susceptibility, or other reasons, even though the wine quality can be excellent when the variety is grown under the climatic conditions of the region it is historically associated with. Examples include Zinfandel and Sangiovese. The relative weaknesses of these varieties need to be pointed out just as the attributes of other varieties are illustrated. In the end, there will always be other varieties that could be added, as experience and research broadens our knowledge and as breeding programs release new material for evaluation and commercial adoption. For example, three new varieties were released by the grape breeding program at the New York State Agricultural Experiment Station alone in 2006. Readers should consult their state resources to keep abreast of variety developments and for local recommendations.

No effort is made to provide specific planting recommendations because many factors determine the suitability of a particular variety to a particular location, not the least of which is grower/vintner preference. The goal is to provide a brief overview of the strengths and weaknesses of each variety as compiled from many sources, including controlled evaluations in research plantings. Because this book is written for a large and geographically diverse region, we encourage readers to consult state or local recommendations—including the market demand for a particular variety—before committing.

Species, Cultivars, Varieties, and Clones

Grapevines are members of the genus *Vitis*, which includes two subgenera, *Euvitis* and *Muscadinia*. *Euvitis* subgenus is the bunch grapes; *Muscadinia*, the muscadine grapes. Muscadine grape varieties (for example, Scuppernong), which are cultivated in the more southern areas of the eastern US, are not described in this text.

Most wine grape varieties belong to the species *Vitis vinifera* (abbreviated *V. vinifera*), which originated in middle Asia, and are called "European grapes" or simply "vinifera." There are more than twenty species of grapes native to North America, but most of these are unsuitable for commercial wine production. Notable exceptions include Norton (*V. aestivalis*) and Concord (*V. Labruscana*). Grape breeding programs have, however, used North American grape species to develop pest-resistant rootstocks and pest-resistant and cold-hardy hybrid grapes. Vidal blanc, Chambourcin, and Vignoles are examples of interspecific hybrid grapes.

The descriptions of grape varieties in this chapter are divided into five classes:

- Nonfruiting rootstocks used for grafting to desired scions
- Vinifera used primarily for red and rosé wines
- Vinifera used for white wines
- Hybrid and American varieties used primarily for red or rosé wines
- Hybrid and American varieties used for white wines

We have used the term "variety" to designate a named grape selection, such as Cabernet Sauvignon or Vidal. Varieties are also called "cultivars" (cultivated variety) in more formal circumstances. In addition, a single variety (for example, Pinot noir) may have many different clones, which may differ from each other in berry size, color, or flavor or in cluster compactness, vegetative characteristics, or other properties. Each clone in this case, however, is still a Pinot noir. Clones have typically been selected from a single plant (called a "mother vine") that expressed the unique, desired property. The factors that can contribute to clonal variation are numerous but have occasionally involved genetic mutations or virus infections. The unique vine can be vegetatively propagated by taking cuttings, and the group of resulting plants can be given a clonal number or name. A nonexhaustive listing of clones available through state and commercial sources is provided in table 3.3 (page 66).

Before planting, it is important to know as much as possible about the land, grape varieties and clones, rootstocks, and the intended markets. Although more attention is being given to matching certain varieties and clones to specific climate and soil conditions, research and experience with many varieties in eastern North America is extremely limited. We have included information from other regions of the US and Europe with the caveat that growers who choose to plant "alternative" or niche-market varieties do so with some degree of additional risk.

How a given variety, clone, or rootstock will perform will vary from one vineyard to another or even within a vineyard. Local experience with a given variety, clone, or rootstock should be sought before planting. For larger plantings (for example, greater than 5 acres), we suggest that two or more clones of a given variety be planted, but in separate blocks, to provide complexity to the resulting wines.

All the variety names and synonyms used in this book are those approved by the US Department of the Treasury's Alcohol and Tobacco Tax and Trade Bureau (TTB) (http://www.ttb.gov).

Pollination

All commonly planted fruiting grapevines in eastern North America, except some muscadines, are self-fertile and, therefore, self-fruitful. They can, therefore, be planted in large contiguous blocks without the need for cross-pollinating varieties. Some muscadine grape varieties (*V. rotundifolia*) do, however, require a pollinator. Growers interested in muscadine grapes should determine in advance if a pollinator is necessary.

Rootstocks

All vinifera grapes and hybrid grapes with 50% or more vinifera in their parentage should be grafted to a rootstock that provides resistance to phylloxera and nematode-transmitted viruses, such as tomato ringspot virus (page 239). Grafting consists of joining the variety that will bear fruit (scion) to the variety that will provide the roots (rootstock). This process is performed by specialized nurseries (listings available from individual state resources) with finished vines sold after a year's growth in the nursery. Phylloxera (page 250) are very small, aphid-like insects, indigenous to eastern North America. Phylloxera feed on susceptible grapevine roots and can eventually weaken or kill the vine. Phylloxera also exist as an aerial form that creates galls on grapevine leaves and can occasionally cause commercial damage. North American species of *Vitis* or their hybrids that have resistance to phylloxera are used as rootstocks. Grafting of hybrid or American varieties to an appropriate rootstock can also increase vigor and improve yields, particularly if the soils are inherently low in nutrients or water-holding capacity. Besides pest resistance, rootstock selection should consider the physical and chemical properties and the water-holding capacity of the soil, growth characteristics of the scion variety, and the intended vine spacing, training system, and other cultural inputs.

Almost all rootstocks are crosses between three North American species: *V. riparia*, *V. rupestris*, and *V. berlandieri* (figure 3.1). Examples include the following:

Vitis riparia* × *V. rupestris (C-3309, C-3306, 101-14): *Vitis riparia* offers excellent phylloxera resistance and good adaptation to moist soils (table 3.1). Rootstocks derived from *V. riparia* × *V. rupestris* prefer fertile, deep, and moist soils. One of the most common rootstocks in Eastern vineyards is C-3309. Scion vigor is moderate., but may be low when vines are grown on dry, shallow, and heavy soils. Experiences in South Africa indicate good resistance to crown gall. Rootstock C-3309 is reportedly susceptible to nematodes, especially root-knot nematodes; however, this has not been extensively examined in eastern North America. Rootstock C-3309 may induce potassium deficiency with overcropped, young vines on clay soils. Another rootstock in this group is 101-14, which is considered less vigorous than C-3309. Situations in which 101-14 may be superior to C-3309 would include sites with deep, fertile soils or with very vigorous scion varieties (for example, Cabernet Sauvignon).

Vitis berlandieri* × *V. riparia (5BB, 5C, SO4, 420A): *Vitis berlandieri* is indigenous to the alkaline soils of the southwest US and is well adapted to limestone soils and drought. Generally, rootstocks from this group are more vigorous than are those from *V. riparia* × *V. rupestris*, especially under plentiful water. However, *V. berlandieri* × *V. riparia* rootstocks are more drought-tolerant than the *V. riparia* × *V. rupestris* rootstocks.

Rootstock 5C was extensively used in the 1980s but was mistakenly propagated and distributed under the name SO4, a distinct rootstock. Rootstock 5C is widely used in California, is well-suited to well-drained, fertile soils, and could be a good choice for heavy soils (clays and clay loams). But it does not perform well in dry soils. It has good resistance to root-knot and dagger nematodes. 5BB and SO4 are commonly available but they tend to produce larger, more vigorous vines than is desirable for

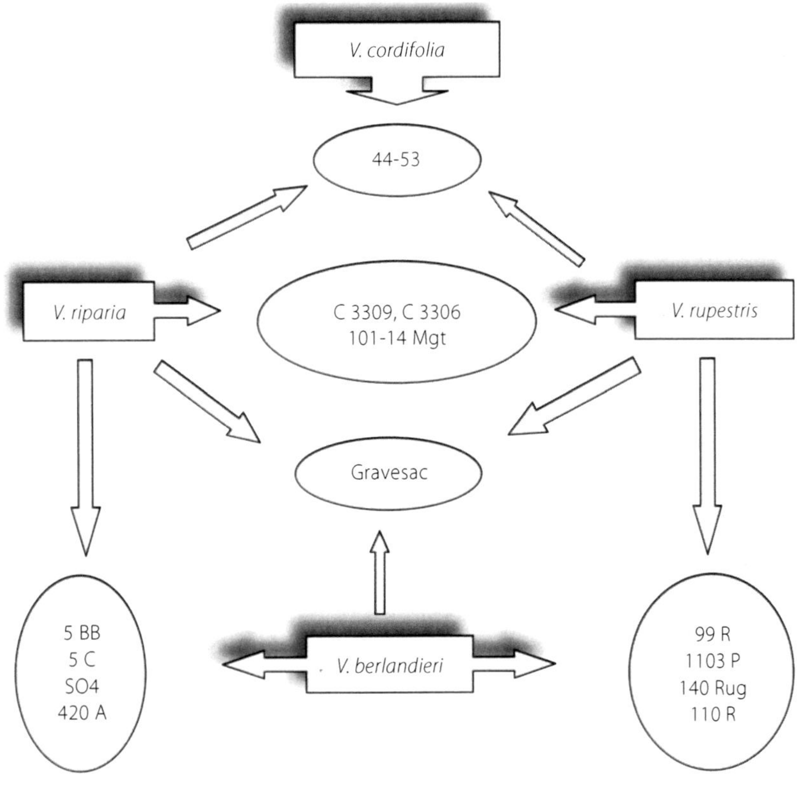

Figure 3-1 ▪ Schematic diagram of the parentage of principal rootstocks used for wine grapes. *Source: Wolf* et al. *(1999).*

Table 3-1 ▪ Characteristics and performance of major rootstocks reported from various viticultural regions

Rootstock	Vitis species parentage	Phyllx[a]	Nematodes[a] R-K	D	Crown gall[a]	Scion vigor[b]	Drought[a]	Soil alkal.[c]	Soil acidity[a]	Water logging[a]	Veg. maturity[d]
Gloire de Montpellier	Riparia	5	3	?[e]	?	1-2	1	1	2	4	+
St. George	Rupestris	4	1	1	?	4	3	3	2	2	-
C-3309	Riparia x Rupestris	4	2	4	4	3	2	2	1	4	+
C-3306		4	3	1	?	3	1	2	2	?	+
101-14		4	3	2	4	2–3	1	2	1	4	+
5 BB	Berlandieri x Riparia	4	3	2	4	4	2	4	2	4	+
5 C		4	4	4	?	4	2	3	?	4	+
SO4		4	4	2	2	4	2	3	1	4	+
420 A		4	2	2	?	1–2	2	4	?	3	+
99 R	Rupestris x Berlandieri	4	4	?	1	4	3	3	3	2	-
110 R		4	3	2	1	3	5	4	3	3	-
140 Rug		4	2	?	1	5	5	5	4	3	-
1103 P		4	4	2	2	3	4	4	3	4	-
44-53 M	Riparia x Rupestris x Cordifolia	4	2	3	?	3	2	2	2	4	+
Gravesac	Riparia x Berlandieri x Rupestris	5	2	?	?	3	3	3	3	?	?

Sources: Howell (1987), Galet (1979), Kasamatis and Lider (1980), Pongracz (1983).

[a] 1 = sensitive (or susceptible); 5 = resistant (or tolerant): 2, 3 and 4 = intermediate ratings. Phyllx = Phylloxera; R-K = Rootknot nematode; D = Dagger nematode

[b] 1 = low vigor; 5 = high vigor; 2, 3 and 4 = intermediate ratings.

[c] Tolerance to % lime, a measure of soil alkalinity: 1 = sensitive to value below 10%; 5 = tolerant to value above 30%; 2, 3 and 4 = intermediate ratings.

[d] Scion vegetative maturity in fall: "+" = advances maturity; "-" = delays maturity

[e] ? = Data not available or unknown

conventional plant spacings and training systems. SO4 does best in light, well-drained soils of low fertility. Reports from France indicate that SO4 is susceptible to magnesium deficiency and that the combination of Cabernet Sauvignon grafted to SO4 is particularly susceptible to late-season bunch stem necrosis. In New York state, vines grafted to SO4 or 5C produced greater crop yields than did C-3309 because of larger vine size; however, vines grafted to SO4 or 5C also sustained greater cold-injury than did those grafted to C-3309.

Vitis rupestris* × *V. berlandieri (99 R, 110 R, 140 Rug, 1103 P): *Vitis rupestris* and *V. berlandieri* are well-adapted to drought stress; thus, rootstocks produced from them are suited to warm regions where water is limited. These rootstocks were developed for Mediterranean-like growing conditions and nonirrigated vineyards. This group has the most vigorous rootstocks and the best adaptability to poor growing conditions, including infertile soils and drought. These may be worth considering in low-vigor sites with poor soils and no irrigation.

Other complex crosses: *Vitis riparia* × *V. cordifolia* × *V. rupestris* (44-53): 44-53 has attributes similar to C-3309 and should be suitable for a wide range of conditions found in eastern North America. It has moderate vigor, performs well under dry conditions, and tolerates somewhat acidic soils; but it has received scant commercial attention in eastern North America.

V. riparia* × *V. berlandieri* × *V. rupestris (Gravesac): This is a relatively new rootstock (circa 1985) developed in Bordeaux, France for tolerance to acidic soils. Gravesac has moderate vigor and is suited for well-drained soils of low fertility. Availability of Gravesac among North American nurseries is limited.

Disease and Disease Resistance

All commonly grown commercial grape varieties in eastern North America are susceptible to one or more foliar, fruit, or systemic, vascular diseases (table 3.2, page 62). The most common diseases include black rot, powdery mildew, downy mildew, Botrytis bunch rot, phomopsis, and crown gall (see chapter 11, page 216). Certain cultural practices can reduce the severity of these diseases; however, most will need to be managed with an effective pesticide program. For the varieties in this chapter, only those diseases to which a variety is unusually sensitive, particularly with fruit infections, are included in the descriptions.

Certified versus noncertified nursery material. Growers often have the option of purchasing "certified disease-free" material versus noncertified material from commercial sources. What does this mean? Buying certified grapevines provides some assurance that the nursery made an effort to maintain vine health and verify the variety as true-to-type. It is not, however, a guarantee that the stock is free of potential pathogens or is true-to-variety. The reasons are complex. The nursery might have obtained certified, true-to-type plant material from a reputable source. That material is propagated in increase blocks that, depending on the state, might have been visually inspected for disease or for rogue plants in a voluntary certification program. Material from increase blocks is used for grafting or other propagation, grown in nursery production rows, and sold to the end-user. Infection by disease-causing pathogens can occur and remain inconspicuous during this production process. Some diseases such as crown gall are not deliberately tested (indexed) for, but rather vines are evaluated only visually for presence or absence of symptoms. While the certification process is not absolute or fail-safe, the purchase of certified material is still considered superior to purchasing noncertified material given the range of pathological problems that can affect grapevines.

Sensitivity to Sulfur and Copper

Sulfur fungicide is injurious (phytotoxic) to the foliage of many red hybrid varieties, such as Chambourcin, and certain other American type varieties such as Norton (table 3.2). Sulfur is an important powdery mildew fungicide, however, particularly for those trying to grow grapes organically and/or for use in fungicide resistance management programs. Those varieties that are sensitive to sulfur should be planted far enough away from other varieties to prevent them from being affected by spray drift.

Copper, which is used in some fungicides such as Bordeaux mix, can also be phytotoxic, particularly to Chancellor and Norton; and similar planting precautions are advised.

Vinifera versus Hybrids

Most vinifera varieties command higher raw product and wine prices than do the hybrid or American-type grapes (see chapter 1, page 1). The higher potential crop values of vinifera varieties must, however, be weighed against their generally inferior cold-hardiness and generally greater disease susceptibility, relative to hybrids or American varieties. Cold-injury is the primary threat to *V. vinifera* production in much of eastern North America. Cold-hardiness is the ability to acclimate to and resist cold-injury, and the degree of hardiness varies throughout the dormant period. While genetics determines the ultimate degree of cold-hardiness expression, the environment, as well as cultural practices and pest management, affects that expression. In the mid-Atlantic region, dormant buds, canes, and trunks usually attain maximum cold-hardiness in late December and normally retain this degree of hardiness through mid-February or late February, depending upon prevailing temperatures. Understanding how well a given variety tolerates cold is critical to making a good decision on which variety to plant. We have provided a relative rating of cold-hardiness in table 3.2. In a similar sense, growers should be aware that vinifera varieties will require a more diligent and expensive spray program than that required for most hybrid or American-type grapes because of their generally greater susceptibility to fungal diseases (table 3.2, page 62) The potential exists to lose entire crops to powdery mildew or black rot with vinifera varieties, and that risk must be weighed in varietal decisions.

Growing Season Length and Heat Summation

Will a variety ripen at your site? Will it fully ripen given the heat summation of your site? Is a variety's wine potential affected by high-temperature exposure during the post-veraison ripening period? We do not know the answers to these questions on such a large scale as intended with the title of this production guide. The answers to the first two questions depend on both the number of frost-free days in the growing season as well as the amount of heat accumulated at the site (see chapter 2, page 14). Grapes require as little as 140 days or more than 200 days to ripen a crop—a difference of two months. Planting a long-season variety such as Norton or Mourvèdre at a site with only 150 frost-free days is a recipe for chronic disappointment. On the other hand, varieties that ripen early and fully in a cool climate (for example, Region I or II of UC classification), such as Pinot noir, might not produce their optimal quality if grown in a hot region (for example, Region V) because of accumulated heat or heat spikes during the ripening period that can negatively alter fruit chemistry. Matching a variety to a particular environment to maximize wine potential occurred over at least two millennia in the Old World. Today that goal is still a very active area of academic research and grower/vintner trial, due in part to the expansion of wine production into novel regions.

The effect of high-temperature exposure on wine potential (question 3) has not been defined or adequately evaluated in eastern North America, although this is a particularly important topic in continental climates. Changes in climate may affect the occurrence of some diseases, such as Pierce's disease; and this may alter local variety recommendations. For these reasons we encourage the reader to consult local resources for current recommendations on variety adaptation to local environments. The resources should include an evaluation of wines produced from the variety of interest and grown under local conditions. For those who wish to venture from the mainstream, we further recommend reading the perspectives of Gladstones, *Viticulture and Environment*, "Past and Future Climatic Indices for Viticulture," and Happ, "Site and Varietal Choices for Full Flavour Outcomes in a Warm Continent" (see reference list, page 70).

Wine Grape Markets

We have not included any market information on the described varieties. Before planting any variety for commercial production, the reader should thoroughly explore the market demand for the variety or the wine that will be produced from it. Some grape varieties are easy to grow but have no markets. On the other hand, a winery might express a strong interest in buying grapes that are consistently difficult to grow in a particular region.

VINIFERA VARIETIES, PRIMARILY FOR RED AND ROSÉ WINES

In addition to the information provided in this section, table 3.2 summarizes potential wine quality and viticultural characteristics for each variety. Table 3.3 (page 66) provides a nonexhaustive list of clones (and their characteristics) for all the varieties discussed in this section.

Barbera

General History and Origin

Barbera, a blue-black variety that ripens midseason, is an important wine grape variety in Italy, particularly the Piedmont area, where it is believed to have originated.

Strengths

- Good basal bud fertility.

Weaknesses

- Leaf roll virus has been prevalent in commercial nursery sources.
- Highly susceptible to Pierce's disease and downy mildew.

Viticultural Considerations

In California, this vigorous variety is trained on bilateral or quadrilateral cordons, depending on the site. Geneva Double Curtain and lyre systems have also been used. If yields are too high and grapes are grown on vigorous sites, the wines can lack concentration and may have high acidity. Vigorous vines in high-nitrogen sites can have excessive flower shatter at bloom. Experience in eastern North America is minimal but Barbera has shown cold-tenderness.

Wine Descriptors

Barbera has high natural acidity, and wines from warm areas are used for blending. In cool regions of California at lower yields, Barbera produces quality varietal wines of varying styles.

Cabernet Franc

General History and Origin

Cabernet Franc produces blue-black berries and ripens late season. This variety is believed to have been established in Bordeaux in the seventeenth century.

Strengths

- Good resistance to fruit rots and splitting.
- Somewhat more cold-hardy than Cabernet Sauvignon.

Weaknesses

- Early bud break.
- Leafroll virus was common in much of the propagative stock prior to the 1990s. Leafroll virus can reduce yields, fruit quality, and perhaps cold-hardiness. Bunch stem necrosis can occasionally reduce yields.
- High bud fertility can lead to overcropping and poor wine quality unless crops are rigorously managed.

Viticultural Considerations

Cabernet Franc can have excessive vegetative growth on fertile sites. The shoots grow strongly upright; therefore, upright shoot-positioned training and trellis systems are preferred. When grown on fertile sites, vines can be very vigorous and may benefit from divided canopies, such as open-lyre, Smart-Dyson, or Smart-Dyson Ballerina. Highly vigorous sites can produce wines with pronounced vegetative aromas.

Wine Descriptors

Cabernet Franc can produce a varietal wine; however, it is usually blended into Bordeaux-style blends. If not blended, the wine from this variety frequently lacks the desired broad spectrum of aromas and flavors, and good tannin structure. Ripe fruit and good canopy management help minimize the concentrations of methoxypyrazines, the compounds partly responsible for the assertive, vegetal flavor. Varietal descriptors include cherry, clove, violet, dill, spice, and berry-like aromas.

Cabernet Sauvignon

General History and Origin

Cabernet Sauvignon produces small, blue-black berries that ripen late. The parent varieties are Cabernet Franc and Sauvignon blanc. Cabernet Sauvignon is a native of the Bordeaux region of France.

Strengths

- Late bud break, usually 10 days after Chardonnay.
- Fruit is more resistant to cracking and rots than many of the commonly grown varieties.

Weaknesses

- Requires at least 180 frost-free days to ripen. Vines have a long vegetative cycle and mature very late in season.
- A temperature of −11°F killed all of the primary buds and caused some trunk damage in Virginia. Very sensitive to poor soil drainage that can exacerbate winter cold-injury problems.
- Susceptible to late-season bunch stem necrosis that can reduce the crop. In bunch stem necrosis the cluster stem (rachis) dies at or shortly after veraison.
- In areas that are too cool, the variety can develop undesirable herbaceous or "green bell pepper" aromas.
- In areas too warm for the variety, fruit will not develop normal varietal characters.
- Grafting onto SO4 rootstock is not recommended.

Viticultural Considerations

Cabernet Sauvignon has an upright growth habit and is often excessively vigorous. This variety is often trained with cordons and spur-pruned in upright training systems; however, Cabernet Sauvignon cordons do produce an abundance of base shoots which increases shoot thinning labor. When grown on fertile sites, vines can be very vigorous and require divided canopies, such as open-lyre, Smart-Dyson, or Smart-Dyson Ballerina. Use of size-restricting rootstocks (for example, 420A or Riparia Gloire) is recommended when high-vigor situations are anticipated. Remedial canopy management is often needed during the growing season to obtain optimal fruit quality. Vines are very susceptible to poor growth, winter injury, and crown gall because of poor soil internal water drainage. Performs best on soils with low water-holding capacities. Cabernet Sauvignon can produce fruit with a high pH, especially by overcropping or during dry seasons. In France, this variety can give good results on well-drained, gravelly, acid soils that are well-exposed.

Wine Descriptors

Cabernet Sauvignon, which is frequently used in Bordeaux-style blends, produces a high-quality wine if the grapes are harvested at full maturity. The wine can be robust and full-bodied with aging potential. Wines from immature or shaded clusters can have assertive herbal tones usually described as green bean, eucalyptus, and green bell pepper. The grapes must be fully matured to provide ripe, supple tannins and to minimize the herbaceous tones. In Bordeaux-style blends, this variety provides structural tannins, rich mouth feel, and body. Varietal descriptors include cedar, mint, plum, black currant (cassis), violet, green bell pepper, eucalyptus, and black cherry.

Limberger (Lemberger)

General History and Origin

Also referred to as Blaufränkisch. A red-wine grape widely planted in Austria, Hungary, Germany's Württemberg region, Washington State, and in limited amounts in the eastern US.

Strengths

- Moderate cold-hardiness.
- Good resistance to Botrytis bunch rot.

Weaknesses

- Early bud break.

Viticultural Considerations

Vigorous, with a tendency for over-cropping. Vines performed well at Winchester, Virginia; however, wines were of mediocre quality.

Wine Descriptors

Wines have a deep red color, rich tannins, and good acidity. Limberger produces dry, fruity wines with Merlot-like tannins and a dark chocolate/raspberry flavor. Varietal characters include raspberry and cracked black pepper.

Merlot

General History and Origin

Merlot produces blue-black berries that ripen midseason. The origin of Merlot is unknown; however, it has been cultivated in the Bordeaux region since the eighteenth century.

Strengths

- Grape and wine quality can be exceptional.

Weaknesses

- Sensitive to cold-injury, crown gall, and Eutypa.
- Susceptible to poor fruit set if cool weather occurs during bloom.
- Susceptible to nonspecific bunch rots.
- Very susceptible to phosphorus deficiency in acid soils.

Viticultural Considerations

Merlot has medium to high vigor and a semitrailing growth habit; it can be trained to low or high cordons, Smart-Dyson, lyre, or Geneva Double Curtain, depending on the vineyard site and propensity for high vigor.

Wine Descriptors

Merlot can be made into a varietal wine or be blended and has supple tannins. Adequate maturity and good vineyard management is essential to minimize herbal or vegetal flavors. The most common blending partner is Cabernet Sauvignon, which adds rich berry-like aroma and flavors, structure, and backbone. Varietal descriptors include herbaceous, leafy, perfumed, cherry, raspberry, fruitcake, and black currant.

Mourvèdre (Mataro)

General History and Origin

Mourvèdre produces blue-black fruit that ripens very late. This variety was possibly introduced into the Barcelona area of Spain by the Phoenicians in 500 BC. Mourvèdre is one of the principal red varieties produced in southeastern France, especially in the Rhône region. In Spain, where it is known as Mataro, Mourvèdre is the second most important variety.

Strengths

- Late bud break. In Virginia, Mourvèdre breaks buds later than Cabernet Sauvignon and about 16 days after Chardonnay.
- Somewhat resistant to Botrytis bunch rot (thick skins) and phomopsis.

Weaknesses

- Very cold-tender.
- Large-clusters and high bud fertility increase the potential to over-crop this variety, leading to mediocre wine quality. Extra attention to crop-thinning is essential.
- Sensitive to downy and powdery mildew, as well as crown gall. Sulfur may cause phytotoxicity.
- Late fruit maturation. Mourvèdre should not be planted in sites with less than 180 frost-free days.
- In France, Mourvèdre is sensitive to drought stress and potassium and magnesium deficiencies.

Viticultural Considerations

Mourvèdre shoots have an upright growth habit and very large clusters. Cordon training with spur-pruning and vertical-shoot-positioned training works well. Well-suited to warm climates and a wide range of soils; however, reportedly susceptable to magnesium deficiency in some situations.

Wine Descriptors

Produces aromatic, balanced wines that age well. Wine styles range from fruity rosés to dark-red wines with strong tannins. Mourvèdre is frequently used in blends, such as in Chateauneuf-du-Pape, for improving structure. Mourvèdre has also been blended with Norton, Syrah, and Tannat in Virginia.

Nebbiolo

General History and Origin

Nebbiolo has violet to blue-black fruit that ripen late-season. Believed to have been cultivated in the Langhe district of Italy before the fourteenth century, Nebbiolo is considered the premier variety of Italy's Piedmont region and is the basis of Barolo wines.

Strengths

- Low sensitivity to bunch rots.
- High potential wine quality if crop level is aggressively controlled.

Weaknesses

- Notoriously difficult to grow and make into fine wine outside of its home region of Piedmont.
- Early bud break.
- Extremely cold-tender.
- Poor basal bud fertility, often necessitating cane-pruning to achieve adequate yields.

Viticultural Considerations

Nebbiolo is vigorous. The shoots are long and trailing. Low basal bud fertility dictates cane-pruning to afford acceptable yields.

Wine Descriptors

In Italy, Nebbiolo wines vary greatly among production sites. Nebbiolo wines are astringent in youth but can evolve into well-structured, richly scented wines. Varietal descriptors include violets, cherry, and clove spice. The wines have abundant tannin but often poor color density.

Petit Verdot

General History and Origin

Petit Verdot originated in southwest France and is one of the five principal red Bordeaux varieties. The midsize clusters of blue-black berries ripen late-season.

Strengths

- Excellent fruit quality in the mid-Atlantic region, with potential for blended or varietal wines.
- Slightly greater cold-hardiness than Cabernet Sauvignon but within the same five-degree injury threshold (table 3.2, page 62).
- Small berries and loose clusters help minimize fruit-rot problems.

Weaknesses

- High acid and pH. Occasionally, when the sugars are adequate, the pH and titratable acidity can be high (more than 3.7 and 8 grams per liter, respectively).
- Requires a long growing season.

Viticultural Considerations

Shoots are semiupright. Upright shoot-positioned training systems are generally recommended, but high training is possible with rigorous shoot positioning. In France, this variety is reportedly well-adapted to gravelly soils. Traditional fruit maturity gauges—such as sugar, acid and pH—may be inadequate predictors of wine quality; and tasting the skins for the degree of tannin polymerization may be a better predictor of fruit maturity (see chapter 16, "Wine Grape Quality: When Is It Time to Pick?" page 282). Seed color is not predictive of fruit ripening with this variety.

Wine Descriptors

The following description is from Wolf et al. (1999). Petit Verdot produces a medium- to full-bodied wine that is more tannic and colored than the other four traditional red Bordeaux varieties. Wines produced from fully ripe fruit are rich; are age-worthy; and, like Syrah, can have a peppery, spicy aroma profile laced with hints of currants and black cherry. In a blended wine, Petit Verdot adds aroma and flavor, alcohol, palate weight (body), tannins, and color. The fruit can attain 23 to 24 °Brix in dry falls. Varietal descriptors include berry, strawberry, cinnamon, raspberry, spice (black pepper), rose, and floral.

Pinot noir

General History and Origin

Pinot noir is a blue-black, early-ripening variety. It is one of the oldest cultivated varieties and is thought to have been cultivated by the Romans. This variety may be a native of Burgundy.

Strengths

- Early harvest.
- Versatile to a range of wine styles, including sparkling wine, light, fruity reds, and deep, full-bodied reds.

Weaknesses

- Early bud break.
- Cool, damp weather during bloom can dramatically reduce fruit set and yield.
- Very susceptible to common fungal diseases, Pierce's Disease, and North American Grapevine Yellows.
- Not well-adapted to warm or hot grape regions, which tend to reduce color and flavor.
- Berries are prone to splitting and nonspecific fruit rots because of compact clusters, and fruit may rot before intended ripeness in wet falls.

Viticultural Considerations

Pinot noir has moderate vigor, and vines are usually trained upright from bilateral cordons. The fruit is susceptible to sunburn, so Geneva Double Curtain or vertically divided canopies, such as Scott Henry or Smart-Dyson, are not recommended. In France, Pinot noir gives the best results in calcareous-clay soils when vigor and yield are strictly controlled.

Wine Descriptors

Pinot noir is used to produce both sparkling and still wines. The latter range in style from light-red varietal wines to full-bodied, tannic reds. Varietal descriptors include strawberry, cherry, and woody aromas.

Sangiovese

General History and Origin

Sangiovese, a blue-black variety that ripens late-season, is thought to have originated from selections made by the Etruscans in the Villanovan Era in what is now Tuscany, Italy. In Italy, there are two types: Sangiovese grosso (or dolce or gentile) and Sangiovese piccolo (or forte or montanino). Both types have multiple clones (table 3.3, page 66). Sangiovese grosso has large berries, and Sangiovese piccolo has small berries with small- to medium-sized clusters.

Strengths

- Potential to produce a range of wine styles including light, fruity reds to deep, full-bodied reds.

Weaknesses

- Tendency to overcrop, particularly with "grosso" clones.
- Difficult to achieve wines with concentrated color and flavors in sites that promote excessive vine growth.
- Very cold-tender.

Viticultural Considerations

Sangiovese vines are vigorous and have a semiupright growth habit. It is usually trained to bilateral cordons. Vertical shoot positioning is commonly used in low-vigor sites. Smart-Dyson, Geneva Double Curtain, or lyre systems can be used on higher vigor sites.

Wine Descriptors

Sangiovese grapes can make varietal wines that range from rosé to full-bodied red wines. This grape is also used for light- to medium-bodied Chianti-style wine. Sangiovese can also be blended with Cabernet Sauvignon, Merlot, and Cabernet Franc. Varietal descriptors include violet, plum, cherry, leather, and tobacco.

Syrah (Shiraz)

General History and Origin

Syrah produces small- to medium-sized clusters of blue-black berries that tend to shrivel when ripe. This variety ripens midseason and is a cross between two French varieties, Dureza and Mondeuse Blanche, both of which are from the Rhône region. Syrah is believed to have been cultivated since Roman times in the northern Côte du Rhône.

Strengths

- Bud break is fairly late.

Weaknesses

- Fruit tends to shrivel during later stages of ripening, often at 21 to 22° Brix.
- Botrytis infection of shoots and clusters can be a problem in wet springs.
- Shoots are very susceptible to wind breakage prior to bloom.
- Sensitive to bud necrosis, which can limit yields.
- Extremely cold-tender.

Viticultural Considerations

The vines are very vigorous, and shoots are upright to trailing. Syrah is suited to vertical-shoot-positioned training in low vigor sites. In more vigorous sites, it does well on horizontal quadrilateral systems such as Geneva Double Curtain or lyre.

Wine Descriptors

Syrah can produce good-quality varietal wines that have high alcohol content and improve with aging, and it may be used for blended wines. Syrah has deep color, is not overly tannic, and has fruity aromas. The tannins are dark and complex; and aromas are delicate (violet, leather, olive). Syrah can also be made into dessert wines and port-style wines.

Tannat

General History and Origin

Tannat is a blue-black grape that has small berries, has large clusters, and ripens late season. This variety is thought to have originated in the Pyrenean area of southwest France. It is important in the Madiran region of France, in Uruguay, and in minor amounts in the US mid-Atlantic.

Strengths

- Late bud-break—about 13 days later than Chardonnay.
- Good fruit quality, high sugar accumulations, and some resistance to fruit rots.

Weaknesses

- Total acidity can remain high in cool years.
- Cold-tender.

Viticultural Considerations

Tannat is extremely vigorous and has a semiupright growth habit. It is often trained to bilateral cordons and spur-pruned, but canopy division might be advised to accommodate the high vigor. Crop control is necessary to produce quality wines.

Wine Descriptors

Tannat is used in varietal and blended red and rosé wines. The wines are deeply colored, perfumed, and have rich, firm tannins and a high alcohol content, high acidity, and relatively low pH. Varietal descriptors include black cherry, plum, tobacco, and chocolate. The bouquet is not particularly elegant with varietal wines but can be improved with blending.

Tempranillo (Valdepeñas)

General History and Origin

This midseason variety has large, long clusters of blue-black fruit. Tempranillo is believed to be either a natural hybrid of Cabernet Franc and Pinot noir that originated in France, or a variety from northern Spain (Rioja).

Strengths

- Potential for high wine quality in hot, dry falls.

Weaknesses

- Very susceptible to Eutypa and Crown Gall in mid-Atlantic trials.
- In warm climates, acid levels can be low and pH marginally high.
- Cold-tender.

Viticultural Considerations

The vine is vigorous, and vertical shoot positioning is recommended. If the site is vigorous, divided canopies can be used. In California, the vine can be head-trained or trained to Geneva Double Curtain. Cane pruning is not recommended. Cluster thinning is often required to promote higher wine quality.

Wine Descriptors

Tempranillo can produce good varietal wine with good color, or it can be blended to produce table wines or port wines. Under favorable growing conditions, the wines are full-bodied and well-colored.

Zinfandel

General History and Origin

Zinfandel has deep blue-black fruit in medium to large clusters. The variety ripens midseason to late-season. The name "Zinfandel" is peculiar to California, but how it arrived there is still debated; one claim is that it was imported by a Massachusetts nurseryman. Regardless of origin, Zinfandel has been produced in California since the 1850s. A clonal subgroup is called Primitivo, although TTB considers Primitivo to be a separate variety, even if DNA fingerprinting does not.

Strengths

- Potential for high wine quality in rare hot, dry falls.

Weaknesses

- Compact clusters are prone to uneven ripening, berry splitting, and bunch rots—particularly in wet weather.
- Cold-tender.

Viticultural Considerations

Vines are trained with spurs on bilateral cordons or head-trained with vertical shoot positioning. This variety needs cluster thinning, as it tends to have large clusters and a deficit of leaf area to crop. In very warm climates, a significant proportion of the clusters may shrivel or produce raisins. Very little positive experience in eastern North America. The Primitivo clones might be slightly less susceptible to berry crack and rots (table 3.3, page 66).

Wine Descriptors

Zinfandel can be used in varietal wines or blends, including white variations. Wine descriptors include raspberry, blackberry, boysenberry, cranberry, black cherry, and cinnamon spice. In warm areas, the varietal character is less obvious. Zinfandel can produce high sugar levels and can be made into high-alcohol table wine or port-style dessert wines when matured in hot, dry falls.

VINIFERA VARIETIES FOR WHITE WINES

In addition to the information provided in this section, table 3.2 (page 62) summarizes potential wine quality and viticultural characteristics for each variety. Table 3.3 provides a nonexhaustive list of clones (and their characteristics) for all the varieties discussed in this section.

Chardonnay

General History and Origin

Chardonnay produces yellow to amber fruit that ripens early to midseason. It is a cross between Pinot noir and Gouais blanc and originated in the Bourgogne region of France.

Strengths

- Good consumer recognition.
- Adaptable to a wide range of site conditions and wine stylistic goals.

Weaknesses

- Early bud break increases risk of spring frost damage.
- Susceptible to winter cold-injury; however, the general experience in eastern North America has been that Chardonnay is one of the more adapted varieties to winter cold.
- Highly susceptible to many common diseases, including powdery mildew, Botrytis bunch rot, and North American Grapevine Yellows. Chardonnay is often the first variety in which growers observe powdery mildew outbreaks. Excellent canopy management, including leaf pulling around the fruit, can reduce the incidence of both powdery mildew and botrytis bunch rot.

Viticultural Considerations

Chardonnay is usually trained to grow upright with bilateral cordons and spur pruning. Other alternatives are head training, Smart-Dyson, Smart-Dyson Ballerina, and open-lyre. High training is not generally recommended because of the difficulty in positioning shoots downward and the increased potential for fruit sunburning in warmer regions.

Wine Descriptors

Chardonnay is considered a premium grape variety and is typically produced as a varietal wine. The fruit is suited to a wide range of wine styles from sparkling wine cuvées (base wines) to semidry wines to Burgundy-style wines with rich, complex bouquets. Wines can be well-balanced, strong, ample, soft, and full.

Gewürztraminer

General History and Origin

Although made into "white" wine, the fruit of Gewürztraminer has a unique tan-pink color when the fruit is fully ripened with some sun exposure. Fruit ripen early and Gewürztraminer should only be grown in cooler regions to optimize the retention of volatile aroma and flavor compounds. Gewürztraminer may be from the Pfalz region of Germany or it may have originated in France.

Strengths

- Excellent wine quality when produced in cool, dry falls.

Weaknesses

- Vulnerable to spring frosts due to early bud break.
- Fruit is very prone to rot, often before the fruit is ripe.
- Varietal fruit character can be lacking when grown in hotter regions.
- Prone to poor fruit set.
- Cold-tender.

Viticultural Considerations

Gewürztraminer is vigorous and has a trailing habit. It is usually cordon-trained and spur-pruned; however, in California, this variety is head-trained and cane-pruned. Good canopy management is critical to minimize fruit rot issues and afford optimum fruit exposure.

Wine Descriptors

Varietal character develops late in the ripening period, and the fruit often has low acidity at ripeness. Gewürztraminer produces distinctive wines with a spicy, floral aroma. table wines are usually slightly sweet to offset a natural phenolic bitterness. Gewürztraminer also produces excellent dessert wines.

Petit Manseng

General History and Origin

Petit Manseng is a white-fruited variety that has small clusters and matures late season. This variety originated in France in the Pyrénées-Atlantiques and is most notable in the Jurançon region of France.

Strengths

- Loose cluster, thick-skinned, very small berries.
- Resistant to bunch rots.
- High concentration of sugar, acid, and flavors when fully ripe.
- Fairly cold-hardy for *V. vinifera*.

Weaknesses

- Late-season ripening—requires at least a 180-day growing season.
- Little commercial experience outside of Virginia.

Viticultural Considerations

Vines are vigorous. Cordon training and spur-pruning with vertical-shoot-positioned training is appropriate.

Wine Descriptors

Petit Manseng can produce very aromatic wines of good quality or dessert wines, or it can be made into dry wines similar to Riesling or Gewürztraminer. This variety can accumulate high sugar levels with high titratable acidity and low pH. The ripe fruit has pronounced flavor of honey with some citrus and pineapple, similar to Vignoles.

Pinot blanc

General History and Origin

Pinot blanc, a mutation of Pinot noir that produced white grapes, is thought to have originated in the Alsace in the sixteenth century.

Strengths

- Generally easier to grow than Pinot noir, with relatively less fruit-rot problems.
- Vine are nearly as cold-hardy as Chardonnay.

Weaknesses

- Lacks consumer recognition of Pinot noir or Pinot gris.
- Very susceptible to all major fungal diseases.

Viticultural Considerations

The vines are vigorous and more productive than Pinot noir and Pinot gris. Growth habit and recommended training systems are comparable to Pinot noir.

Wine Descriptors

Pinot blanc is made into a varietal wine and can be blended to produce sparkling wines. The wines are pleasant, if not memorable.

Pinot gris (Pinot Grigio)

General History and Origin

Pinot gris is a variant of Pinot noir that is used to produce white wines. The clusters are small and berries range in color from pink, to coppery-gray, to brownish-pink, even within the same cluster. Pinot gris was first described in the fourteenth century and is grown throughout Europe.

Strengths

- Somewhat cold-hardy.
- Secondary buds are fruitful.

Weaknesses

- Cool, damp weather can cause poor fruit set.
- Slightly low acidity.
- Very susceptible to all major fungal diseases.

Viticultural Considerations

Moderately vigorous. In France, Pinot gris is well-adapted to deep, calcareous, and dry soils that have good exposure. Growth habit and recommended training systems are comparable to Pinot noir.

Wine Descriptors

Wines made from Pinot gris can be very fine, strong, robust, and aromatic. Aromas are typically more pronounced than are those of Pinot blanc.

Riesling (White Riesling)

General History and Origin

Riesling has small clusters of green to golden-yellow fruit that ripen midseason This variety is from the Rhine and Moselle regions of Germany, where it has been grown possibly since Roman occupation.

Strengths

- Relatively cold-hardy.
- Relatively late bud break.

Weaknesses

- Fruit tends to split and rot before fully ripe in the warmer areas of the mid-Atlantic.
- Flavors and aromas are not consistently obtained in warmer regions.
- Yields are often poor because of primary bud necrosis.

Viticultural Considerations

Riesling is moderately vigorous. Traditionally, Riesling is head-trained and cane-pruned. Cordon training and spur pruning have been used successfully in many areas of eastern North America. Cane pruning can increase fruitfulness. For sites with high vigor, Smart-Dyson, Scott Henry, or Geneva Double Curtain and open lyre have been used.

Wine Descriptors

Riesling is used to produce varietal wines or late-harvest dessert wines. Styles can range from dry to very sweet. This variety can produce high-quality, dry white wines that are aromatic, pleasant, and lively and can have a good keeping quality.

Sauvignon blanc (Fumé blanc)

General History and Origin

The yellow-green fruit of Sauvignon blanc ripens early in the season. The origin of Sauvignon blanc is unknown; however, it has been grown for several centuries in Bordeaux and the Loire Valley.

Strengths

- Ripens in early season.
- Wine quality can be excellent in the rare dry fall.

Weaknesses

- Highly susceptible to Botrytis bunch rot and berry splitting.
- Vines take a long time to mature wood for winter and are prone to winter injury.
- Cordon-trained, spur-pruned vines may be of low fruitfulness; therefore, cane-pruning recommended.

Viticultural Considerations

Sauvignon blanc is very vigorous. Head training with canes and renewal spurs is recommended because of low basal bud fertility. On vigorous sites, a lyre or U-divided canopy is often used. Geneva Double Curtain is not recommended because the fruit can sunburn. In California, berry drop at set can occur if vegetative growth is excessive during the period. Sauvignon blanc is recommended for weak to moderately fertile soils, grafting to a low-vigor rootstock, or using both to manage the high-vigor potential.

Wine Descriptors

Sauvignon blanc can produce a varietal wine, or it can be used in blends. Styles range from dry to slightly sweet. If it is harvested late or if noble rot occurs, Sauvignon blanc can produce very pleasant, sweet wines. Dry wines made from this variety can be pleasant, fine, and well-balanced.

Grassy aromas and flavors can be pronounced when fruit is matured in cool falls. Varietal descriptors include citrus, peach, apricot, bell pepper, asparagus, and green olive.

Viognier

General History and Origin

Viognier produces small clusters of yellow and amber fruit that ripen early. Viognier is from the Rhône districts of Condrieu and Château-Grillet. The vine may have been brought from the Dalmatia region by the Romans.

Strengths

- Excellent fruit quality.
- Viognier tends to produce fruit with high concentrations of sugars.
- Good resistance to berry splitting and bunch rots.

Weaknesses

- Early bud break increases concerns about spring frosts.
- Susceptible to winter cold-injury. A temperature of −11°F killed all primary buds in Virginia.
- Weak growth. Vines can be spindly and slow to fill trellis.
- Modest yields (3 tons per acre on average) in mid-Atlantic trials.
- Bud necrosis is common and severe with Viognier.

Viticultural Considerations

Viognier has moderate vigor and is usually trained to grow upright with bilateral cordons and spur training. This variety may not be vigorous enough for divided-canopy systems, such as Smart-Dyson, or open-lyre. In eastern North America, cane pruning is used in cooler areas.

Wine Descriptors

Viognier is typically produced as a varietal wine that can have distinctive aromas of honeysuckle, melon, orange, muscat, pears, cloves, honey, and tropical fruits. Viognier can also make sparkling or sweet wines. Aromas and flavors develop in the fruit only at high sugar levels. As sugar approaches 23 °Brix or higher, there is a large increase in varietal aroma and flavor, and skin phenols and acidity drop rapidly. The change in Brix corresponds to a change in fruit from green to yellow. Shaded berries or clusters have limited varietal character. Wines may lack acidity and often do not age well.

HYBRID AND AMERICAN VARIETIES FOR RED AND ROSÉ WINES

In addition to the information provided in this section, table 3.2 (page 62) summarizes potential wine quality and viticultural characteristics for each variety.

Baco noir (Baco #1)

General History and Origin

The jet-black berries of Baco noir are small and ripen early to very early. Baco noir is a cross between Folle Blanche (*V. vinifera*) and *V. riparia* that was developed by Francois Baco in 1902.

Strengths

- Ripens very early.

Weaknesses

- Tight clusters are susceptible to bunch rots.
- Early bud break increases risk of spring frost damage.
- Susceptible to black rot, ringspot viruses, and crown gall.
- High titratable acidity in otherwise ripe fruit.
- Prone to bird damage and foliar phylloxera.

Viticultural Considerations

Baco noir is very vigorous and has a semitrailing growth habit. The shoots resemble those of *Vitis riparia*. The canes must be pruned long to be productive. This variety may be excessively vigorous, which can increase the risk of winter injury. Tomato and tobacco ringspot virus are common, and these viruses weaken and often kill infected vines. Graft to a pest-resistant rootstock to decrease the likelihood that vines will be infected with virus through nematode feeding.

Wine Descriptors

Baco noir produces deeply colored wines that are low in tannin and have high acidity. Others describe the wine as "Rhone-style" or "Beaujolais-style" and state that Baco noir can be very good when well-ripened fruit are vinified with good cellar technique. The wines can also be herbaceous and bitter and require long aging to improve. Not generally recommended for quality wine production.

Catawba

General History and Origin

Catawba is an American variety that produces large, reddish blue berries that ripen late season. Catawba is believed to have been introduced by John Adlum, who took cuttings from a vineyard in Montgomery County, Maryland, in 1819. Other references suggest that the variety was found along the Catawba River of North Carolina in 1802. In either event, it was widely distributed in Ohio, Virginia, and New York by the mid-1800s. Catawba is either a cross of *V. labrusca* and *V. aestivalis* or a cross of *V. labrusca* and *V. vinifera*.

Strengths

- Vines are vigorous, productive, and cold-hardy.

Weaknesses

- Pronounced *V. labrusca* aromas and flavors.
- Sensitive to black rot, downy mildew, and phomopsis.

Viticultural Considerations

Catawba is a vigorous vine, with a procumbent growth habit that lends itself to high-wire, cordon training such as Hudson River Umbrella or Geneva Double Curtain.

Wine Descriptors

Catawba is primarily used for white or pink dessert wines. High acidity is a problem in some years. Not recommended for quality wine production.

Chambourcin

General History and Origin

Chambourcin (Joannes-Seyve 26-205) is a blue-black grape that ripens midseason to late season.

Strengths

- Loose clusters provide good resistance to fruit rots, particularly bunch rots.
- Good resistance to powdery mildew and, to a lesser extent, downy mildew.

Weaknesses

- Late ripening, which can be a problem in short-season sites.
- Sensitive to nutrient deficiencies in high-pH soils and oxidant stipple caused by atmospheric pollution. Grower experience suggests that Chambourcin has a high nitrogen fertilizer demand relative to other varieties.

Viticultural Considerations

Chambourcin has semitrailing growth and is usually trained on nondivided trellises, including high or low cordons with spur pruning. This hybrid may have poor vigor, which can be improved by grafting to pest-resistant rootstocks. The nutrient status of Chambourcin needs to be closely monitored, and routine applications of nitrogen are often needed to maintain vigor, vine size, and productivity. Low nitrogen status may also exacerbate foliar expression of Rupestris leaf spot, a common disorder on vines with *V. rupestris* parentage. Others report that the vines can be vigorous with rampant growth and that they have a tendency to overcrop, particularly on coarse-textured, well-drained soils.

Wine Descriptors

Chambourcin is used to make varietal wines and is blended with other mid- to full-bodied red wines, such as Cabernet Sauvignon. Chambourcin has a distinct aroma and herbaceous flavors that are more vinifera-like than most red hybrids. Chambourcin can produce rosé-style, Beaujolais-style, or other medium- to full-bodied, fairly complex wines, or ports. The fruit frequently has immature tannins and high acid content that can cause unbalanced wines. Varietal descriptors include raspberry, cloves, cherry, plum, and tobacco.

Chancellor

General History and Origin

Chancellor (Seibel 7053) is a blue-black-fruited variety that ripens midseason. This hybrid is a cross between Seibel 5163 and Seibel 880 and was named by the Finger Lakes Wine Growers Association in 1970.

Strengths

- Moderately cold-hardy.
- Highly productive.

Weaknesses

- Early bud break makes it susceptible to spring frosts; however, this variety can have a good secondary crop after a spring frost.
- Fruit is highly susceptible to downy mildew, and foliage to powdery mildew.
- Fruit is borne on large, compact clusters that can contribute to overcropping and bunch rots.
- Crown gall may develop when vines are grown on heavy, water-retentive soils.

Viticultural Considerations

Growth is semiupright. Chancellor requires cluster thinning, particularly of fruit on young vines and vines planted in heavy, high-water-content soils.

Wine Descriptors

The wine quality can occasionally be good. Chancellor is used to produce varietal wines with notes of plum and cedar, or it can be used in blends. Not generally recommended for quality wine production.

Chelois

General History and Origin

Chelois (Seibel 10878) is a cross between Seibel 5163 and Seibel 5593, its parentage a mix of American species and *V. vinifera*. The berries are small and blue-black and may have high titratable acidity at harvest. Chelois ripens early to midseason.

Strengths

- Late bud break.

Weaknesses

- Berry splitting and subsequent bunch rots may be severe in some years.
- Reportedly susceptible to winter damage.
- Susceptible to tomato and tobacco ringspot virus.

Viticultural Considerations

The vines are vigorous and may require cluster thinning to prevent overcropping.

Wine Descriptors

Produces medium-bodied, fruity wines with berry, leather, and earthy aromas. Can be used for varietal wines but is probably best for blending, as with Chambourcin, Baco noir, Chancellor, or vinifera varieties.

Concord

General History and Origin

Concord grapes are thought to have originated as an unintentional seedling of *V. labrusca* possibly crossed with *V. vinifera*. Concord was introduced in 1854 by E. W. Bull of Concord, Massachusetts. Liberty Hyde Bailey coined the term *Vitis Labruscana* to distinguish Concord from the wild *V. labrusca* species. The mild fruit flavor of Concord and Niagara (relative to the wild labrusca) and the appearance of perfect flowers (not found on wild labrusca), coupled with what's known about their ancestry, perhaps lead Bailey to believe that these cultivated vines had some vinifera in their ancestry. Bailey, therefore, termed Concord and Niagara "cultigens" and assigned the classification *V. Labruscana* (with a capital "L") for these varieties.

Strengths

- Cold-hardy.
- Very productive.
- Moderately resistant to downy and powdery mildew.

Weaknesses

- Early bud break increases spring frost-damage concerns.
- Foxy flavor limits market appeal.
- Susceptible to black rot and phomopsis.

Viticultural Considerations

Vigorous vines with procumbent shoot growth. Vines are usually trained on a high cordon and long spur-pruned. Fruit may not color uniformly when Concord is grown in hot regions. The related variety, Sunbelt, was developed and released by the Arkansas Agricultural Experiment Station in 1993. Sunbelt is reported to color more uniformly than Concord does when grown in hot regions.

Wine Descriptors

Usually made into sweet wines with a pronounced American grape flavor owing largely to methyl anthranilate, a natural flavor compound common with many American grape species. Not recommended for quality wine production; more often used for juice and jelly production.

De Chaunac

General History and Origin

De Chaunac, also called "Seibel 9549," is a cross between Seibel 5163 and Seibel 793 (*V. labrusca, V. lincecumii, V. riparia, V. rupestris, V. vinifera*) and is named for Canadian enologist Adhemar DeChaunac of Bright's Winery. This variety produces medium to large, loose clusters of thick-skinned, blue-black berries that ripen early.

Strengths

- Very cold-hardy and very productive, including fruitful secondary buds.
- Resistant to bunch rots.
- Resistant to black rot and Botrytis bunch rot.

Weaknesses

- Ringspot viruses are common in this variety and can weaken and often kill infected vines.
- Very early bud break increases concern for spring frost damage.
- Susceptible to anthracnose.

Viticultural Considerations

De Chaunac has a tendency to overcrop and requires cluster thinning. Although once widely planted, the acreage of De Chaunac is decreasing.

Wine Descriptors

De Chaunac makes only a fair-quality wine, although color can be intense. Not generally recommended for quality wine production.

Frontenac

General History and Origin

Frontenac, a cross between Landot 4511 and *V. riparia* 89, was named and released by the University of Minnesota in 1996. It produces small, blue-black berries on medium to large, moderately loose clusters. As with other varieties from this program, excellent cold-hardiness is a primary attribute.

Strengths

- Very cold-hardy.
- Highly resistant to downy mildew. Good resistance to powdery mildew and Botrytis bunch rot.

Weaknesses

- Fruit can have very high acidity at otherwise ripe condition.

Viticultural Considerations

Frontenac is a vigorous, semitrailing variety that is usually trained to high cordons. This variety may need cluster thinning.

Wine Descriptors

Produces deep-colored wines with cherry, blackberry, black currant, and plum notes. Can also be used in production of port-style wines.

GR 7

General History and Origin

Released by the grape breeding program at the New York State Agricultural Experiment Station in Geneva, New York in 2003, GR 7 is a cross between Buffalo and Baco noir. The "GR" refers to Geneva (New York) Red, although the name has not yet been approved by TTB.

Strengths

- Very good winter cold-hardiness.
- Tolerates moderate 2,4-D herbicide exposure without damage.
- Good disease resistance.

Weaknesses

- Breaks bud very early and can be susceptible to spring frosts.

Viticultural Considerations

The vines can be very vigorous, and fruit ripens midseason, occasionally with both high acidity and high pH. Commercial experience outside of the Finger Lakes region of New York is limited, but it should perform well in other cool-climate grape regions.

Wine Descriptors

GR 7 can produce good-quality red wines with soft tannins and cherry aromas and is often used in blends with other hybrid and vinifera varieties.

Léon Millot

General History and Origin

Léon Millot (Kuhlmann 194-2) produces small, loose clusters of blue-black fruit that ripen early. Léon Millot originated from a cross of *V. riparia* × *V. rupestris* × *V. vinifera* cultivar Gold Riesling, as did the variety Maréchal Foch.

Strengths

- Very winter-hardy.

Weaknesses

- Susceptible to bird damage.
- Lacks consumer recognition/appeal.

Viticultural Considerations

Léon Millot has a semitrailing growth habit, and is more vigorous and, arguably, more productive than Maréchal Foch. This variety must be pruned long.

Wine Descriptors

The deeply pigmented wines have fresh berry aromas and low tannins but are not generally recommended for quality wine production.

Maréchal Foch (Foch)

General History and Origin

Maréchal Foch (Kuhlmann 188-2) was developed by Eugene Kuhlmann of Alsace. The variety produces small, tight clusters of blue-black fruit that ripen very early. This hybrid has the same parentage as Léon Millot.

Strengths

- Vines are cold-hardy to very cold-hardy.
- Ripens very early.
- Relatively good fungal disease resistance.

Weaknesses

- Birds are attracted to the early-maturing, small berries.
- Early bud break and poor secondary bud fruitfulness pose a frost concern.

Viticultural Considerations

Vines are moderately vigorous and should be grafted to a phylloxera-tolerant rootstock.

Wine Descriptors

Wine styles can range from fruity and light to hearty and full-bodied, but are typically thin unless blended with a more substantial wine. Good wines of a blush style have been produced. Wines may have strawberry or raspberry character. Not generally recommended for quality wine production.

Norton (Cynthiana)

General History and Origin

Norton is thought to have originated in Virginia as a hybrid of *V. aestivalis*. Small blue-black berries are borne on small clusters, and fruit ripens very late.

Strengths

- Good cold-hardiness, but not as hardy as Concord or many other American varieties. Norton expressed 23% bud kill after exposure to a −11°F field episode in Virginia.

- Resistant, but not immune, to common fungal pathogens.

Weaknesses

- Early bud break can increase risk of frost damage.
- Fruit can mature with both high titratable acidity and high pH.
- Wine quality can be negatively affected by high acidity and pronounced American grape-like flavors.

Viticultural Considerations

Norton shoots have a trailing habit, which facilitates high cordon training, such as Geneva Double Curtain. High training and otherwise good canopy management are required to promote fruitfulness and balanced fruit chemistry at harvest. Non-grafted vines have low to moderate vigor in first few years but usually increase in vigor after three or four years, but then may decrease in vigor again over time. Proper nutrition is imperative with this variety to maintain vigor and foliage health, particularly with respect to nitrogen, magnesium, and iron.

Wine Descriptors

Norton can be used in varietal wines, including port style, but may also be blended with other reds. Depending on fruit maturity, winemaking, and wine aging, wine descriptors include spicy, fruity (labrusca grape-like to raspberry character), black pepper, tobacco, and chocolate. Wines have intense color density. High titratable acidity and persistent, herbaceous characteristics can be problematic.

Rougeon

General History and Origin

Rougeon, also called "Seibel 5898," ripens midseason.

Strengths

- Cold-hardy.
- Will produce some crop after being frosted in spring.

Weaknesses

- Can exhibit a biennial pattern of fruit bearing.

Viticultural Considerations

Rougeon is vigorous with semiupright shoots. Bud break is fairly early.

Wine Descriptors

The wine is ordinary and is used primarily for blending to provide color. Not recommended for quality wine production.

HYBRID AND AMERICAN VARIETIES FOR WHITE WINES

In addition to the information provided in this section, table 3.2 (page 62) summarizes potential wine quality and viticultural characteristics for each variety.

Aurore

General History and Origin

Aurore (Seibel 5279) is a very-early-ripening variety that produces large bunches of amber-colored berries used for white wine.

Strengths

- Good cold-hardiness.

Weaknesses

- Bird damage and fruit rots are frequently problems.

Viticultural Considerations

The vines are vigorous and trailing.

Wine Descriptors

Wine quality is mediocre at best. This variety has been used for bulk wine production and is not recommended for quality wine production.

Cayuga White

General History and Origin

Cayuga White is a hybrid that originated from a cross of Seyval and Schuyler. It was released by the grape breeding program at the New York State Agricultural Experiment Station in Geneva, New York in 1972. Early ripening fruit are greenish-yellow.

Strengths

- The vines are hardy and productive.
- Good disease resistance to most rots and mildews.

Weaknesses

- Wines from fully ripened grapes can have a strong *labrusca* character and lack refinement.

Viticultural Considerations

The vines are vigorous and semitrailing.

Wine Descriptors

Cayuga White produces pleasant, Germanic-style wines if fruit is harvested before full ripeness. The wines have medium body and good balance. Cayuga White is versatile: it can be made into semisweet wines emphasizing the fruity aromas, as well as dry, less fruity wine with oak aging. When harvested early, it can produce a sparkling wine with good acidity, good structure, and pleasant aromas.

Chardonel

General History and Origin

Chardonel is a hybrid cross of Seyval and Chardonnay and was released by the grape breeding program at the New York State Agricultural Experiment Station in Geneva, New York in 1991. This variety ripens early to midseason, and its clusters are moderate to large.

Strengths

- Excellent fruit and wine quality.
- Clusters are less compact than those of either Seyval or Chardonnay.
- Cold-hardiness is better than Chardonnay and equal to or slightly inferior to that of Seyval. In Virginia, only 26% of the buds were killed at field exposure of −11°F with no effect on the subsequent year's crop.
- Resistance to bunch rots in the mid-Atlantic was superior to that of Chardonnay or Seyval, perhaps because of somewhat loose cluster architecture.

Weaknesses

- Susceptible to crown gall, particularly in wet sites.
- Susceptible to root form of phylloxera. Chardonel should be grafted to a phylloxera-tolerant rootstock.

Viticultural Considerations

Cordon training with spur-pruning and vertical, upright shoot positioning has worked well in Virginia, but Chardonel can also be successfully trained on high cordons with single curtains. Chardonel is similar in hardiness to Seyval, while yields, vegetative growth, and bunch rot resistance are more similar to Vidal blanc.

Wine Descriptors

Chardonel is typically produced as a varietal wine and finished dry to semidry. Chardonel shows the attributes of its parents, Chardonnay and Seyval, but can have high alcohol levels. Chardonel also has the potential for fine-quality, dry still wines produced with barrel fermentation and/or barrel aging. Chardonel is also used as a sparkling wine base. This variety can attain relatively high Brix while maintaining high acid and low pH. The wines made from mature fruit have the fruit aroma characteristics of both parents. Unripe fruit will contribute herbaceous tones to wine.

Delaware

General History and Origin

The original wood was taken from a New Jersey garden. This variety was propagated in Delaware, Ohio, beginning in 1849. The pink berries form medium to small clusters that ripen early.

Strengths

- Cold-hardy.
- Early ripening.

Weaknesses

- Early bud break increases concern for spring frost damage.
- Berries tend to crack during ripening if it rains.
- May need grafting to improve vigor.
- Very susceptible to downy mildew and phomopsis.

- Susceptible to phylloxera.
- Foxy flavor limits market appeal.

Viticultural Considerations

Delaware has low vigor and a trailing growth habit. This variety requires deep, fertile, well-drained soil for satisfactory growth. The clusters do not need thinning.

Wine Descriptors

Mild American flavor. Once prized for sparkling wine production and continues to be used in dessert wine production.

Diamond

General History and Origin

Diamond is thought to be a cross between Iona (a labrusca-vinifera hybrid variety) and Concord, credited to Jacob Moore of Brighton, New York in 1885. Diamond ripens midseason.

Strengths

- Good winter cold-hardiness.

Weaknesses

- Skins may crack during wet seasons.
- Very susceptible to powdery mildew and black rot.

Viticultural Considerations

Procumbent growth habit facilitates high training.

Wine Descriptors

Used to produce dry table wines and sparkling blends, somewhat similar to Niagara wines, or its Concord parent. A limited but appreciative market, but not generally recommended for quality wine production.

Niagara

General History and Origin

Niagara grapes were introduced in 1882 by the Niagara Grape Company. This early to midseason white grape is a cross between Concord and Cassady. See the description of Concord (page 54) regarding Niagara's taxonomic classification.

Strengths

- Moderately cold-hardy.

Weaknesses

- Early budbreak.
- Very susceptible to black rot, downy mildew, and phomopsis.
- Strong flavor limits market for table wines.

Viticultural Considerations

Moderately vigorous, trailing vines do not generally need cluster thinning.

Wine Descriptors

Niagara produces wines with a pronounced American or labrusca flavor and are usually finished semisweet. Can also be used for desert wines (dry and cream sherry) and is popular as nonalcoholic, white grape juice. Not generally recommended for quality wine production.

Seyval (Seyval blanc)

General History and Origin

Seyval, also called Seyve-Villard 5276, is frequently marketed as Seyval blanc. This variety ripens early and produces large, compact clusters.

Strengths

- Good hardiness when not overcropped.
- Fruitful secondary or base buds ensures a modest crop in event of primary crop loss to frost.

Weaknesses

- Very susceptible to Botrytis, nonspecific bunch rots and powdery mildew, and moderately susceptible to downy mildew, black rot, and crown gall.
- Fruitful base buds can lead to substantial overcropping, even with severe dormant pruning.

Viticultural Considerations

Vines can easily be overcropped, causing vine size and subsequent crop yields to suffer. Seyval needs cluster thinning and benefits from grafting to vigor-inducing rootstocks.

Wine Descriptors

Wine quality can be good. When grapes are harvested at optimal maturity, wines have attractive aromas of grass, hay, and melon. Wine body tends to be thin. Others describe the wine as clean and fresh. This versatile variety can be finished fresh and dry, barrel-fermented with malolactic fermentation and *sur lie* aged (on the lees or spent yeast cells), or made into sparkling wines.

Traminette

General History and Origin

Traminette, a cross between Joannes Seyve 23-416 and Gewürztraminer, was named and released in 1996 by the New York State Agricultural Experiment Station in Geneva, New York. This yellow-green variety ripens midseason.

Strengths

- Excellent wine quality. This variety produces fruit and wines similar to those of Gewürztraminer but with better balance of pH, Brix, and acidity, especially in warmer regions.
- Good cold-hardiness. Traminette is hardier than Gewürztraminer, comparable to Seyval. A field exposure to −17°F in New York killed 15% of Traminette's primary buds.
- Good disease resistance. The foliage and fruit are moderately resistant to powdery mildew, black rot, and Botrytis bunch rot.
- Large, loose clusters.

Weaknesses

- Trunks were reportedly susceptible to winter damage in Michigan, particularly on wet soils.

Viticultural Considerations

Very vigorous, semitrailing vines require proper canopy management. Vines grafted to C-3309 rootstock are still very vigorous and require a divided canopy, such as Geneva Double Curtain. Vines grafted to C-3309 produce higher yields than do non-grafted vines.

Wine Descriptors

Traminette is typically produced as a varietal wine which has some of the aroma and flavor characteristics of Gewürztraminer. Traminette's relatively high acid and low pH help complement its fresh-fruit aromas and flavors. Typically, wines made with some skin contact have strong spice and floral aromas, a full structure, and long aftertaste. The wine can be made dry or sweet but is usually finished with some residual sweetness. Varietal descriptors include floral, spicy, perfume, and lavender.

Vidal blanc

General History and Origin

Vidal blanc (Vidal 256) is a hybrid variety with long, loose clusters that ripen midseason to late season. Vidal blanc is a cross between Ugni blanc and Seibel 4986. The fruit is greenish-white with pronounced, darkened lenticels at fruit maturity.

Strengths

- Vidal blanc breaks buds very late and has good cold-hardiness but is not as hardy as Seyval.
- This variety is well suited to a wide variety of climatic and soil conditions.
- Relatively resistant to fruit rots, including Botrytis, downy mildew, and black rot.
- Can produce a commercial crop from the basal buds when all buds retained at pruning have been winter-killed.

Weaknesses

- Tomato and tobacco ringspot virus are common in this variety. These viruses weaken and often kill infected vines. Grafting to a pest-resistant rootstock may decrease the likelihood that vines will be infected with virus through nematode feeding.
- Foliar phylloxera infestations are a common occurrence.
- Only moderate (for hybrids) winter cold-hardiness.

Viticultural Considerations

Cordon training with spur-pruning and vertical, upright shoot positioning has worked well in Virginia. Some growers have reported better results with high cordons and downward shoot positioning. If the spurs become sparse in the mid-cordon area, some growers have resorted to cane pruning. High fruitfulness can lead to overcropping. This variety can easily produce more grapes than it can adequately ripen. If overcropped, the vines are stunted, and the yields and wine quality both suffer. This variety must be properly dormant-pruned or shoot-thinned after the threat of spring frost to adjust crop. Additional crop control—through shoot thinning, fruit cluster thinning, or a combination of the two—is often necessary. Regular nitrogen fertilization is generally required to maintain vigor, particularly with non-grafted vines.

Wine Descriptors

Vidal blanc is typically produced as a varietal wine. Like Chardonnay, Vidal blanc is versatile and can be used to make a variety of wine styles, including off-dry Germanic-style wines, sparkling wine cuvées (base wine), dry barrel-fermented table wines, and complex Burgundy-style products. Vidal can be a successful blending component with high terpene varieties, such as Riesling, Muscat Ottonel, and Malvasia bianca. Vidal blanc has also been used to produce late-harvest-style wines and ice wines (wines that have been made from fruit naturally frozen on the vine and still frozen at the time of pressing). Varietal descriptors include melon, pineapple, lead pencil, pears, and figs.

Vignoles (Ravat 51)

General History and Origin

This white-fruited variety originated from a cross between Siebel 6905 and Pinot de Corton and was named in 1970 by the Finger Lakes Wine Growers Association. Vignoles ripens early to midseason.

Strengths

- Late bud break.
- Ripens early.
- Vines are moderately to very cold-hardy.

Weaknesses

- Small, tight clusters are highly susceptible to Botrytis bunch rot and other cluster rots.
- Problems with fruit cracking.
- Low yields because of low bud fruitfulness.
- Very susceptible to powdery mildew.

Viticultural Considerations

Vignoles has an upright growth habit and low vigor when young, but vigor improves as vines establish.

Wine Descriptors

Fruit develops high sugar content while retaining acidity. Vignoles can produce many different styles of wine, including dry, barrel-fermented, *sur lie* aged wine, and sparkling wine cuvee. Vignoles is frequently used for dessert wines, particularly when picked late in the season. The wines from ripe fruit have tropical fruit, citrus, and pineapple flavors.

Table 3.2 ▪ Potential wine quality and viticultural characteristics of selected wine grape varieties

		Characteristic						
Variety	More detail	Potential wine quality [a,b]	Berry color	Harvest time [c]	Vigor [d]	Growth habit	Winter hardiness [e]	
Aurore	• page 57	Poor	White	Very early	Vigorous	Trailing	4	−10°F to −20°F
Baco noir (Baco #1)	• page 52	Poor to mediocre	Blue-black	Very early	Very vigorous	Semitrailing	4	−10°F to −20°F
Barbera	• page 42 • page 66 (clones)	Good to excellent	Blue-black	Midseason	Vigorous	Semiupright	1	+5°F to −5°F
Cabernet Franc	• page 43 • page 66 (clones)	Excellent	Blue-black	Late	Vigorous	Upright	3	−5°F to −15°F
Cabernet Sauvignon	• page 43 • page 66 (clones)	Excellent	Blue-black	Very late	Very vigorous	Upright	2	0°F to −10°F
Catawba	• page 53	Mediocre	Red/blue	Late	Vigorous	Trailing	5	−15°F to −25°F
Cayuga White	• page 57	Good	White	Early to midseason	Vigorous	Semitrailing	4	−10°F to −20°F
Chambourcin	• page 53	Good to excellent	Blue-black	Mid- to late-season	Low to moderate	Semitrailing	3	−5°F to −15°F
Chancellor	• page 53	Mediocre	Blue-black	Midseason	Moderate	Semiupright	4	−10°F to −20°F
Chardonel	• page 58	Excellent	White	Early to midseason	Moderate to vigorous	Semiupright	4	−10°F to −20°F
Chardonnay	• page 49 • page 66 (clones)	Excellent	White	Early to midseason	Vigorous	Upright	3	−5°F to −15°F
Chelois	• page 54	Mediocre	Blue-black	Early to midseason	Vigorous	Semiupright	4	−10°F to −20°F
Concord	• page 54	Poor to mediocre	Blue-black	Late	Vigorous	Trailing	5	−15°F to −25°F
De Chaunac	• page 55	Poor to mediocre	Blue-black	Early	Vigorous	Semiupright	5	−15°F to −25°F
Delaware	• page 58	Mediocre to good	Red-rose	Early	Low	Trailing	5	−15°F to −25°F
Diamond	• page 59	Mediocre to good	White	Midseason	Vigorous	Trailing	4	−10°F to −20°F
Frontenac	• page 55	Good	Blue-black	Midseason	Vigorous	Semitrailing	6	−25°F to −35°F
Gewürztraminer	• page 49 • page 66 (clones)	Excellent	White	Early	Vigorous	Trailing	2	0°F to −10°F
GR 7	• page 55	Good	Blue-black	Midseason	Very vigorous	Semitrailing	5	−15°F to −25°F
Léon Millot	• page 56	Mediocre	Blue-black	Early	Vigorous	Semitrailing	5	−15°F to −25°F
Limberger (Lemberger)	• page 44 • page 66 (clones)	Mediocre to good	Blue-black	Midseason	Vigorous	Semiupright	2	0°F to −10°F
Maréchal Foch (Foch)	• page 56	Mediocre	Blue-black	Very early	Moderate	Semitrailing	5	−15°F to −25°F
Merlot	• page 44 • page 67 (clones)	Excellent	Blue-black	Midseason	Vigorous	Semitrailing	1	+5°F to −5°F

Table 3.2 ▪ Potential wine quality and viticultural characteristics of selected wine grape varieties (*cont.*)

		Characteristic						
Variety	**More detail**	**Potential wine quality [a,b]**	**Berry color**	**Harvest time [c]**	**Vigor [d]**	**Growth habit**	**Winter hardiness [e]**	
Mourvèdre (Mataro)	• page 45 • page 67 (clones)	Good to excellent	Blue-black	Very late	Vigorous	Upright	2	0°F to −10°F
Muscat Ottonel		Good to excellent	White	Very early	Vigorous	Upright	2	0°F to −10°F
Nebbiolo	• page 45 • page 67 (clones)	Good	Blue-black	Late	Vigorous	Trailing	1	+5°F to −5°F
Niagara	• page 59	Poor to mediocre	White	Early to midseason	Moderate	Trailing	4	−10°F to −20°F
Norton (Cynthiana)	• page 56	Good to excellent	Blue-black	Very late	Moderate to vigorous	Trailing	4	−10°F to −20°F
Petit Manseng	• page 50 • page 67 (clones)	Excellent	White	Late	Vigorous	Semiupright	3	−5°F to −15°F
Petit Verdot	• page 46 • page 67 (clones)	Good to excellent	Blue-black	Late	Vigorous	Semiupright	2	0°F to −10°F
Pinot blanc	• page 50 • page 67 (clones)	Good	White	Early	Vigorous	Semiupright	2	0°F to −10°F
Pinot gris (Pinot Grigio)	• page 50 • page 67 (clones)	Excellent	White	Early	Moderate	Semiupright	2	0°F to −10°F
Pinot noir	• page 46 • page 68 (clones)	Good to excellent	Blue-black	Early	Moderate	Upright	2	0°F to −10°F
Riesling (White Riesling)	• page 51 • page 68 (clones)	Excellent	White	Midseason	Moderate	Semiupright	3	−5°F to −15°F
Rougeon	• page 57	Poor	Blue-black	Midseason	Vigorous	Semiupright	4	−10°F to −20°F
Sangiovese	• page 47 • page 68 (clones)	Good	Blue-black	Late	Vigorous	Semiupright	1	+5°F to −5°F
Sauvignon blanc (Fumé blanc)	• page 51 • page 68 (clones)	Excellent	White	Early	Very vigorous	Semiupright	1	+5°F to −5°F
Seyval (Seyval blanc)	• page 59	Good	White	Very early to early	Low to moderate	Semiupright	4	−10°F to −20°F
Syrah (Shiraz)	• page 47 • page 68 (clones)	Good	Blue-black	Midseason	Very vigorous	Semitrailing	1	+5°F to −5°F
Tannat	• page 47 • page 68 (clones)	Good to excellent	Blue-black	Late	Very vigorous	Semiupright	1	+5°F to −5°F
Tempranillo (Valdepeñas)	• page 48 • page 68 (clones)	Good	Blue-black	Midseason	Vigorous	Semiupright	1	+5°F to −5°F
Traminette	• page 60	Good to excellent	White	Midseason	Very vigorous	Semitrailing	4	−10°F to −20°F
Vidal blanc	• page 60	Good to excellent	White	Midseason to late	Moderate	Semiupright	3	−5°F to −15°F
Vignoles (Ravat 51)	• page 61	Good to excellent	White	Early to midseason	Moderate	Upright	5	−15°F to −25°F
Viognier	• page 52 • page 69 (clones)	Excellent	White	Early	Moderate	Upright	3	−5°F to −15°F
Zinfandel	• page 48 • page 69 (clones)	Good	Blue-black	Midseason to late	Vigorous	Upright	1	+5°F to −5°F

Table 3.2 ▪ Potential wine quality and viticultural characteristics of selected wine grape varieties *(cont.)*

Variety	Susceptibility to Diseases							Sensitivity	
	Black rot	Botrytis bunch rot	Downy mildew	Powdery mildew	Phomopsis	Crown gall	Eutypa	Copper [f]	Sulfur [f]
Aurore	High	High	Moderate	Moderate	Slight	Moderate	High	Some evidence	Uncertain
Baco noir (Baco #1)	High	Moderate	Slight	Moderate	Slight	High	Moderate	Uncertain	Uncertain
Barbera	Uncertain	Uncertain	High	High	Uncertain	Moderate	Uncertain	Uncertain	Uncertain
Cabernet Franc	Moderate	Slight	High	High	Moderate	High	High	Uncertain	Uncertain
Cabernet Sauvignon	High	Slight	High	High	High	High	High	Uncertain	Uncertain
Catawba	High	Slight	High	Moderate	High	Slight	Slight	Some evidence	Uncertain
Cayuga White	Slight	Slight	Moderate	Slight	Slight	Moderate	Slight	Some evidence	Uncertain
Chambourcin	High	Moderate	Moderate	Slight	Slight	Moderate	Uncertain	Uncertain	Some evidence
Chancellor	Slight	Slight	High	High	High	High	Slight	Some evidence	Some evidence
Chardonel	Moderate	Moderate	Moderate	Moderate	High	Moderate	Uncertain	Uncertain	Uncertain
Chardonnay	Moderate	High	High	High	High	High	Moderate	Uncertain	Uncertain
Chelois	Slight	High	Slight	High	High	Moderate	Moderate	Some evidence	Uncertain
Concord	High	Slight	Slight	Moderate	High	Slight	High	Some evidence	Some evidence
De Chaunac	Slight	Slight	Moderate	Moderate	High	Moderate	High	Some evidence	Some evidence
Delaware	Moderate	Slight	High	Moderate	High	Slight	Slight	Some evidence	Uncertain
Diamond	High	Moderate	Slight	High	Uncertain	Uncertain	Moderate	Uncertain	Uncertain
Frontenac	Moderate	Moderate	Slight	Moderate	Slight	Uncertain	Uncertain	Uncertain	Uncertain
Gewürztraminer	High	High	High	High	Uncertain	High	Uncertain	Some evidence	Uncertain
GR 7	Slight	Moderate	Moderate	Moderate	Slight	Slight	Slight	Uncertain	Uncertain
Léon Millot	Slight	Slight	Moderate	High	Slight	Uncertain	Slight	Uncertain	Some evidence
Limberger (Lemberger)	High	Slight	High	High	Uncertain	High	High	Uncertain	Uncertain
Maréchal Foch (Foch)	Moderate	Slight	Slight	Moderate	Uncertain	High	Uncertain	Some evidence	Some evidence
Merlot	Moderate	Moderate	High	High	Slight	High	High	Some evidence	Uncertain
Mourvèdre (Mataro)	Uncertain	Slight	High	High	Slight	High	High	Uncertain	Some evidence
Muscat Ottonel	High	Moderate	High	High	Uncertain	High	High	Uncertain	Uncertain
Nebbiolo	Uncertain	Slight	Moderate	High	Uncertain	Uncertain	Uncertain	Uncertain	Uncertain
Niagara	High	Slight	High	Moderate	High	Moderate	Slight	Some evidence	Uncertain

Table 3.2 ▪ Potential wine quality and viticultural characteristics of selected wine grape varieties (*cont.*)

Variety	Susceptibility to Diseases							Sensitivity	
	Black rot	Botrytis bunch rot	Downy mildew	Powdery mildew	Phomopsis	Crown gall	Eutypa	Copper [f]	Sulfur [f]
Norton (Cynthiana)	Slight	Slight	Moderate	Slight	Slight	Slight	Slight	Some evidence	Some evidence
Petit Manseng	Uncertain	Slight	Moderate	Moderate	Uncertain	Uncertain	Uncertain	Uncertain	Uncertain
Petit Verdot	High	Slight	High	High	Uncertain	High	High	Uncertain	Uncertain
Pinot blanc	High	Moderate	High	High	Uncertain	High	Moderate	Some evidence	Uncertain
Pinot gris (Pinot Grigio)	High	Moderate	High	High	Uncertain	High	High	Uncertain	Uncertain
Pinot noir	High	High	High	High	Uncertain	High	Moderate	Some evidence	Uncertain
Riesling (White Riesling)	High	High	High	High	Moderate	High	Moderate	Uncertain	Uncertain
Rougeon	Moderate	Moderate	High	High	High	Moderate	Slight	Some evidence	Some evidence
Sangiovese	Moderate	Slight	Moderate	Moderate	Moderate	High	Moderate	Uncertain	Uncertain
Sauvignon blanc (Fumé blanc)	High	High	High	High	Slight	High	Moderate	Uncertain	Uncertain
Seyval (Seyval blanc)	Moderate	High	Moderate	High	Moderate	Moderate	Slight	Some evidence	Uncertain
Syrah (Shiraz)	Moderate	Slight	Moderate	Moderate	Moderate	High	Moderate	Uncertain	Uncertain
Tannat	Moderate	Slight	Moderate	Moderate	Moderate	High	Moderate	Uncertain	Uncertain
Tempranillo (Valdepeñas)	Moderate	Slight	Moderate	Moderate	Moderate	High	High	Uncertain	Uncertain
Traminette	Slight	Slight	Moderate	Slight	High	Moderate	Slight	Uncertain	Uncertain
Vidal blanc	Slight	Slight	Moderate	High	Slight	Moderate	Slight	Uncertain	Uncertain
Vignoles (Ravat 51)	Slight	High	Moderate	High	Moderate	Moderate	Moderate	Uncertain	Uncertain
Viognier	Moderate	Slight	Moderate	Moderate	Moderate	Moderate	Uncertain	Uncertain	Uncertain

Sources: Adapted from Reisch et al. 2000, Bordelon 2001.

Note: Rankings are relative; and some, such as potential wine quality, are subjective.

[a] "Wine quality" is a subjective generalization about market appeal of well-made, representative wines, starting with high-quality fruit. The rating is not necessarily reflective of the experience with the variety in eastern North America. For example, Zinfandel can make an excellent wine under the right growing conditions, but the experience in the eastern North America has been poor.

[b] "Excellent"—can be made into varietal wines and blended. "Good"—can be made into varietal wines and blended. "Mediocre"—used for blending. "Poor"—used only for blending or specialty wines, or to be avoided.

[c] Harvest season may span two ratings, depending upon wine stylistic goals.

[d] Vigor, as grown on moderately fertile soils, with adequate moisture, and not overcropped. All species will, in general, be more vigorous if grafted to a pest-tolerant rootstock than if grown on their own roots. A dual rating here (for example, "low to moderate") reflects potential vigor differences due to grafting.

[e] Winter hardiness is a relative index (1 = least hardy; 6 = most hardy) based primarily on performance in the mid-Atlantic. Temperatures shown are the approximate warmest temperature ranges where 50%–100% primary bud kill might be expected to occur in midwinter, under optimal acclimation and cold hardiness conditions. The ranges could be decreased (colder) approximately 5°F in more northerly regions and may be slightly higher (warmer) in more southerly areas.

[f] Where noted, there is some evidence or experience that the variety is sensitive to either sulfur- or copper-fungicide phytotoxicity used at label rates. High temperatures (warmer than 90°F) may exacerbate sulfur phytotoxicity, while prolonged wet weather may exacerbate copper phytotoxicity. Uncertain generally means that evidence of phytotoxicity is either lacking or has not been evaluated.

Table 3.3 ▪ Nonexhaustive List of Available Clones of *V. vinifera* Varieties

Variety ▪ page references to more detail	Clone number or name (synonyms are in parentheses; see note for explanation of abbreviations)	Clone characteristics and details (see note for explanation of abbreviations)
Barbera ▪ page 62 (table 3.2) ▪ page 42	FPS 02 (Rauscedo 6), 03 (CVT 171), 04 (CVT 84), 06, 07, and 08.	▪ Leafroll virus was pervasive with the original Barbera FPS 01. FPS 06 is similar to FPS 01 but apparently free of leafroll virus. ▪ Under California conditions, Barbera FPS 02 was more fruitful and productive and had more compact clusters than did FPS 01, while Barbera FPS 06 had better fruit composition than that of FPS 02. ▪ Barbera FPS 03 (CVT 171) and FPS 04 (CVT 84) were selected in the Piedmont region of Italy.
Cabernet Franc ▪ page 62 (table 3.2) ▪ page 43	Clones FPS 01 (France); FPS 03 (ISV 1, Italy); FPS 04 (France 332); FPS 05 (France 331); FPS 09 (VCR 10, Italy); FPS 11 (France 214); FPS 12 (France 327); French 312; ENTAV 214; ENTAV 327; and ENTAV 623.	▪ Cabernet Franc FPS 01 was higher-yielding than were the other European clones under California conditions. ▪ Of the Italian clones, FPS 03 is from Conegliano, Italy; and FPS 09 is from Rauscedo. ▪ ENTAV 214, ENTAV 327, and ENTAV 623 are all considered to produce superior wines.
Cabernet Sauvignon ▪ page 62 (table 3.2) ▪ page 43	At least 33 clones of Cabernet Sauvignon are available in the U.S.: ▪ Clones FPS 02, 06, 07, 08, 11, 22, 23, 24, 29, 30, and 31 were selected from California sources. ▪ Clones FPS 04 and 05 are from Argentina. ▪ Clones FPS 12, 13, 14, 15, 19, 20, and 21 are from Chile. ▪ Clone FPS 10 is from Germany. ▪ FPS 38 is from Italy. ▪ Others include: FPS 33 (France 191), FPS 34 (France 191), FPS 35 (France 585), ENTAV 15EV, ENTAV 169, ENTAV 170, ENTAV 191, ENTAV 337, ENTAV 338, ENTAV 341, and ENTAV 412.	▪ In Virginia trials, FPS 06 produced wine with better color and phenol structure and with more varietal aroma, compared with FPS 08. ▪ The yield of FPS 06 is about 60% that of FPS 08. ▪ In France, ENTAV 169 can experience fruit-set problems; however, it produces balanced wines with round tannins. ENTAV 170, from the Loire, has small clusters; and ENTAV 412, from Espiguette, has an earlier maturing date.
Chardonnay ▪ page 62 (table 3.2) ▪ page 49	▪ Foundation Plant Services (FPS) currently lists 52 Chardonnay clones, including three ENTAV clones (76, 96, and 548). ▪ Other ENTAV clones available in the U.S. include 75, 95, 124, 131, 277, and 809.	▪ Nearly 100 clones and subclones are available. The clones will differ in yield, vigor, fruit intensity, and flavor profiles. ▪ ENTAV 77, ENTAV 809, FPS 79, and FPS 89 have a muscat flavor and are sometimes called "Chardonnay musqué." ▪ In California, FPS 04 and 05 are the most widely planted clones. ▪ Clones FPS 04, 05, 25, and 352 have relatively large clusters with large berries. FPS 04 is a high-yielding clone that was prone to rot and made a poor wine. In the same trials, FPS 06, FPS 15, and FPS 17 produced the highest quality wines; FPS 06 had small clusters and high yields. Both FPS 15 and FPS 17 were low-yielding with small, loose clusters of small, "shot" berries that ripened early. In preliminary tests in New York and Virginia, FPS 06 and FPS 15 had small clusters, and FPS 17 had average-size clusters.
Gewürztraminer ▪ page 62 (table 3.2) ▪ page 49	▪ FPS 01, 02, 03, 11, and 12. ▪ ENTAV 47.	▪ FPS clones 01, 02, 11, and 12 were selected in Alsace, France. ▪ FPS 03 originated in California.
Limberger (Lemberger) ▪ page 62 (table 3.2) ▪ page 44	FPS 01 and FPS 02.	Both Limberger clones originated in Germany.

Table 3.3 ▪ Nonexhaustive list of available clones of *V. vinifera* varieties (*continued*)

Variety ▪ page references to more detail	Clone number or name (synonyms are in parentheses; see note for explanation of abbreviations)	Clone characteristics and details (see note for explanation of abbreviations)
Merlot ▪ page 62 (table 3.2) ▪ page 44	At least 23 clones of Merlot are available in the U.S.: ▪ Clones FPS 01, 03, 06, and 18 are of U.S. origin. ▪ Clones FPS 09 (Rauscedo 3), FPS 10 (ISV-V-F-2), FPS 11 (ISV-V-F-5), FPS 12 (ISV-V-F-6), FPS 21 (ISV-V-F-5), FPS 23 (VCR 1), and FPS 24 (VCR 101) are from Italy. ▪ Merlot FPS 08 is from Argentina. ▪ The remainder of the clones are from France: FPS 15 (France 181), FPS 19 (France 343), France 347, FPS 20 (France 348), FPS 25 (France 314), ENTAV 181, ENTAV 314, ENTAV 343, ENTAV 346, ENTAV 347, and ENTAV 348.	▪ In California, FPS 03 is the standard due to its consistent fruit set, yield, and fruit composition. FPS 01 (Inglenook) and FPS 06 (Monte Rosso) have similar performance to FPS 03. ▪ FPS 08 reportedly has a low yield and poor fruit set, particularly in cool weather. ▪ FPS 09 seems to be similar to clones FPS 01 through 08. ▪ ENTAV 181 has small clusters, and ENTAV 348 reportedly produces wines with high polyphenol content.
Mourvèdre (Mataro) ▪ page 64 (table 3.2) ▪ page 45	▪ FPS 04 (Mataro FPS 01) and FPS 03 were selected from California sources. ▪ ENTAV 233, 369, and 450 are also available from some sources.	In California, ENTAV 369 has relatively small clusters and can have problems with fruit set.
Nebbiolo ▪ page 64 (table 3.2) ▪ page 45	▪ Nebbiolo FPS 01 and Nebbiolo Fino FPS 02 (FPS 09) were selected from California sources. ▪ FPS 02, FPS 03, FPS 04, FPS 06 (CVT 142), FPS 06 (CVT 142), FPS 07 (CVT 036), FPS 08 (CVT 230), Nebbiolo Lampia FPS 01 (FPS 10), and Nebbiolo Michet S1 (FPS S1) all originated from the University of Torino, Italy.	▪ Nebbiolo clones are subdivided into morphologically distinct subtypes or subclones—Rosé, Lampia, Michet, and Bolla—which Manini (1995) described as follows: The subtypes vary in leaf and cluster shape and color, yields, and wine quality. ▪ The more vigorous vines tend to produce lower-quality wines. In Italy, Michet is considered to have the best enological aptitude; Lampia and Rosé, good; and Bolla, medium-low. ▪ Rosé has blue-violet berries, and the others have blue-black berries. ▪ Michet has moderate vigor and yield. Rosé has slightly lower vigor and yield. Lampia has good vigor and yield. Bolla has the highest yields. ▪ In keeping with U.S. wine labeling rules of TTB, all Nebbiolo subclones were recently renamed. Lampia FPS 01 became FPS 10. Fino FPS 02 became FPS 09. Michet FPS S1 became FPS S1.
Petit Manseng ▪ page 65 (table 3.2) ▪ page 50	ENTAV clones 440 and 573.	▪ In France, two ENTAV clones (440 and 573) are available, and both are from the Pyrénées-Atlantique (Toulouse). ▪ ENTAV 440 reportedly has a smaller cluster weight, and ENTAV 573 has denser vegetation and makes a particularly good sweet wine. ▪ The Petit Manseng evaluated in Virginia was obtained from the New York State Agricultural Experiment Station in Geneva, New York.
Petit Verdot ▪ page 63 (table 3.2) ▪ page 46	At least two clones are available in the U.S.: ▪ ENTAV 400 from France and FPS 02 from the U.S.	Growers in the mid-Atlantic have reported occasionally poor fruit set with FPS 02.
Pinot blanc ▪ page 63 (table 3.2) ▪ page 50	At least five clones are available in the U.S.: ▪ FPS 05 (Rauscedo 9), FPS 06 (Rauscedo 5), FPS 07 (France 55), ENTAV 54, and ENTAV 55.	▪ ENTAV 54 and ENTAV 55 were obtained from Colmar, France. ▪ Clones FPS 05 and 06 originated from Rauscedo, Italy.
Pinot gris (Pinot Grigio) ▪ page 63 (table 3.2) ▪ page 50	At least nine Pinot gris clones are available: ▪ FPS 01 (California); FPS 04 (France 53); FPS 05 (France 53); FPS 06 (Germany); FPS 08 (VCR 5); FPS 09 (France 52); FPS 10 (SMA 505, Italy); FPS 11 (Alsace, France); and ENTAV 52.	

Table 3.3 ▪ Nonexhaustive list of available clones of *V. vinifera* varieties *(continued)*

Variety ▪ page references to more detail	Clone number or name (synonyms are in parentheses; see note for explanation of abbreviations)	Clone characteristics and details (see note for explanation of abbreviations)
Pinot noir ▪ page 63 (table 3.2) ▪ page 46	There are more clones of Pinot noir than any other wine grape variety. At least 59 are available in the U.S.	▪ Pinot noir clones are divided into four main groups: standard (Pinot fin), highly fruitful (Pinot fructifier), upright shoots (Pinot droit), and loose-clustered (Mariafeld). ▪ In clonal trials in New York, Clones 29 (Jackson), Mariafeld, and Pernard all produced good-quality wines. Clone 29 was cold-hardy but susceptible to bunch rots; Pernard had above-average winter hardiness but was susceptible to bunch rots; Mariafeld was resistant to bunch rots but was cold-tender. Yields for these clones were moderate.
Riesling (White Riesling) ▪ page 63 (table 3.2) ▪ page 51	At least five Riesling clones are available in the U.S.: ▪ FPS 04; FPS 09 (Geisenheim 110, Germany); FPS 10 (Martini #107, California); FPS 12 (Neustadt 90, Germany); and FPS 17 (Geisenheim 198, Germany).	▪ Geisenheim 239 is the most widely planted in Germany. ▪ In California, FPS 09 has an extremely fruity, slightly muscat flavor and, in warmer sites, is not regarded as typical of German Riesling wines; FPS 17 has lower yields and produces wines of elegant fruitfulness and pronounced flavor.
Sangiovese ▪ page 65 (table 3.2) ▪ page 47	At least 14 clones are available in the U.S.: ▪ FPS 15 and 17 were selected from California sources. ▪ The other clones are from Italy: FPS 02, 04 (Rauscedo 10), 05, 06, 07 (VCR 6), 08 (VCR 19), 09 (VCR 30), 10 (VCR 23), 12 (B-BS-11), 13 (VCR 102), 14, and 18 (VCR 221).	▪ Sangiovese clones are broadly divided into grosso (large-berried) and piccolo (smaller-berried) clones. ▪ Sangiovese grosso has less vigor, and the must is usually less sweet and acidic than Sangiovese piccolo; however, the grosso form has more consistent yields. ▪ In California trials, clone FPS 02 was more fruitful than FPS 04. FPS 04 tends to have heavier berries, more bunch rot, and inferior fruit composition compared to FPS 02.
Sauvignon blanc (Fumé blanc) ▪ page 68 (table 3.2) ▪ page 51	At least 19 clones of Sauvignon blanc are available in the U.S.: ▪ FPS 03, 22, 23, and 26 are from California heritage sources. ▪ Clones FPS 06 (ISV 5), FPS 07 (ISV 2), FPS 17 (ISV 1), and FPS 24 (ISV 3) are from Italy. ▪ Clones FPS 14 (French 316), FPS 18 (French 317), FPS 20 (French 242), FPS 21 (French 378), FPS 25 (French 378), and FPS 27 are from France. ▪ The ENTAV clones 241, 297, 317, 376, and 530 are also available.	The following description is from Ministry of Agriculture, Fisheries, and Food: ▪ ENTAV 297 produces balanced wines and is somewhat sensitive to Botrytis. ▪ ENTAV 241 produces light wines that are slightly aromatic. ▪ ENTAV 317 is from Gironde and produces typical dry or sweet wines. ▪ ENTAV 530 is from Cher and produces dry, aromatic, complete wines that can be heavy. ENTAV 530 also ripens slightly earlier than the other clones. ▪ Sauvignon blanc FPS 27 has a muscat flavor.
Syrah (Shiraz) ▪ page 63 (table 3.2) ▪ page 47	At least 25 clones of Syrah (Shiraz) are available in the U.S., including: ▪ FPS 04 (France 300), 05 (France 174), 06 (France 100), 07 (France 877), 08, 09, and 10. ▪ ENTAV INRA clones 99, 100, 300, 471, 525, and 877.	▪ At least 25 clones of Syrah (Shiraz) are available in the U.S. ▪ The French distinguish between petit Syrah and grosse Syrah. Grosse Syrah has better berry set, is more productive, and produces a wine of lesser quality than petit Syrah. ▪ Syrah 08 (Durell clone) originated in California. ▪ Most of the other Syrah clones originated in France.
Tannat ▪ page 63 (table 3.2) ▪ page 47	▪ FPS 01 was selected in California. ▪ ENTAV 474 and ENTAV 794 are from France.	
Tempranillo (Valdepeñas) ▪ page 63 (table 3.2) ▪ page 48	At least five Tempranillo clones are available in the U.S.: ▪ FPS 02 and FPS 03 (Clone 43) are from Spain. ▪ FPS 07 originated in Italy. ▪ Valdepeñas FPS 03 (Tempranillo) originated in California. ▪ ENTAV 770 is from France.	

Table 3.3 ▪ Nonexhaustive list of available clones of *V. vinifera* varieties *(continued)*

Variety ▪ page references to more detail	Clone number or name (synonyms are in parentheses; see note for explanation of abbreviations)	Clone characteristics and details (see note for explanation of abbreviations)
Viognier ▪ page 63 (table 3.2) ▪ page 52	FPS 01, FPS 02, FPS 03, FPS 04, and ENTAV 642.	▪ Viognier FPS 01 was selected from INRA (France). ▪ ENTAV 642, from the Rhône Valley, is a productive clone that produces wine with good varietal character. ▪ Viognier FPS 02, 03, and 04 were selected from California sources.
Zinfandel ▪ page 63 (table 3.2) ▪ page 48	At least seven Zinfandel clones are available in the U.S., all from California sources: ▪ FPS 01A, 02, 03, 06 (FPS 01A), 08, 13, and 16.	▪ In California, Zinfandel FPS 01, 02, 03, and 06 have a poor reputation under California conditions because they produce large clusters and berries, have poor fruit color, and lack varietal character. ▪ Primitivo FPS 03, 05, and 06 (all of Italian origin) are commercially available and have more clusters with smaller berries. Primitivo (a clonal subgroup) ripens before Zinfandel and is less prone to bunch rot.

Note: Table 3.2 (page 62–65) lists characteristics of these and other *V. vinifera* varieties.

Note: Clonal designations are subject to revision. Performance characteristics will vary depending upon the environment, cropping goals, and seasonal variance. Readers should consult local or state sources to determine if more specific recommendations are available for clones.

Note: The following abbreviations are used in clone numbers and names and originate as follows:

- B-BS. Brunello-Biondi-Santi (Italy)
- CVT. (Italy)
- ENTAV. L'Etablissement National Technique pour l'Amélioration de la Viticulture (France). INRA (l'Institut National de la Recherche Agronomique) is the French National Institute for Agricultural Research. Clones listed as "ENTAV" are ENTAV-INRA® proprietary clones. See HTTP://WWW.ENTAV.FR/ for further details. ENTAV-INRA® is a registered trademark used for distribution of ENTAV-INRA–derived clones through internationally licensed distributors. (France)
- FPS. Foundation Plant Services (HTTP://FPMS.UCDAVIS.EDU/GRAPE/GRAPEPROGRAMINDEX.HTML).
- ISV (IS-V-F) (Italy)
- SMA (Italy)
- TTB. U.S. Department of Treasury's Alcohol and Tobacco Tax and Trade Bureau (HTTP://WWW.TTB.GOV).
- VCR. Vivai Cooperativi Rauscedo (Italy)

References

Bettiga, L. J., et al. 2003. *Wine Grape Varieties in California.* University of California Agriculture and Natural Resources Pub. 3419. 188 p.

Bordelon, B. 2001. *Grape Varieties for Indiana.* Purdue Cooperative Extension Service Commercial HO-221-W, West Lafayette. Available HTTP://WWW.HORT.PURDUE.EDU/EXT/HO-221.PDF

Delas, J. J. 1992. Criteria used for rootstock selection in France. Pp 1-14. In: J.A. Wolpert, M.A. Walker and E. Weber (Eds.) Rootstocks: A Worldwide Perspective, Reno, NV. American Society for Enology and Viticulture, Davis, CA.

Galet, P. 1979. A practical ampelography. Grapevine identification. Translated and adapted by Lucie T. Morton. Cornell University Press, London. 248 p.

Gladstones, J. 1992. *Viticulture and Environment.* Winetitles, Adelaide 310 p.

GRIN. National Genetic Resources Program. Germplasm Resources Information Network (GRIN). [Online Database].USDA/ARS, Beltsville, Maryland. Available: HTTP://WWW.ARS-GRIN.GOV/.

Happ, E. 2000. Site and varietal choices for full flavour outcomes in a warm continent. *The Australian and New Zealand Wine Industry Journal* 15:54-62.

Howell, G.S. Vitis Rootstocks. 1987. In: *Rootstocks for Fruit Crops.* R.C. Rom and R.F. Carlson (Eds). pp 451–472. John Wiley and Sons, New York.

Kasimatis, A.N. and L.A. Lider. 1980. *Grape Rootstock Varieties.* Leaflet #2780, Division of Agriculture and Natural Resources Publications, University of California.

MAFF. 1997. Catalogue of selected wine grape varieties and certified clones cultivated in France. Ministry of Agriculture, Fisheries and Food. Onivins. Imprimerie Maraval St. Pons de Thomieres. France. 269 p.

Manini, F. 1995. Grapevine Clonal Selection in Piedmont (Northwest Italy): Focus on Barbera and Nebbiolo. Proceedings of the International Symposium on Clonal Selection. American Society of Enology and Viticulture. June 1995. Pp. 20–32.

Pongracz, D.P. 1983. *Rootstocks for Grapevines.* David Philip, Cape Town, South Africa.

Pool, R. M., et al. 1995. Pinot noir clonal research in New York. Proceedings of the International Symposium on Clonal Selection. American Society of Enology and Viticulture. June 1995. Pp. 45–51.

Reisch, B. I., R.M. Pool, D.V. Peterson, M.-H. Martens, and T. Henick-Kling. 2000. *Wine and Juice Grape Varieties for Cool Climates.* Adaptation of Information Bulletin 233, Cornell Cooperative Extension. HTTP://WWW.NYSAES.CORNELL.EDU/HORT/FACULTY/REISCH/BULLETIN/WINE

Robinson, J. 1986. *Vines, Grapes and Wines.* Knopf, New York. 280 p.

Southey, J. M. 1992. Grapevine rootstock performance under diverse conditions in South Africa. In : J.A. Wolpert, M.A. Walker and E. Weber (Eds.) *Rootstocks : A Worldwide Perspective,* Reno, NV. pp 27–51. American Society for Enology and Viticulture, Davis, CA.

Wolf, T. K., et al. 1999. *Commercial Grape Varieties for Virginia.* Virginia Cooperative Extension Publication 463-019. 34 p. HTTP://WWW.EXT.VT.EDU/PUBS/VITICULTURE/463-019/463-019.PDF

Additional Resources

Cold Climate Cultivars. Iowa State University, Department of Horticulture. 2008. Available: HTTP://VITICULTURE.HORT.IASTATE.EDU/CULTIVARS/CULTIVARS.HTML.

National Grape Registry. University of California, Davis, CA. Available: http://ngr.ucdavis.edu/.

Vitis International Variety Catalogue (VIVC). Institute for Grapevine Breeding—Geilweilerhof, Germany HTTP://WWW.VIVC.BAFZ.DE/INDEX.PHP.

—4—

Vineyard Design and Establishment

Author

Thomas J. Zabadal, Michigan State University

When a site has been selected and the decision has been made to plant a vineyard, the next step is vineyard design. The design may be straightforward if the field is square or rectangular and has uniform topography. The process becomes much more challenging with undulating topography and an irregularly shaped parcel of land. Below are guidelines for designing a vineyard.

ROW ORIENTATION

The preferred orientation of rows with standard vertical trellising systems is north to south, with slight deviations (less than 10°) from that axis to maximize sunlight interception on the east and west sides of the grapevine canopies. However, other row orientations also can result in highly productive vineyards with quality fruit. If the dimensions of a field would result in either numerous, short north-to-south oriented rows or fewer, much longer, east-to-west oriented rows, then choose the latter. Or, if north-to-south rows would run up-and-down a slope, then plant east-to-west rows across that slope to reduce the risk of soil erosion (see sidebar, page 72).

Vineyard rows are almost always designed to be straight. However, on rare occasions they may follow topographic contours, as in the case of terraced vineyards. Such an unusual vineyard design mandates special considerations for the design, installation, and management of a trellis.

ROW SPACING

Vineyard row spacing is heavily influenced by the vineyard equipment that is to be used. Modern vineyard equipment in eastern North America typically has a width of 60 to 65 inches. Therefore, vineyard row widths from 8 to 9 feet are common when using standard two-dimensional trellising. Exceptionally narrow or over-the-row equipment can be used to reduce row width. The use of three-dimensional trellises, such as those with Geneva Double Curtain or Lyre training systems, will require wider rows. The selection of vineyard row spacing involves a trade-off between yield and fruit quality. Excessively wide rows do not optimally intercept sunlight; therefore, vineyard productivity is unnecessarily compromised. Narrow rows, on the other hand, may shade each other. Although yields may be increased with narrow rows, fruit quality may be compromised if the row width to canopy height ratio is less than 1:1 due to shading by adjacent rows.

SIDEBAR

Preferred Orientation of Rows on a Slope

The preferred orientation for rows on a slope has typically been perpendicular to the direction of the slope, with the rows running across the slope. This arrangement minimizes the likelihood of soil erosion. Erosion is more apt to occur with rows that run up and down slopes because surface water can accelerate down the weed-free strip under the trellis. On steep slopes—those greater than about 15%—rows planted across the slope present their own problems: Towed machinery such as mowers and sprayers can drift into the downhill row, and over-the-row machinery such as hooded-boom sprayers or canopy trimmers, can be difficult to align with the vertical axis of the trellis. Furthermore, equipment operation becomes increasingly susceptible to roll-over as the slope approaches 25%. For these reasons, some vineyards are being installed with rows that run up and down steep slopes. Two requirements in these situations are that the vineyard has a tractor (usually four-wheel drive or fitted with tracks) with sufficient horsepower to navigate the incline while pulling a load, and that elaborate erosion control measures are used in the vineyard. The latter might simply consist of allowing sod or weeds to develop under the trellis on all or a portion of each row to check the downhill movement of water and soil. A modest, potential advantage of rows that run up and down the hill is the improved cold air drainage out of the vineyard afforded by this orientation.

TRELLIS HEIGHT

Standard trellis height is 5.5 to 6 feet. The appropriate maximum height of a trellis is related to vineyard row spacing. To avoid excess shading of one row of vines by another, the height of the grapevine canopy should not exceed the vineyard row width. For example, with row spacing of 9 feet, grapevine canopies up to 9 feet high might be used to enhance sunlight interception. There are additional factors to consider. The effective distance between canopies in two adjacent rows is not the distance between the two planes of the trellis but rather the distance between the outer edges of the canopies. Canopies 18 inches wide in a vineyard with a 9-foot row spacing would have only 7.5 feet from the wall of one canopy to the other. Also consider the location of the functional canopy on the trellis. If the vine canopy begins 3 feet off the ground at a cordon, shade below that is relatively unimportant. Finally, consider how far a canopy will be allowed to develop above the top wire of a trellis. A vineyard with a 6-foot high trellis could have canopies extending 7.5 to 8.0 feet above the ground. Each of these factors should be considered when adjusting the rule-of-thumb guideline that the trellis (actually canopy) should not be higher than the row width.

ROW LENGTH

Vineyard row length depends upon the engineering of the trellis, topography, dimensions of the vineyard site, and the extent of access desired to vineyard rows. Most eastern North American vineyards use a rigid end-post assembly to maintain tension of the trellis wires. Vineyard rows 1,500 feet or more in length can be efficiently managed in this manner, but, in other regions trellises are constructed so that wires are secured to each line post or even individual stakes at each vine. There is no tensioning of the wires at end-post assemblies. This is typical in warm climates when frost heaving of posts is not a factor. With this type of trellis construction, the length of rows is unlimited. When a vineyard is situated on rolling topography, a swale area is a logical place for an alley between two vineyard blocks. Long vineyard rows reduce the cost of end-post assemblies and increase the efficiency of vineyard tasks involving equipment such as pesticide spraying, mechanical harvesting, mowing, and disking row middles; however, manual tasks such as pruning, tying of vines, crop adjustment, suckering, and hand harvesting may be more easily managed with relatively short rows. Consider these factors to determine the appropriate vineyard row length.

VINE SPACING

A wide range of vine spacing has been used in vineyard plantings. From the viticultural perspective, optimum vine spacing allows vines to fill an entire trellis by midseason with a grapevine canopy that has one-and-one-half to two layers of leaves. All leaves in this ideal canopy would be at least partially exposed to the sun and, as a result, would be photosynthetically functional.

Several factors influence grapevine canopy development. These include: the fertility of the soil; the inherent vigor of the variety being grown; the relative vigor of the rootstock, if one is used; fertilization practices; vineyard floor management practices; vine pruning practices; the choice of a vine training system; the cropping level; the choice of canopy management practices; the particular growing season relative to heat and precipitation; the possible use of irrigation; and the overall health of the vine relative to pressures of pests (insects and disease). With all of these factors influencing vine development, it should be apparent that there is no definitive vine spacing recommendation. If the vine space for a vineyard is too wide, the mature grapevines will make incomplete use of the vineyard trellis, display less than the ideal vine canopy, and produce less than the full crop potential. Very narrow in-row vine spacing (less than 4 feet) increases vineyard establishment costs and can lead to dense canopies by excessive lateral shoot development. Additional vineyard tasks including leaf removal, shoot thinning, summer pruning, and shoot positioning will be needed to manage the excess vigor (chapter 6, page 124). Therefore, a grower's choice of vine spacing at the time of planting is a best estimate of the mature size of these vines as a result of all the factors listed above. The most common vine spacings for wine grape varieties in eastern North America ranges from 5 to 7 feet. So-called high-density plantings use an in-row spacing as narrow as 3 feet (approximately 1 meter) and require careful attention to maintain vine manageability.

HEADLANDS, ALLEYWAYS, ACCESS ROADS

Portions of a vineyard site must be developed for the movement of equipment and people in and out of the vineyard. Areas at the end of the vineyard rows are called headlands. Headlands must be wide enough to allow turning of equipment in and out of rows. Depending on the equipment used, a 20-foot-wide headland can be a tight fit; generally, a 30-foot-wide headland provides a more comfortable turning radius. When planning the width of a headland, keep in mind that the first vine in a vineyard row often will be situated 3 to 4 feet from the end post. An anchoring mechanism for that end post might be 2 to 3 feet beyond the end post itself. Because of this, the first vine in a row may need to be 36 to 38 feet from the edge of a field to end up with a 30-foot-wide headland.

Alleyways are breaks between vineyard blocks. They may be logical breaks in swale areas, or they may simply break up long rows to facilitate vineyard access for hand work. Often, they are not as wide as headlands to conserve land because large pieces of equipment such as sprayers and harvesters are typically not turning around in these areas. A 20-foot-wide alleyway is often adequate. The proximity of access roads to a vineyard is often determined by the lay of the land and the characteristics of the surrounding land. For example, consider a vineyard that is planted with rows parallel to adjacent woodland. A 30-foot-wide access road along the woodland edge might be the best use of that land. If grapevines were to be planted in that strip, they would be shaded.

Access roads may also make use of the least desirable parts of the field such as low-lying or swale areas. Low-lying areas of a field often are undesirable for planting, but they may be appropriate as staging areas for equipment during harvest. Some vineyards may have special design requirements for establishing perimeter electric fencing for deer or raccoon control or modification of headlands and access roads for specialized equipment such as mechanical harvesters or bird-netting application apparatus.

MAPPING THE VINEYARD

When the details of the vineyard design have been resolved, it's time to put stakes in the ground at the corners of the vineyard. A grower is fortunate when a vineyard is a straightforward rectangular area that can be marked by four corner stakes. If, however, the land dictates an irregularly shaped vineyard with rows of uneven length, this is not a problem. Simply place a stake at each point along the periphery of the vineyard at which the direction changes. Then measure the length and direction of each segment of the vineyard periphery from stake to stake.

Calculate the area of a rectangular vineyard by multiplying the length times the width. For irregularly shaped vineyard blocks, sketch an outline of the vineyard block on paper. Then subdivide that area into a series of rectangles and right triangles (figure 4.1). Compute the total area of the vineyard block by adding the areas of the rectangles and right triangles (figure 4.1). If the stakes are placed in the ground so that they mark the positions of outside rows of the vineyard, the actual vineyard will be somewhat larger because the vines in the outside rows will occupy a half-row of space outside the calculated area. To account for this, add to the calculated area an additional area, calculated by multiplying the length of the longest side of the vineyard by one row-width. This will provide a very close estimate of the total vineyard area. Use this value to calculate the number of vines to order for planting. The sketch of the vineyard should then be transferred to a permanent map. Details such as the direction of rows, row numbers, and alleyways can be added as the vineyard is developed.

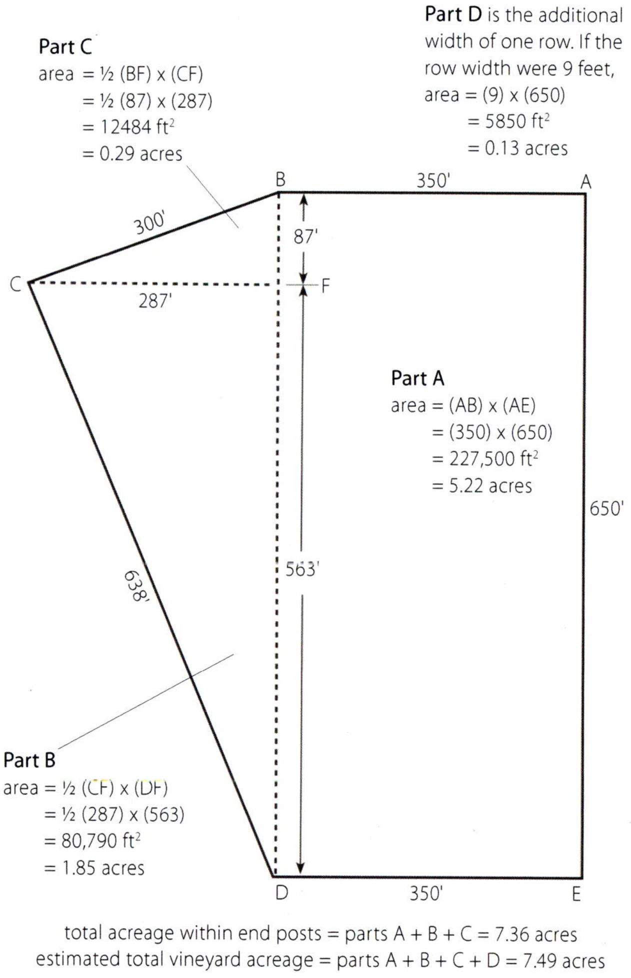

Figure 4.1 ▪ A schematic showing how to calculate the area of a new vineyard block.

Geographic Information Systems (GIS) have been used for vineyard design for several years. This technology integrates different types of vineyard information such as soil types and topography into a spatial design. Those interested in this approach to vineyard establishment will find information at several internet web sites.

PREPARING THE VINEYARD SITE

After the vineyard design is set, it is time to begin preparing the site for planting. Too often, individuals decide during the winter, or even in the early spring, to plant grapevines. Hastily prepared ground for a new vineyard may result in long-term poor vine performance. Good vineyard site preparation often takes one or two years or, in special instances, longer.

The first action that often comes to mind is soil testing and soil amendments. Indeed, one or more soil tests—including soil pits to evaluate soil physical properties—should be conducted prior to vineyard planting and as part of the vineyard design process. Soil evaluation is discussed in chapter 2, page 14, and chapter 8, page 141.

Lime, potassium, or other nutrient amendments indicated by soil tests should be incorporated into the soil profile by plowing, disking, or rotovating at an early stage of site preparation. Removal of brush, rocks, and other debris from the vineyard site must be done before the establishment of a permanent sod or vine planting. The management of soil surface water flow should also be addressed during site preparation. Sodded waterways can be used to channel the flow of surface water. Stand pipes that are plumbed into underground drainage are possibilities in swale areas (figure 4.2). Internal soil drainage is another aspect of site preparation. Presumably, a site has been selected with moderate to well-drained soils. Nevertheless, even with those soils there may be a need to improve internal soil drainage in portions of the field. Tile lines have traditionally been set at 25- to 50-foot intervals. In recent years, some vineyards on heavy clay

Figure 4.2 ▪ Swale area.

A swale area on sloping ground that will be planted to a vineyard. The bales of straw were used to stabilize the soil immediately after plowing and disking. The orange stand pipes are located in catch basins and are connected underground to a 6-inch tile to carry water downhill without causing surface soil erosion.

soils have had tile lines placed alongside every vineyard row. Professional, agronomic consultation on both internal and surface water management is appropriate if the grower wishes to maximize grape and wine potential.

Weed control is often the most important aspect of vineyard site preparation (see chapter 13, page 262). Weed control involves not only the obvious removal of trees and shrubs but also the control of herbaceous perennials and annuals. Weeds are much more difficult to control after vines have been planted, so there should be a dedicated effort to control weeds during site preparation. Any plant growth that cannot be controlled within several feet of planted vines will significantly reduce vine growth. The goal should be to eradicate herbaceous perennials and to minimize the population of annual weed seeds. This is achieved through a combination of cover cropping and herbicide use. Several strategies may be employed. A systemic herbicide such as glyphosate may be used to kill all existing vegetation on a vineyard site in the late summer or early fall of the year before planting (figure 4.3). If there is a heavy sod and the site is not vulnerable to surface erosion, killed sod may be rough disked or plowed in the fall to facilitate the breakup of the sod with freeze/thaw cycles through the winter. Otherwise, the sod may be left intact until the following spring. Cover cropping is very helpful for managing weed populations. The use of Roundup-ready™ soybeans as a cover the year before planting a vineyard makes it possible to simultaneously cover-crop the site, increase soil nitrogen, and spray all existing weeds with herbicides (figure 4.4). It would seem that annual weed seeds are ubiquitous and plentiful. However, the number of annual weed seeds produced just the year before planting vines may greatly influence the weed populations that develop immediately after the planting of vines. The practical impact for the grower is that a good pre-emergence herbicide program applied to a new vineyard immediately after the planting of vines may perform quite differently depending upon the weed

Figure 4.3 ▪ A field in early spring that will be planted to a vineyard.

The field was sprayed with Roundup (glyphosate) in the previous fall. In the distance, drain tile is being installed in a low spot of the field prior to the planting of vines.

Figure 4.4 ▪ A field being prepared for a vineyard.

The brown plants are Roundup-resistant soybeans that have died back naturally. During the previous growing season, Roundup (glyphospate) was sprayed to eradicate weeds from the field. In late summer, a fescue cover was direct-seeded on top of the soybean planting.

seed population present. Cover-cropping the year before planting of vines, rather than allowing the site to grow freely to weeds, can greatly reduce the weed pressure in the first year of the vineyard. Annual rye is effective at reducing the carryover of annual weed seeds.

Several species of nematodes—very small, worm-like organisms in the soil—feed on the roots of grapevines. Some can be the source of serious grapevine virus diseases such as fan leaf, tomato ring spot, and tobacco ring spot. It is common practice in some regions to reduce nematode populations in the soil by actively managing the soils of a vineyard site before planting. Practices include cover cropping and periods of fallowing and/or fumigation. In other viticultural areas, the hazard of nematode infestations in the soils of vineyard sites is seldom addressed. In all viticultural regions, the hazard of nematode and other root pests is greatly increased when the grapevines are planted on a site that was previously planted to grapevines or to several other woody plant species. Seek local expertise to determine the merits of managing the vineyard site with regard to nematode populations prior to planting (see sidebar).

SIDEBAR

Preplant Renovation and Soil Conditioning for New Vineyards

Joseph A. Fiola, Ph.D.
Specialist in Viticulture and Small Fruit
University of Maryland

POOR PERFORMANCE OF NEW PLANTINGS

Many newly planted vineyards seem to do well for some years, while others show signs of early plant decline or never seem to reach their full potential. The latter situation often occurs on sites where fruit had been previously planted and also on sites that are left fallow for many years. There are many factors that can lead to poor performance of new plantings; however, two reasons that are frequently overlooked are high populations of plant-parasitic nematodes and low organic matter.

Plant-Parasitic Nematodes

Plant-parasitic nematodes are tiny, eel-like roundworms that feed in various ways on plant roots. There are three species of nematodes that can cause damage and reduce wine grape production, either directly or through induced disease. Lesion nematodes (*Pratylenchus penetrans*) enter roots and travel longitudinally within them, killing many cells and opening the roots to secondary invasion by root rotting fungi. Dagger nematodes (*Xiphinema americanum*) feed from outside the root but can cause severe stunting since they feed just behind the growing tips. Dagger nematodes can also vector tomato ringspot and tobacco ringspot viruses (see chapter 11, page 216). Root knot nematodes (*Meloidogyne spp.*) also feed from within the roots, inducing large gall-like deformities that impair overall root function.

Soil Organic Matter

Many of the upland agricultural soils in eastern North America contain only marginal amounts of organic matter (often less than 2%). Soils with high organic matter content (3% to 4% or higher) are generally teeming with microbial organisms (bacteria, fungi, protozoa, nonparasitic nematodes), many of which are antagonistic to soil-borne plant pathogens.

SITE CONDITIONING PROGRAM

A preplant, soil-conditioning program is aimed at improving native soil conditions and revitalizing old sites by encouraging the establishment of a soil ecosystem that will support long-term productivity. This is done by incorporating organic matter and reducing resident populations of soil-borne, plant-parasitic nematodes using green manure crops. The "complete" program is recommended for soils that have tested positive for specific plant parasitic nematodes and that are also low in organic matter. The program

Irrigation of grapevines immediately after planting is desirable in some situations. However, many successful vineyards receive no irrigation. Contact local viticultural authorities regarding the importance of irrigation to your newly planted vines. Irrigation throughout a vineyard's first growing season often results in significantly increased vine growth (see chapter 9, page 169).

Good vineyard site preparation will make it possible to work the site as early in the spring as the weather allows, ensuring early planting and an early start for the vines.

PURCHASING VINES

The purchase of grapevines is a critical step in the establishment of a productive vineyard. The choice of planting stock would seem complicated enough when deciding which variety to grow, but the choice is complicated with other considerations such as the selection of a clone within a variety and rootstock selection for grafted vines and the health of grapevines. Viral, fungal, and bacterial diseases can be present in grapevine tissues and can be transmitted to a new vineyard. To protect against these,

works best if initiated two years prior to planting. If the soil tests negative for nematodes or if the level of the organic matter is sufficient (see chapter 2, page 14), the program is probably not needed. Timing is critical, as each has specific timing and temperature requirements for maximum effectiveness.

Two Years Before Planting

March to May. Collect and submit soil samples taken from the top 12 to 16 inches of soil for pH and basic fertility determinations. Incorporate lime if needed (see chapter 8, page 141). In mid-May, broadcast 50 pounds of actual nitrogen (N) per acre and the recommended amounts of potassium (K) and phosphorus (P) needed for forage crops based on the soil test. Plant Sudex (sorghum x Sudan grass hybrid) at 20 to 25 pounds of seed per acre. Sudex is the crop of choice because it produces large amounts of biomass in a short time and its roots deeply penetrate the soil. It is also very good for both accenting the presence of and eliminating residual herbicides that may be present from the previous planting.

Mid-July through August. Mow Sudex just before seed heads mature and apply an additional 30 pounds of actual nitrogen per acre using ammonium sulfate. In mid-August, mow again to reduce the bulk of the plant residue before plowing it down thoroughly with a moldboard plow. Two weeks after plowing, plant rape variety 'Dwarf Essex' at 8 to 10 pounds of seed per acre. In addition to adding more organic matter to the soil, rape produces chemicals that are highly toxic to plant-parasitic nematodes. Retest soil and adjust pH if needed. Add 15 to 20 pounds additional nitrogen per acre as ammonium sulfate. (Note: The sulfur contributes to the amount of toxicant produced within the rape plants.)

One Year before Planting

Mid-April. Mow rape using a flail mower **and PLOW down plant residue within one to two hours of mowing.** The toxicants within the plants are released with wounding and must be incorporated quickly to avoid loss to the atmosphere. Two weeks after plowing broadcast 50 pounds of actual nitrogen per acre (ammonium sulfate) and plant a second crop of Dwarf Essex (same rate as above).

Mid-August. Mow and plow down the second rape crop as previously described. Make final lime adjustments for a pH of 6.5.

Early September. Plant fescue ground cover. Endophyte-infested, Kentucky 31 tall fescue (30 pounds per acre) will suppress nematode populations and exclude broad leaf weeds but will require frequent mowing. Dwarf fescue grasses will reduce the need for mowing, but are less tolerant of heavy equipment travel and may not suppress nematode populations.

Year of Planting

April. Two weeks before planting, apply glyphosate herbicide in 3- to 4-foot wide strips where rows are to be planted. Plant vines through the killed sod.

Source: Halbrendt, J. (2006), Personal Communication.

there has been considerable effort to develop disease-free grapevines. The purchase of so-called "certified" grapevines is highly desirable because it reduces the risk of planting unhealthy vines that could reduce both vineyard productivity and fruit quality. The buyer should understand, however, that certified grapevines are not guaranteed to be free from all diseases or from even the diseases for which they were tested. Rather, certification ensures only that the original vine, from which material was obtained for propagation purposes, has been tested for specific grapevine diseases and that those tests were negative. Growers in cool climates should note the likelihood that certified grapevines purchased from nurseries in warm climates have not been tested for the presence of the bacterial disease known as crown gall. Despite the limitations of the certification process, the purchase of certified grapevines, whenever possible, is recommended as an additional step to promote the establishment of a quality vineyard.

Nurseries often sell grapevines according to grades of quality such as 2-year #1, 1-year-extra, 1-year #1, 1-year-medium, or 1-year #2. These growing standards are applied according to the size and branching patterns of the root systems. Unfortunately, there is no uniform application of these grading standards among all nurseries. Because of this, the actual quality of vines may more reliably be determined by a description of the size and branching pattern of the root systems of the vines. Purchasing vines of the highest grade (typically 1-yr "extra" or 1-year #1) of vines from a specific nursery is often a good investment for the long-term health of the vineyard.

The number of grapevines required for a planting can be calculated based on the area of the planting and the row and vine spacing. For example, in figure 4.1 (page 74) a total effective planting area for the vineyard was calculated to be 7.49 acres. If this vineyard were to be planted on a 9-foot row by 6-foot vine spacing, or 54 square feet per vine, we could calculate the approximate number of vines required for planting as follows:

7.49 acres x 43,560 = 326,264 square feet
326,264 square feet / 54 = 6,042
vines required for this planting

This calculation will provide a close but not exact estimate of the vines required. Moreover, there often may be breakage of weak grafts at the time of planting and some vines are likely to fail during the first year of growth. With this in mind, it is often helpful to add 1% to 2% to this estimate for replacements. Because the availability of grapevines is unpredictable, especially when a specific clone and rootstock combination is needed, order vines a year or more in advance of planting. Grapevines may be purchased from nurseries which are located in several viticultural regions. A listing of nurseries may be obtained from your state or local extension service or through a search of the internet.

PREPARING FOR PLANTING

Grapevines can be planted successfully in the fall if the vines can be obtained from the nursery prior to freezing of the soil; however, there are hazards associated with fall planting in a cool climate. Frost heaving of soils during the winter may damage new root systems, and there may be winter injury to grapevine tissues that have not been fully acclimated to low temperatures. Most vineyards are planted in the spring, preferably as early as possible after the risk of a severe freeze has passed and it is possible to prepare the site. In many regions, this will be from mid-April to mid-May. Typically, a field is prepared through a combination of plowing, disking, and dragging (figure 4.5). If a permanent sod has been established on erodible, sloping ground, vines can be planted into strips that have been killed with a systemic herbicide and then rotovated (figures 4.6a, 4.6b). There are several satisfac-

Figure 4.5 ▪ Grapevines being planted with a modified tree planter.

The field has been plowed, disked, and then marked for planting.

a. This hillside vineyard site had been sown to tall fescue the previous August. In May, strips are being rotovated; strips were previously sprayed with glyphospate herbicide. Vines will be planted into those rotovated strips.

b. Vines were planted into the rotovated strips. A pre-emergence herbicide was then applied over these strips to keep them weed-free. Photo taken in late July.

Figure 4.6 ▪ Hillside vineyard site preparation and planting.

tory techniques for the actual planting of vines. Precise planting is possible for small plantings. Planting can be made into augered holes or into a trench using a modified tree planter. In recent years laser-guided equipment (figure 4.7) has made extremely accurate planting possible. With the exception of laser-guided planting systems, there is a basic requirement for marking the field prior to planting. This typically begins with a baseline for the planting of the first row. The baseline might be a line parallel to an adjacent woodland or a line across the slope of uniformly sloping ground.

The most common way of marking a fully cultivated field for planting is to develop a grid of shallow furrows in the soil from this baseline (figure 4.8, page 80). Furrows in one direction identify vine rows. Furrows perpendicular to the row furrows create intersections that identify the location of vines to be planted. A long tool bar on a 3-point hitch that slightly exceeds two vineyard row widths in length may be used to mark this grid. For example, if a vineyard were to be planted to 9-foot wide rows, then a tool bar about 20 feet long would be needed. Adjustable cultivating teeth are clamped along this tool bar. One cultivator tooth is placed directly in the middle of the tractor and two others are placed at the appropriate row spacing at the ends of the tool bar. The lifter arms of the 3-point hitch of the tractor must not sway when using this tool bar to mark the field. This can be done by using a combination of stabilizer bars and/or by rigidly chaining the unit to one side of the tractor. Even with these precautions the marker may shift slightly. This may result in alternating wider and narrower-than-desired rows of vineyard when marking the furrows in both directions. One way to minimize the marking error associated with a shifting of the lifter arm assembly on a tractor is to do all marking in one direction. On a large planting this can be a lengthy process. A more efficient approach is to start marking in the mid-section of the field so that all marking of one side of the vineyard is marked in one direction and the other half of the vineyard in the other direction. Check your marking strategy with a measuring tape as you proceed. After the first pass with this marking tool, three rows of the new vineyard will have been marked. Subsequent passes can be made in two ways. One of the outer markings of the

Figure 4.7 ▪ Laser-guided grapevine planter.

A laser-guided grapevine planter being demonstrated to growers.

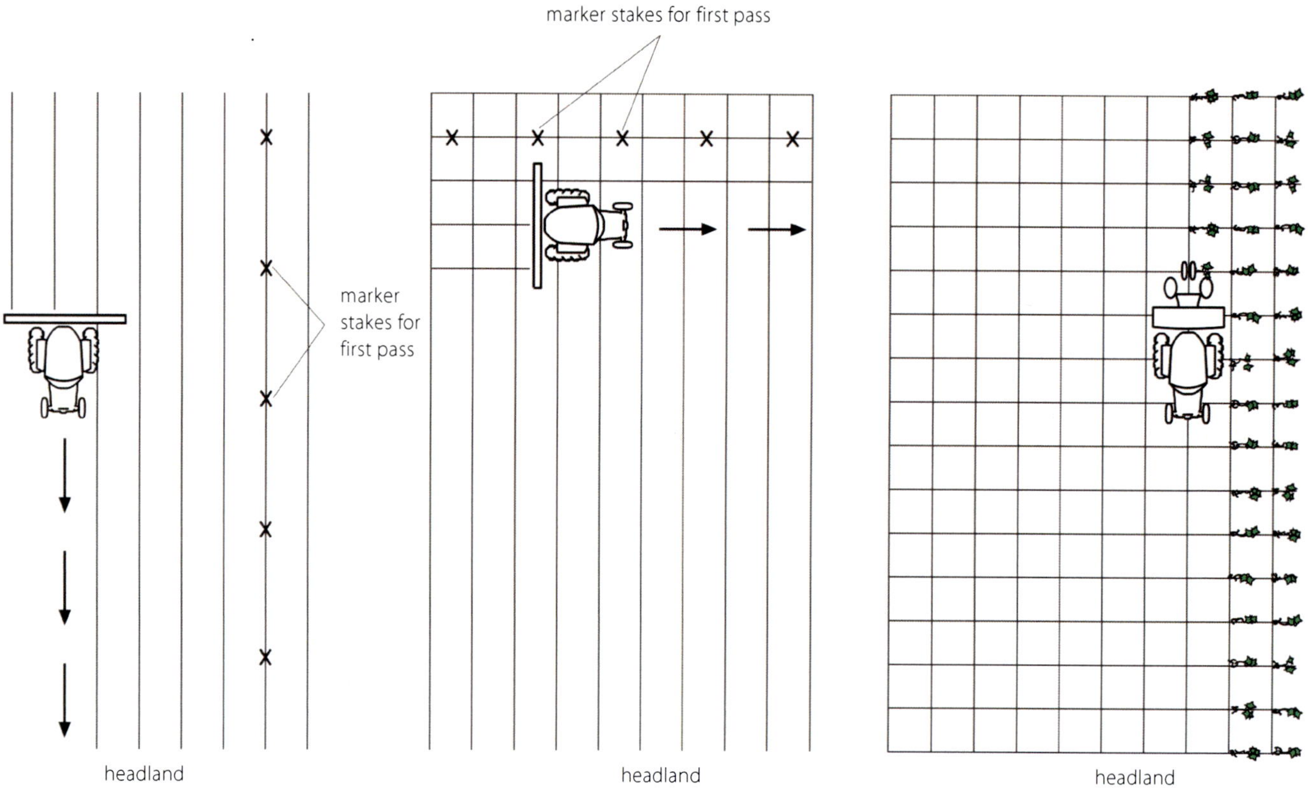

Figure 4.8 ▪ **Marking a fully cultivated field for planting.**

A schematic showing the marking of vineyard rows *(left)*, the cross-marking of vine spaces *(middle)*, and the planting of vines using the marked grid as a reference *(right)*.

first pass may be used as a guide for one of the outer marking teeth for the next pass of the marking tool. In this way, two additional rows are marked each time the field is traversed. Another variation of marking uses precise staking of every third vineyard row (as discussed below) as a guide for the tractor to mark three new independent rows with each pass. When all the rows have been marked, adjust the cultivating teeth on the tool bar to the appropriate vine spacing. The process of marking furrows is then repeated perpendicular to the row markings. When the entire process is complete, a grid of shallow furrows on the whole field identifies the location of each grapevine to be planted at the intersections of the furrows. One grower variation of this marking scheme, which has been used with a tree planter, is to mark only the first row to be planted and then the perpendicular furrows to identify the vine spacing. Then as the first row is planted, a boom with a small disc on its end on the front of the tractor marks the location of the next row to be planted (figure 4.9).

When planting manually or with laser-guided equipment, stakes are needed to mark the ends of each row. These are established along a line perpendicular to the baseline at both ends of the field. These perpendicular

Figure 4.9 ▪ **Planting one row of grapevines and marking the next row.**

While one row of grapevines is being planted, a bar on the front of the tractor, with a small disk on its end, is marking the next row for planting.

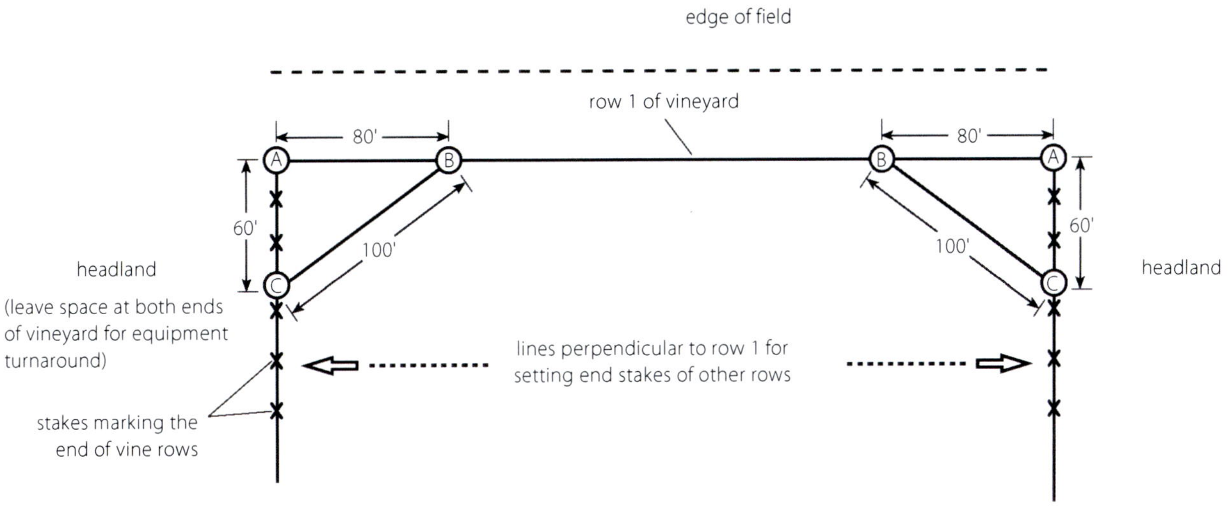

Figure 4.10 ▪ **Marking the ends of vine rows.**

A schematic showing the use of a 3-4-5 triangle to mark off lines at right angles to a baseline. Stakes are then placed along these perpendicular lines to mark the spacing of rows for the vineyard.

lines can be easily established without sophisticated instrumentation by using the dimensions of a 3-4-5 right triangle as described and illustrated in figure 4.10. First, place stakes at the ends of the baseline at the anticipated locations of the end posts (stake A in figure 4.10). Then place stakes along the baseline 80 feet from each end stake towards the middle of this line (stake B in figure 4.10). Using two measuring tapes, hold the end of one tape on an end stake of the baseline (stake A) and extend that tape to the 60-foot mark. Hold the end of the other tape on the stake at the 80-foot location (stake B) and extend that tape to its 100-foot mark. Where the 60-foot and 100-foot marks on those tapes intersect, place a stake (stake C in figure 4.10). This new stake (C) and the end stake (A) will define a line that is perpendicular to the baseline. Extend the length of this new, perpendicular line for the entire width of the new planting. Place stakes along this line at the appropriate row spacing interval. Repeat this process at the other end of the field. It is helpful to also repeat this process at intervals along the baseline for plantings with long rows. In this way, stakes will identify the location of rows at intervals along a line of sight.

Care and patience are required to properly mark a field for planting. It may take as long to mark a field as it takes to actually plant the vines. Effort spent to properly mark a field for planting will provide benefits over the life of the vineyard.

SETTING OF THE VINES

Quality grapevines should come from a nursery in a fully dormant condition. Coordinate the delivery of vines so that they arrive just before planting. Keep them cool (refrigerated, if possible) and moist. If you have any doubt about the moisture status of the vines, soak them for a few hours before planting. Transport vines to the field in a way that keeps them shaded and moist (figure 4.11). Do not let the roots of vines dry out during any part of the planting process. Most grapevines will come

Figure 4.11 ▪ **Keeping vines shaded and moist.**

Growers have just put a new supply of grapevines on a planter from under blue tarps, which are keeping a trailer load of grapevines moist during the planting process.

from nurseries bundled and ready to plant. Pruning roots before planting will often depress vine growth. At the same time, a large vine root system jammed into a small planting hole will not allow for use of that extensive root system and will probably hinder vine development. To guard against this, ensure that root pruning is performed only if the root system is too large to be distributed well in the soil volume. The goal should be to plant the largest possible root system in the greatest possible soil volume. Therefore, regardless of the method for opening the soil for the grapevine, be sure the opening is large enough to effectively distribute the roots of the grapevine. For a small planting, this means using a shovel rather than a manual post-hole digger.

When planting with an auger, you may want to use a tool as large as 18 inches in diameter.

When planting with a tree planter, many growers have modified the flanges that open the trench to make a furrow that is wider and deeper than normal (figure 4.12).

When planting vines, distribute the root system uniformly within the 360° circumference of the hole. As the soil is filled in around the grapevine, gently pull the grapevine upward to establish a root orientation that is somewhat vertical. When most of the soil has been placed around the grapevine, make the final height adjustment of the grapevine. For grafted grapevines, the ideal placement of the graft union is approximately 2 inches above the level ground surface (figure 4.13). Shallow planting of vines such that the roots are just below the soil surface may result in extensive vine mortality.

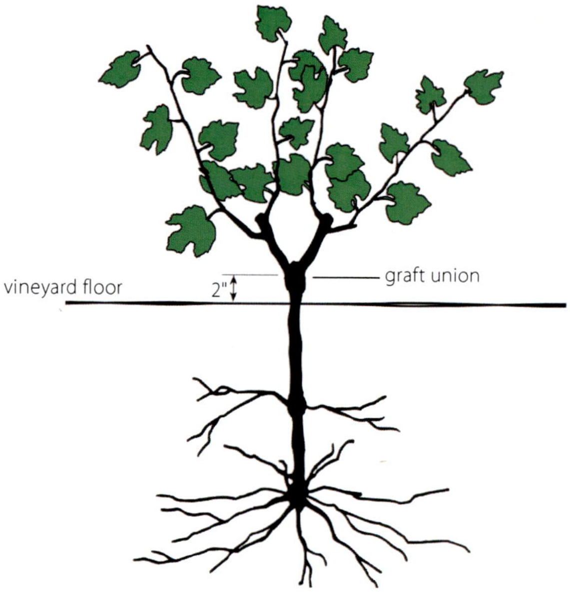

Figure 4.13 ▪ Placement of graft unions.

Graft unions on vines should be placed approximately 2 inches above the surface of the soil.

When planting manually with a shovel or an auger, tamp the soil firmly around the grapevine to hold it in a precise vertical orientation and to prevent graft unions from settling below the soil line. In heavy soils, an auger may create a heavily compacted cylinder of soil that hinders the grapevine root penetration. In such a case, it may be necessary to chip off the surface of the soil in the hole during the planting process. When planting with a tree planter, the person setting the vines should key on the

Figure 4.12 ▪ Modified tree planter.

A modified tree planter makes a planting trench in the soil that is deeper and wider than normal.

Figure 4.14 ▪ Grapevine being placed in the trench created by a tree planter.

The person on the planter is looking to the side to line up the placement of the vine with the shallow furrow marked in the soil at right angles to the direction of planting.

vine-spacing furrows to release the vines in the planting troughs at those marks (figure 4.14). Hand follow-up after the planter is usually needed to ensure a vertical orientation of the vine.

If vines are planted early, before the last chance of spring frost, it is advisable to gently mound loose soil over the graft union and a portion of the scion immediately after planting. While this increases the potential for scion rooting, the soil provides additional protection against desiccation of the graft union and late-spring freeze. The mound can be pulled down with hand-hoes later in the summer to minimize the potential for scion rooting.

STEPS IMMEDIATELY AFTER PLANTING: HILLING; FERTILIZING; WEED CONTROL

Several tasks should be performed immediately after grapevines have been planted. If a tree planter was used, the soil around the vines will be either level or in a trough. When vineyards are planted on sloping ground with rows perpendicular to the slope, it is desirable to create a ridge of soil along the rows of vines (figure 4.15). These ridges serve two purposes. First, they interrupt the down-slope flow of surface water that can cause severe erosion during periods of heavy rain. Such ridging is especially helpful in the swale areas of undulating ground. Second, they eliminate depressions around the grapevines that can accumulate concentrations of pre-emergence herbicides during times of surface water run-off. When ridging is performed, it should be the first step after planting. The second step is fertilization. Fertilize grapevines at the time they are planted unless the site has highly organic fertile soils. The development of both the above- and below-ground parts of the vine is the principal goal of the first growing season. Fertilization with 20 to 30 pounds of actual nitrogen per acre will contribute to that goal. Because the roots of newly planted grapevines exploit such a small volume of soil as they begin to grow, apply by ringing fertilizer manually in a circle with a radius of 1 foot around each vine. If that isn't possible, band the fertilizer along the vine rows. For typical vine spacing, 30 pounds of actual N per acre translates to about 2 ounces of ammonium nitrate or 6 ounces of a blended 10-10-10 fertilizer per vine. This should be done soon after planting and should be done only once in the first season. An alternative is to use a dilute, liquid formulation of nitrogen, applied to the soil by hand or through the drip irrigation system. Be patient. First-year vines may grow slowly at first. Adding supra-optimal rates of fertilizer could cause excessive shoot vigor and increase the risk of winter injury to vines.

Weed control is the single most important facet of vine management in a newly planted vineyard, and poor weed control is the most common cause of poor vineyard establishment. Weeds are highly efficient competitors with vines for water and nutrients. Inadequate weed control around newly planted grapevines can hinder vine size development so that optimum vineyard productivity is delayed one or more years (figure 4.16). Weed control

Figure 4.15 ▪ Ridge of soil along a row of vines.

Ridging of soil around young vines controls surface water movement and prevents concentration of herbicides at the base of grapevines.

Figure 4.16 ▪ Lack of weed control.

Lack of weed control in this vineyard planting resulted in a delay of several years in the establishment of this vineyard.

Figure 4.17 ▪ **Size of vegetation-free areas affects vine growth.**

Niagara grapevines were planted with a 4-foot wide vegetation-free band *(left)* and an 8-foot wide vegetation-free band *(right)*. This mid-summer picture shows the vines are already noticeably larger in the wider band *(right)*.

in a new vineyard is a challenging task that should begin during site preparation. If the site has been properly prepared, seed germination will be the only source of plant growth that might develop within several feet of the new grapevines. The goal is to prevent all plant growth within a minimum radius of 24 inches of the grapevines. When the characteristics of the vineyard site allow it without creating a hazard of surface soil erosion (figure 4.17), an increase in the vegetation-free (weeds and sod) radius around vines to 48 inches or more will result in significant increases in vine growth compared to less weed control. Increasing the vegetation-free area around vines will be rewarded with greater vine size development. Several strategies may be followed to achieve weed control around newly planted grapevines. All require a focused commitment. The simplest of these strategies is hand-hoeing or cultivating, which must be faithfully performed at two-week intervals to be successful. Otherwise, the settling and compacting of the soil along with the germination and rapid growth of weeds will make it difficult, or impossible, to keep the area around vines free of weeds. It is relatively easy to keep 80% of the vineyard floor weed-free. However, it is within the remaining 20% of the vineyard floor immediately around vines that weed control is most important and most difficult. Therefore, cultivation strategies for weed control often need supplemental manual or chemical efforts to control weeds immediately around vines. Faithful adherence to a biweekly program of cultivating and/or hoeing can stimulate the development of vine size just as well as chemical weed control (figure 4.18). Several forms of mulching have been effectively used around newly planted grapevines. Plastic mulch, landscaping cloth, stone, newspaper, leaf litter, wood chips, and straw have been used. A drawback to some of these approaches is the cost of labor and materials. An additional drawback of mulching is the refuge that mulch provides for some vine-damaging insects, such as climbing cutworms (see chapter 12, page 241) and, depending on materials, the survival of grape root-borer larvae. Incomplete mulching with any of these materials will result in "monster" weeds growing through the mulch. Nevertheless, with a sufficient commitment, mulching is a viable alternative for establishing weed control around newly planted vines.

The use of herbicides is the most common approach to weed control in new vineyards. Successful herbicide control of weeds in a new vineyard requires planning and attention to detail (see chapter 13, page 262).

The past decade has seen a surge of interest in the use of "grow tubes" with newly planted grapevines. Grow tubes are placed around newly planted grapevines and come in many types and sizes. They dramatically promote rapid elongation of early shoot growth. Side-by-side comparisons of new vine growth with and without grow tubes consistently produce impressive results with regard to shoot extension growth (figure 4.19). How-

Figure 4.18 ▪ **Cultivation strategies for weed control.**

The cultivation of a new vineyard can be a very effective means of vine establishment. Hand hoeing and weeding around the vines in this vineyard were used as a complement to the machine work shown.

Figure 4.19 ▪ Grow tubes on grapevines.

Grow tubes on White Riesling grapevines, seven weeks after planting. Although the vines with grow tubes *(foreground)* are considerably taller than those without grow tubes *(background)*, those in the background developed more leaf area, developed more root growth, and had somewhat hardier tissues than those grown with the grow tubes.

ever, there is research evidence that grow tubes do not increase the rate of vine size development for either above- or below-ground portions of the vine and they do not alter long-term vine productivity. Moreover, the hardiness of trunk and cane tissues may be reduced by the use of grow tubes. Presumably this is related to the frequent defoliation of portions of shoots within the grow tubes. One advantage of grow tubes is that herbicides can be safely sprayed around the base of newly planted vines. Some growers feel that this herbicide-spraying capability alone warrants the cost and effort of installing grow tubes. Growers are urged to carefully weigh what can be a relatively high cost per acre for the tubes, their installation, and removal versus the benefits that will be obtained from their use.

The installation of trellis posts and one or more wires at the time of planting grapevines will be helpful in the management of newly planted grapevines. Nevertheless, numerous vineyards are successfully managed in their first year without the installation of a trellis.

MANAGING YOUNG VINES

The principal goal of vine management in the first two years of a new vineyard is to develop mature-sized vines (see chapter 5, page 98). Adequate vine size to fill the entire trellis with canopy by the end of the third growing season will be the basis for long-term productivity. Healthy, functional leaf area is the basis for that growth. Promoting the early development of a vine training system by limiting the number of shoots on a newly planted vine to one or two is a traditional, orderly approach, especially if those shoots are meticulously tied to individual stakes at each vine. However, such management limits the leaf area on newly planted vines in comparison to those managed with greater numbers of shoots. If weeds are controlled around vines and vines are fertilized at the time of planting, newly planted vines will support the vigorous growth of several shoots. Shoots should be tied to training stakes or trellis wire(s) in order to keep the foliage protected from fungal pathogens and to encourage straight trunks. If grow tubes are used, the tubes will likely foster a dominant shoot that will suppress development of other shoots within the tube.

Standard vineyard establishment procedures involve defruiting vines in their first year of growth so that cropping doesn't compete with the vegetative development of the vine. If it seems necessary to verify the variety in the new planting, one cluster per vine can be left on vines in the second growing season. Under optimal conditions, some growers have been able to take a very small crop (for example, 0.5 tons/acre) in the second year of growth in the vineyard; however, this practice is not recommended for inexperienced growers.

Because vine development depends on healthy, functional leaves, disease and insect pests must be controlled on newly planted vines. Diseases of greatest significance in new plantings are typically powdery and downy mildew, which attack the leaf area. Vines should be scouted weekly to detect emerging insect problems. Insects such as Japanese beetles, rose chafers, and potato leafhoppers can quickly become abundant and destructive to new vines. Frequent scouting and the application of appropriate insecticides when needed are part of good first-year vine management. If a trellis is installed at the time of planting of new vines, it will be necessary to tie the shoots loosely on one or more wires. Individual staking of vines is one option for managing young vines. A less costly option is to attach twine to the vine in a way that will not cause girdling (figures 4.20a, 4.20b, page 86). Extend this twine vertically up to one or more trellis wires. Shoots can then be tied loosely to this twine and then on to the trellis wires. All tying in the first year

a. Grapevines in a newly planted vineyard that are being supported by twine.

b. The twine is tied to a non-growing spur at the base of the vine and to a wire at the top of the trellis. The shoots of the vine are then gently twisted around the twine, perhaps bundled together with an additional loop of twine, and eventually loosely tied to the wires of the trellis.

Figure 4.20 ▪ Grapevines supported by twine.

must be done loosely so there is no girdling of shoots as they develop.

The row middles of a newly planted vineyard can be managed either as a sod or cultivated. Sod is perhaps more common in eastern North America. Sodded row middles should be mowed several times through the growing season to minimize competition of the sod with the grapevines and to facilitate air movement through the vineyard as an aid in disease control.

Grapevines have indeterminate growth. That is, they continue shoot extension growth as long as there are conditions that will stimulate that growth. If rapid growth continues late into the growing season, vine tissues may not develop adequate hardiness for over-wintering. Therefore, when winter hardiness is of concern due to the variety being grown and the location of the vineyard, take steps to slow vine growth in the latter part of the growing season. This means ceasing mowing of

Figure 4.21 ▪ Rye cover crop sown in the row middles.
A new vineyard in the spring the year after vines were planted. A rye cover was sown in the row middles in the latter part of the first growing season. Posts are in the process of being installed in this vineyard.

sodded row middles or sowing a cover crop in row middles that have been cultivated (figure 4.21). Common cover crops include oats and rye. Oats do not require a precise seeding pattern because they will not overwinter to become a weed problem the following season. Rye must be sown carefully to avoid its establishment in the vine rows (figure 4.21). The establishment of permanent sod covers may be appropriate in vineyard row middles, especially if vine size development is proceeding satisfactorily. Check local recommendations for perennial grass species that do well in your area. Turf-type tall fescue (*Festuca arundinacea*), blends of tall fescue and perennial rye, and other blended species mixtures are all used in eastern North America with good success.

TRELLIS CONSTRUCTION

Building a good vineyard trellis is very important, not only because it can be the single largest cash expense in the establishment of a vineyard but also because it can significantly influence the long-term productivity and profitability of a vineyard. The installation of trellis posts and one or more wires at the time of planting grapevines will be helpful in the management of newly planted grapevines. Even so, numerous vineyards are successfully managed in their first year without the installation of a trellis.

The durability of a trellis is an important issue because the real cost of a vineyard trellis is determined by its years of service rather than initial cost. Considerable annual maintenance of trellises and frequent replacement of trellis components are inefficient and unnecessary with today's technology. A good trellis promotes good canopy management with well-exposed leaves and, when desired, well-exposed fruit. It facilitates efficient performance of vineyard tasks. The trend toward increased mechanization of vineyard tasks requires precise vine structures, which begin with a proper vineyard trellis. Crooked vineyard rows, sagging trellis wires, and bowed vine trunks jeopardize the precise management of vines.

When a trellis is installed in stages for reasons of labor management and/or cash flow, install posts and two wires either at planting or in the fall/spring between years 1 and 2, then install end-post anchors and the full complement of trellis wires in the fall/spring between years 2 and 3.

Types of Posts

Selection of trellis post materials will be influenced by the types of posts available, post installation equipment available, choice of a vine training system, cost, and personal preference.

Metal Posts

Metal vineyard trellis posts are an attractive option because they are relatively easy to handle and install, their cost can be competitive with wooden posts, some metal posts have wire slots in their design so they don't require staples or clips, they may minimize damage from lightning strikes on trellis wires, and they are more consistent in quality than wooden posts. However, one possible major drawback to metal posts is their lateral strength. Lateral strength is the amount of force that must be applied perpendicular to a post (usually 48 inches above ground) to cause it to break or permanently bend. Lateral strength values for different gauges and construction of metal posts range from 175 to 325 pounds, compared to an estimated 648 pounds for a 3.5-inch wooden line post. Therefore, a grower should carefully assess the strength of metal posts for vineyards with large vines and crops, especially in windy locations, which are likely to concentrate the weight of the vine on the downwind or lee side of the trellis. Most certainly avoid high-gauge metal posts, (that is, those with relatively thin metal walls) in those situations. (See also "Post Strength and Durability").

Wooden Posts from Native Tree Species

A half-century ago vineyard trellis posts were cut from native tree species. These posts occasionally were subjected to on-farm preservative treatments but often were untreated. Most native tree species are not rot-resistant. Random selection of trellis posts from woodlots is likely to result in a high percentage of post failure in ten or fewer years. The annual cost of a post is its cash cost plus the labor required to install it divided by its years of service. Inexpensive posts with a short life are costly on a per-year-of-service basis. Black locust and white cedar have natural resistance to decay and are still used for vineyard trellis posts. When these posts have 90% or more of their cross-sectional area composed of heartwood, they have a life expectancy of twenty years or more.

Pressure-Treated Wooden Posts

Wooden posts that have been commercially pressure-treated with a preservative are the predominant type of trellis post used in eastern North American vineyards today. Red pine or southern yellow pine is commonly used. These posts would fail in four to five years but with proper preservative treatment they have a twenty- to thirty-year life expectancy. Characteristics that will influence post life expectancy include the diameter of the post, the type of preservative used, the amount of preservative used per unit volume of wood, and the vineyard site. Pressure-treated wooden posts are sold in sizes according to the minimum diameter at the smaller end of the post. The cost of a pressure-treated post increases rapidly as its minimum diameter increases (table 4.1, page 88). Therefore, important information is presented below to address the question "What minimum diameter is adequate for a trellis post?"

Post Strength and Durability

The strength of a post is proportional to its cross-sectional area. For example, posts 2.5 and 3 inches in diameter have cross-sectional areas that are only 39% and 56% and lateral strengths that are only 25% and 42% of a 4-inch diameter post, respectively (table 4.1). Post life expectancy is also greatly influenced by diameter. The rate

Table 4.1 ▪ Costs, Cross-Sectional Area, and Lateral Breaking Force of Pressure-Treated Pine Posts

Post	Cost		Cross-sectional area		Lateral breaking force[a]	
Diameter (inches)	Per post[b] (US$)	Percentage of a 4-inch post	Square inches	Percentage of a 4-inch post	Pounds of tension	Percentage of a 4-inch post
2.5	—	—	4.91	39	238	25
3.0	4.15	58	7.07	56	408	42
3.5	4.75	66	9.62	77	648[c]	67[c]
4.0	7.20	100	12.57	100	970	100
5.0	9.90	138	19.64	156	1,893	195
6.0	12.00	167	28.27	225	3,268	337

[a] Average pressure applied 4 feet above ground to cause the post to fail. Adapted from "How to Build Fences with USS Max-Ten 200 High-Tensile Fence Wire," United States Steel Catalog T-111575.

[b] Values are a supplier price for chromated-copper-arsenate impregnated posts as of August 2006.

[c] Interpolated values from the values reported for the other sizes of posts.

of leaching and weathering of preservative from a post is related to its surface area. As post diameter decreases, the ratio of surface area to volume increases, the percentage of preservative leaks from the wood each year increases, and the rate of post decay increases. Because a post with a 3.5-inch minimum diameter often will be guaranteed for thirty years of service, many growers use this size for line posts because it will provide the lowest cost per year of service.

When choosing end posts, 4-inch-diameter posts with lateral breaking forces of 970 pounds (table 4.1) will be adequate when they are new and if they are anchored so that tension on load-bearing wires is transferred to the anchor. However, as 4-inch posts decay or if anchoring is inadequate, or both, lateral forces in excess of 970 pounds are likely to cause post failure. Therefore, choosing end posts with diameters larger than 4 inches can be cost effective over the life of the post. For example, posts with a 5- or 6-inch diameter will have lateral breaking strengths that are about twice, to more than three times, that of a 4-inch-diameter post (table 4.1), but their current cost is only 38% or 67% greater than a 4-inch diameter post, respectively (table 4.1).

The most common chemical preservative currently used for treating vineyard posts is chromated copper arsenate (CCA). Other preservatives using combinations of copper and fungicides or completely organic formulations are future trends in wood preservation. The American Wood Preservers Association (HTTP://WWW.AWPA.COM) establishes standards for the minimum amount of CCA to be impregnated into wood to ensure long-term resistance to decay. The standard for vineyard posts is 0.4 pounds of material per cubic foot of wood. Certificates of treatment, service life guarantees, or both may be available with pressure-treated post products. Growers should work with their suppliers to obtain these assurances for this costly part of vineyard establishment. If there is a doubt about the extent of preservation treatment when purchasing large quantities of pressure-treated posts, it is possible to have a post sample analyzed by a private laboratory.

Large knots are a major defect in pressure-treated pine posts. They can cause posts to break during installation or before the end of their projected life expectancy.

Quantities of Posts Required per Acre of Vineyard

Row spacing and the distance between posts are the principal factors that determine the number of posts required for an acre of vineyard. The force of gravity causes wires and vines to sag in the middle of a post space. A grower's choice on the distance between posts will depend on the grower's tolerance for sagging, which can be reduced but not eliminated by increasing tension on trellis wires. Increasing tension on the trellis wires beyond a certain point (see "Installing Wires") will not further reduce sagging and will lead to excessive tension and permanent stretching of wires. Post spacing commonly ranges from 18 to 21 feet, and it should never exceed 24 feet. A rough estimate of the number of posts required per acre of vineyard can be calculated as follows:

Posts/acre =
(43,560 square feet/acre) / (row spacing (feet)
x vine spacing (feet) x vines per post space)
+ number of vineyard rows per acre

For example: a vineyard will be planted with 9-foot rows, 6-foot vine spaces, 3 vines per post space; and there will be 12 rows/acre.

$$\begin{aligned}\text{Posts/acre} &= 43{,}560 / (9 \times 6 \times 3) + 12\\ &= 269 + 12\\ &= 281\end{aligned}$$

Divide the total amount of posts into end posts and line posts—that is, those installed within the vineyard rows. By allowing two end posts per row of vineyard, the above example would be divided into 257 line posts and 24 end posts.

Installing Line Posts

A typical 8-foot line post is set 24 to 30 inches into the ground so that the top of the post is 66 to 72 inches above the vineyard floor. Posts may be aligned with techniques as sophisticated as the use of surveying equipment to mark locations prior to installation or simply by "eyeballing" down and across the rows. Installing posts first on the outside rows and then on one or more interior rows can help establish a grid for sighting. Placing posts in the plane of the vineyard row is functionally more important than the aesthetics of post alignment across the rows. Installation of posts is best accomplished when there is good but not excessive soil moisture. Posts should be installed with their larger diameter end down in the soil because this is the portion of the post that will decay and ultimately cause the post to fail. Some vineyard posts are sharpened at their bottom end in some viticultural areas, either prior to purchase or on the farm to facilitate pounding posts into heavy soils. Hydraulic pounding of posts (figure 4.22) is advantageous because it is relatively rapid, the posts can be set to a precise desired depth and they are immediately firm in the ground. A small hand-held hydraulic post pounder provides quick installation of metal posts (figure 4.23). Augering of post holes is also common and requires less expensive equipment than post pounding. Light soil tends to backfill around and firm up posts installed in augered holes relatively quickly, but posts on heavier soils can remain loose for years. Bucket loaders

Figure 4.22 ▪ Hydraulic post pounder.

Vineyard trellis posts are often installed with a hydraulic post pounder like the one shown here.

Figure 4.23 ▪ Hand-held hydraulic post pounder.

A hand-held hydraulic post pounder can quickly install metal line posts.

and other hydraulic equipment have also been used to push posts into the ground. An ingenious technique for vibrating posts can install posts up to 5 feet into the ground in seconds (figure 4.24), but such equipment is not generally available.

A hand-held post hole digger is the common method of installing posts in small backyard plantings. Another easy method for installing a small number of posts or replacing posts uses a pinch bar and post maul (figure 4.25). Begin with the pinch bar to punch a hole and then with a circular motion "auger" the hole deeper and wider. Put the post in this hole and finish the job with a post maul. A post maul is a specialized sledgehammer with a head face about 3.5 inches in diameter (figure 4.25). When using this venerable tool, stand on a platform such as the trailer that carried the posts into the vineyard.

Figure 4.24 ▪ Vibrating posts into the ground.
(Above) This unit bolted to the fork lift on this tractor will rapidly vibrate posts into the ground. This grower-invented device can install end posts 5 feet in the ground in just a few seconds.

Figure 4.25 ▪ Post maul.
This post maul has a head with a 3.5-inch diameter. It is used for installing or replacing a small number of posts.

Installing End Posts

All material preparation and trellis engineering skills will be wasted if the end posts are not installed well. End posts function differently from line posts because they are subjected to the tension of the wires and, therefore, a lateral breaking force. An end post that is not anchored or braced will react to this tension by creating a fulcrum at the soil line. Failure to resist this tension may result in curvature of the post or its being pulled through the soil (figure 4.26). End posts that are strong enough and set deep enough into the soil can resist this tension and perform well without additional engineering. For example, large-diameter poles or railroad ties set 4 feet or more into the ground function in this manner. More commonly, however, an end post is chosen for adequate breaking strength when used in combination with an endpost anchoring or bracing system.

End posts that will be anchored should be installed at an angle to a minimum depth of 3 feet (figure 4.27), either by augering or post pounding. To ensure that end posts will line up properly, first install the end posts on the outside rows. Then lightly tension two wires between these posts, one at their tops and one at the soil line. These wires guide the installation and angling of end posts for the interior rows. When installing end posts in a large number of rows, first install end posts at 15-row intervals to provide guides for installation on the other rows. Angling end posts 60° from horizontal (figures 4.27 and 4.28) will transfer tension from load-

Figure 4.26 ▪ Inadequately anchored end posts.
The end posts in this vineyard were not adequately anchored. They have moved after installation to become this random arrangement.

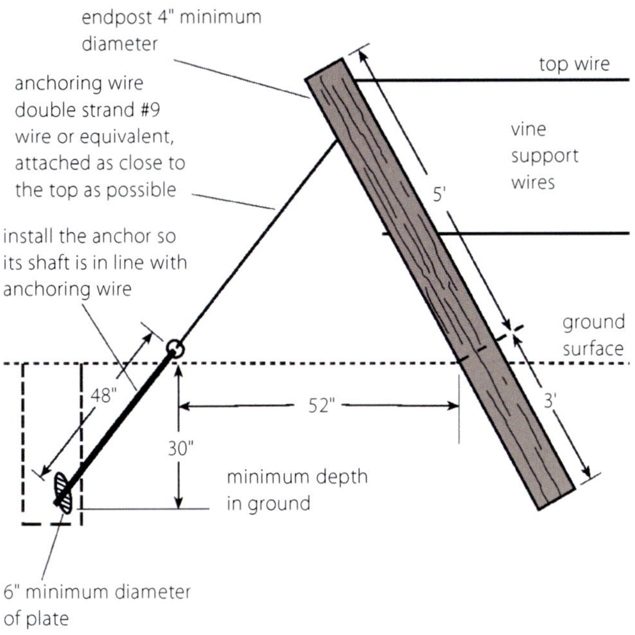

Figure 4.27 ▪ End post with external anchor.

Characteristics of an end post anchoring system that uses an external anchor.

Figure 4.28 ▪ Anchored end post installed at an angle.

This end post was set at a 30° angle from vertical (in other words, at a 60° angle from horizontal). Reinforcement rod connects the post to the anchor, and an L-shaped crank is being used to tighten the one wire attached to this post.

bearing wires to the anchor at a wider angle than setting posts upright. The more vertical an end post and the closer anchors are placed to the end post, the greater the tendency for the tension on trellis wires to pull the end post inward.

There are several good methods for anchoring end posts. Bracing end posts within the row (figure 4.29) is advantageous because it avoids conflict with equipment in the headlands. However, this approach is generally more complex and costly than external anchoring. Therefore, anchoring externally to the end post is the most common method of constructing an end-post assembly. There are several characteristics of good anchor installation:

Figure 4.29 ▪ Bracing an end post.

Bracing an end post avoids conflict between equipment and external anchors in headlands but is more difficult and costly to install.

- Attach the anchoring wire as close as possible on the post to the main crop-bearing wire. This directly transfers tension from the crop-bearing wire to the anchor and reduces the tendency for the point of attachment of the anchor to act as a fulcrum.
- Angle the anchoring wire to avoid a narrow angle between the post and this wire. If the anchor is installed too close to the post, it may keep the post from rising up but not from being pulled into the row.
- Install the anchor so that its shaft rests in line with the anchoring wire (figure 4.27). Otherwise there will be a tendency to pull or bend the anchor shaft through the soil when tension is applied until it achieves this in-line orientation.
- Install the anchor deep enough and with an anchor plate of adequate surface area. The anchor plate should be at a minimum 30-inch vertical depth and have a minimum diameter of 6 inches. Screw-in anchors may or may not provide a satisfactory shortcut to this procedure.
- Make sure that the anchoring plate is in contact with firm, undisturbed soil. Auger holes for anchors vertically. Then use a crowbar to make a narrow slit in the soil, angling the slit from the bottom of the hole up to the point of anchor attachment to the post. The shaft

of the anchor is placed in this slit, and the plate of the anchor sits firmly against undisturbed soil on the side wall of the augered hole (figure 4.27, page 91).
- Use a double-stranded #9 wire or multiple strands of high-tensile wire for the anchoring wire to ensure that it will not be the weakest point in the assembly.

End-post anchoring systems should be in place no later than the start of the third growing season. Final tensioning of trellis wires is done after anchoring is completed.

Wire Characteristics of Importance for Use in Vineyard Trellises

Growers often focus on the gauge and corrosive resistance of wire for vineyard trellises. They intuitively know that the gauge of a wire determines its ability to support a load. Because most wire used for vineyard trellises is made of steel, it is also readily apparent that wire coated with zinc galvanizing or other materials resists rusting and greatly extends its life.

In addition to these factors, there are improved materials and technology that make it possible to install vineyard trellis wires that are cost competitive with traditional materials while improving vine management, reducing trellis maintenance costs, and lengthening the life expectancy of the vineyard trellis. The following information will assist growers to properly choose and install trellis wires.

The tensile strength of a wire measures how much pulling (tension) is required to break the wire. It is determined by the alloy mixture used to make the wire. Formerly, vineyard wire was relatively soft, so-called low-carbon wire. Today, vineyard trellises are increasingly constructed with stiffer, harder, high-tensile (high-carbon) wire. Tensile strength is rated as the pounds of tension that would be required to break a wire if it had a cross-sectional area of 1 square inch. Traditional soft vineyard wire has a tensile strength rating of about 77,000 pounds per square inch (psi), whereas high-tensile wire has a tensile strength of 200,000 psi or more. The ability of a wire to resist breaking is directly proportional to its cross-sectional area. Therefore, the breaking point of a particular gauge of wire can be calculated by multiplying its tensile strength by the cross-sectional area of the wire. For example, a low-carbon, 9-gauge wire with a tensile strength of 77,000 psi and a cross-sectional area of 0.0172 square inches has a breaking point of 77,000 x 0.0172 = 1,324 pounds. Large-diameter (lower gauge number) low-carbon wires with high breaking-point values are useful for the major load-bearing wires of the trellis.

Wire, however, does not simply resist all tension until it reaches its breaking point. Rather, as tension is put on a wire, it begins to stretch. If a small amount of tension is put on a wire, it will return to its original length when the tension is released. As tension on a wire is increased, stretching continues and eventually a portion of that stretching becomes irreversible. When that happens, the wire has reached its yield point, which occurs at 65% to 85% of the tension required to break the wire. The yield point for a wire is very important when constructing a trellis. Relatively modest stretching of trellis wires can cause them to sag considerably between posts. When vines sag, vine trunks bow, and it is more difficult to move equipment through the vineyard or perform tasks like mechanical pruning or hilling and removing soil around vines. Trellis maintenance is also increased when wires stretch because those wires need retensioning. Vines with cordon training systems may be so embedded into wires that retensioning may not be possible. Stretching wire also slightly reduces the cross-sectional area of a portion of the wire. This lowers the yield and breaking points of the wire so that stretching will occur at a lower tension. That can lead to repeated cycles of wire stretching and eventual breakage.

If a wire never reaches its yield point and end posts are firmly anchored, none of the above problems will occur. Therefore, trellis wires should be installed with consideration of their yield points, not their breaking points. A grower can avoid reaching the yield point with low-carbon, soft wires by increasing the diameter of the wire. However, purchasing a large-diameter, soft wire to obtain a higher yielding point wire may not be cost effective. High-tensile wires have yield points that are approximately double that of soft wires for the same diameter (gauge) of wire, and at half the cost per foot (table 4.2). For example, a high-tensile 12.5-gauge wire and a soft 9-gauge wire have approximately the same yield points—1,063 and 1,118 psi, respectively—but their costs per foot are 2.0 cents and 4.0 cents, respectively (table 4.2). High-tensile, crimped wire (figures 4.30 and 4.31) is an excellent choice for a

Table 4.2 ▪ Characteristics and costs of wire typically used to construct vineyard trellises

Wire type	Wire gauge	Diameter (inches)	Cross-sectional area (square inches)	Tensile strength[a] (PSI)	Breaking point[b] (pounds)	Yield strength[c] (PSI)	Yield point[d] (pounds)	Feet per pound	Cost per pound (US$)[e]	Cost per foot (US$)[e]
Low-carbon	12.0	.106	.0088	77,000	678	65,000	572	33.7	0.69	0.020
Low-carbon	11.0	.121	.0115	77,000	886	65,000	747	26.3	0.69	0.026
Low-carbon	10.0	.135	.0143	77,000	1101	65,000	929	20.6	0.69	0.033
Low-carbon	9.0	.148	.0172	77,000	1324	65,000	1118	17.1	0.69	0.040
High-tensile	10.0	.135	.0143	200,000	2860	138,000	1973	20.6	0.75	0.036
High-tensile	11.0	.121	.0115	200,000	2300	138,000	1587	26.3	0.75	0.029
High-tensile	12.0	.106	.0088	200,000	1760	138,000	1214	33.7	0.75	0.022
High-tensile	12.5	.099	.0077	200,000	1540	138,000	1063	38.2	0.75	0.020
High-tensile	14.0	.080	.0050	200,000	1000	138,000	690	58.6	0.78	0.013
High-tensile	16.0	.062	.0030	180,000	540	124,200	373	119.6	0.78	0.070

[a] Minimum tensile strength rating supplied by manufacturer.
[b] Breaking point = tensile strength x cross-sectional area in inches.
[c] Values based on yield and tensile strength data.
[d] Yield point = yield strength x cross-sectional area in inches.
[e] Values based on an average supplier prices per 100 pounds as of August 2006.

Figure 4.30 ▪ High-tensile crimped wire in a notch.

High-tensile crimped wire placed in a notch on top of a line post. The notch was made with a chain saw that had a guard on the blade to allow only a 1-inch depth of the cut.

Figure 4.31 ▪ High-tensile crimped wire wrapped around an end post.

High-tensile crimped wire wrapped around an end post and fastened with a crimping sleeve.

load-bearing wire. It acts as a long spring, which helps it to avoid exceeding its yield point. When properly installed, it can be maintenance-free for many years.

Because vineyard wires are made of steel, as they rust, their tensile strength is reduced. Small-diameter wires without corrosion protection can lose more than half of their tensile strength in fewer than five years. Therefore, corrosion resistance of trellis wire is very important to the grape grower. The highest-quality zinc galvanization for wire corrosion resistance is a type 3 galvanization, which, according to the American Society of Testing Materials (ASTM Designation A116-57), specifies the ounces of zinc required per square foot of wire surface so that wires may resist the onset of rusting for up to thirty years. Other types of wire, such as aluminum or plastic, and other types of wire coatings are also used. Growers should obtain assurances of the resistance to corrosion and ultraviolet light for the wire they purchase.

In summary, wire for vineyard trellises should be purchased on the basis of yield point, cost per foot, and corrosion resistance rather than breaking point and cost per pound.

Amount of Wire Required

The number of wires required to construct a trellis will range from one to ten, depending on the choice of a vine training system (see chapter 5, page 98). Trellis wires generally serve two functions. They are either load-bearing wires, which support most of the weight of the vines and crop, or so-called catch wires, which help orient shoots but do not support much weight. Load-bearing wires should be chosen with emphasis on their yield points. Catch wires have less demanding specifications and can be chosen on the basis of cost per foot of wire and durability. Tables 4.3 and 4.4 will help the grower determine the amount of wire needed per acre of vineyard.

Table 4.3 ▪ Feet of wire per 100 pounds for several gauges of wire used in trellis construction

Wire gauge	Feet per 100 pounds
9.0	1,724
10.0	2,070
11.0	2,617
12.0	3,300
12.5	3,846

Installing Wires

Trellis wires should be tensioned to 270 to 300 pounds. It is quite common to overtension wires. This puts excessive strain on end posts and can stretch wires irreversibly beyond their yield point. A simple, effective technique for proper tensioning of wires involves suspending a bucket (figure 4.32) containing the appropriate weight (table 4.5) from the top wire. The bucket and chain assembly should be 6 inches shorter than the height of the wire and hung on the wire in the middle of a post space. When the wire is pulled to the desired tension, the bucket will lift off the ground.

This technique allows a grower to "get the feel" for the appropriate amount of tension to place on a wire. Then other wires can be tensioned without this technique.

It is difficult to properly tension wires on relatively short trellis rows because very little movement of end posts can result in considerable sagging of wires, even when good end-post anchoring is installed. A couple

Table 4.4 ▪ Length of trellis per acre and pounds of wire, one wire per acre, for five gauges of wire and seven vineyard row spacings

Vineyard row spacing (feet)	Trellis length/acre (feet)	Pounds of wire for one wire per acre[a]				
		9	10	11	12	12.5
6.0	7,260	425	352	276	215	190
7.0	6,223	364	302	237	185	163
8.0	5,445	318	264	207	162	143
8.5	5,125	300	249	195	152	134
9.0	4,840	283	235	184	144	127
9.5	4,585	268	223	174	136	120
10.0	4,356	255	211	166	129	114

[a] Five different wire gauges are shown.

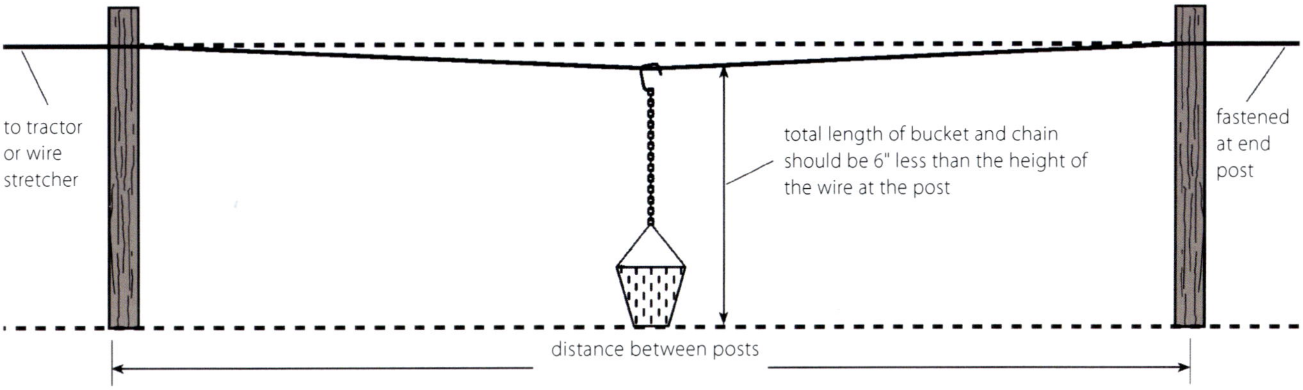

Figure 4.32 ▪ Technique for proper tensioning of wires.

Suspend a bucket containing the appropriate weight (see table 4.5) from the top wire. The bucket and chain assembly should be 6 inches shorter than the height of the wire and hung on the wire in the middle of a post space.

Figures reproduced courtesy of Washington State University Extension.

Table 4.5 ▪ Total test weight of chain, bucket, and contents indicating 270 or 300 psi tension on wire for three post spacings

	Test weight (pounds) for 6-inch sag for three post spacings (feet)		
Desired wire tension (pounds)	24	21	18
300	25.0	28.6	33.3
270	22.5	25.7	30.0

Note: Data indicates results when used as indicated in figure 4.32.

of options for vineyard rows 150 feet long or less are to use specially designed springs, which are placed in line with trellis wires to keep them tight year round (figure 4.33), or to release tension on wires each fall so they will not exceed their yield point when they contract during the winter. Remember to retension these wires in the spring.

Several tips can guide the proper installation of wires on trellis posts. U-shaped, galvanized, or otherwise coated staples approximately 1½ inches long are used commonly for this vineyard task. They should be applied on the windward side of the posts for standard trellis wires. When paired catch wires are installed, staples will also be applied on the leeward side of the trellis. Staples should have a slightly downward orientation when nailed into the posts on level ground. Apply staples to posts in knoll and valley areas at a somewhat exaggerated downward or upward angle, respectively. Nail staples so they are not directly oriented with the grain of the wood in the posts, which often indicates a slightly off-vertical alignment.

A variation for attaching a load-bearing wire at the top of a line post is to place the wire on top of the post rather than on the side. If periodic post pounding is anticipated to reset line posts moved by frost heaving, a groove ¾ inch deep can be made with a chain saw in the top of the line posts in line with the vineyard row (figure 4.30, page 93). Then place the wire in the groove and place a staple on top of the groove. This permits the post to be pounded without damage to the wire.

Figure 4.33 ▪ Springs keep trellis wires tight year-round.

This spring absorbs changes in the trellis wire tension due to temperature fluctuations or crop load, so that the wire never reaches its yield point.

Regardless of which of the many ways you use to attach wires to end-post assemblies, these wires should be installed with gentle bends. Abruptly bending or stretching a wire reduces its yield and breaking points, rendering it the weakest link in the trellis assembly. The simplest method of attaching a wire to an end post is to wrap it around itself several times (figure 4.34). Some prefer to place the staples on each post on the end or back side, while others prefer to place a pair of staples on each side to help maintain the horizontal orientation of closely spaced wires. High-tensile wires may be difficult to attach to end-post assemblies. Soft-metal crimps frequently are used in these situations (figure 4.31). Several excellent in-line tensioning devices are available (figure 4.35) and will work well as long as they do not cause abrupt bending of the wire.

Moveable catch wires are used with some training systems. These typically are installed on upward-slanting nails (or downward-slanting nails in dip areas), on line posts, or on specialized staples or plastic clips (figure 4.36). These wires may be lightly tensioned and permanently attached at the end posts. Some growers have used short (24-inch) lengths of chain on the ends of these wires, which are pulled tight and hooked over nails on the end posts to retension wires after they have been moved.

Tools and Gadgets for Installing Trellis Wires

Basic hand tools for installing vineyard trellis wires include fencing pliers for pulling misguided staples and cutting lengths of wire, a standard claw hammer, a ratcheting wire tightener (figure 4.37) or a chain grab wire puller. A wire gripper (Klein Tools, Chicago, Ill., model #1613-30F) can be added to the ratchet tightener at one or both ends to hold even high-tensile wire quite well (figure 4.37). A wire reel, which can be purchased or fabricated in several designs, is essential for unwinding rolls of wire. Crimps (figure 4.31) and a crimping tool, a post maul (figure 4.25), a pinch bar, and a carpenter's apron to hold staples are also basic tools for trellis construction. There are also several specialized tools that are used to tension wires with their respective tensioning gadgets.

Figure 4.34 ▪ Attaching a wire to an end post.
Low-carbon trellis wire has been attached to an end post by wrapping the wire around the post and then making a series of gentle bends back on the wire itself.

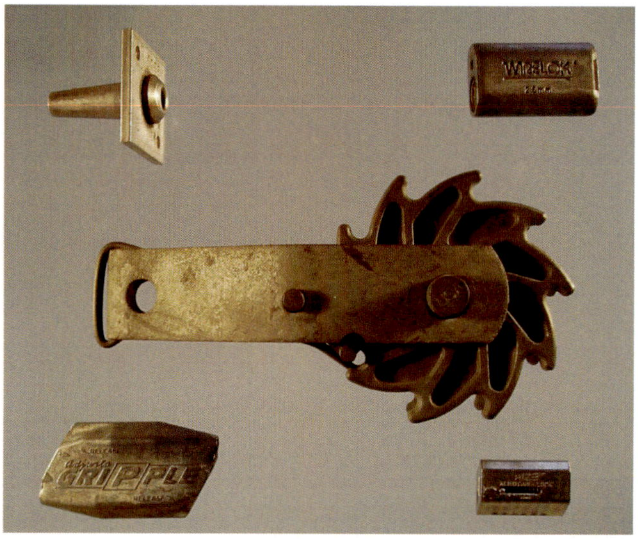

Figure 4.35 ▪ In-line wire-tensioning devices.
These are some of the in-line wire-tensioning devices that can make the task of tightening of trellis wires efficient.

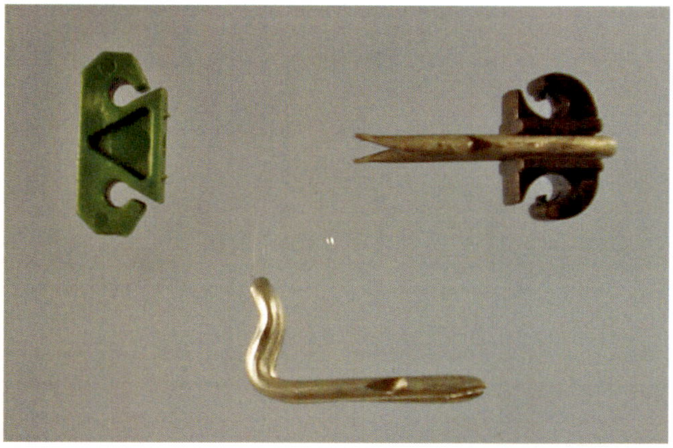

Figure 4.36 ▪ Specialized plastic clips and staples.
Specialized plastic clips and staples can be used to fasten moveable catch wires to wooden line posts.

Figure 4.37 ▪ A ratcheting wire tightener.

A wire gripper has been welded on one end to reliably hold high-tensile wire.

They are a big improvement in the efficiency of trellis maintenance.

Finally, but importantly, a note on safety: high-tensile-strength wire is tightly wound in spools that weigh approximately 100 pounds. Care and prudence are required in handling these rolls to avoid physical injury. Furthermore, the wire has a tendency to recoil when pulled from the spool, and shatter-proof eye protection is a must to avoid eye injury from the sharp end of the wire. Do not allow the wire to play off the wire reel in an uncontrolled fashion—it can be extremely difficult to untangle if this happens. Gloves should be worn when handling wire and posts to minimize contact with heavy metals, wire lubricant, wood preservatives, and splinters. Both hearing and eye protection are desirable when pounding posts. Please follow any additional safety precautions provided by your materials suppliers.

In summary, the careful design, site preparation, planting, and care of a new vineyard are investments for a long, healthy, productive future for that planting. Remedial efforts to correct the neglect of a new vineyard are often more costly and less effective than managing the vines properly from the beginning. Rushing to plant a vineyard improperly may set back the project by more than the year that might have been used for proper preparation. There are many variations for establishing a good vineyard. All of them require attention to the topics presented above.

Further Reading

Anonymous. 1982. How to build orchard and vineyard trellises. U.S.S. Catalog No. T-111578. Pittsburgh, PA: US Steel.

Smart, R. and M. Robinson. 1991. *Sunlight into Wine.* Winetitles, Adelaide.

Zabadal, T.J. 1997. *Vineyard Establishment II—Planting and Early Care of Vineyards.* Michigan State University Extension Bulletin E-2645. E. Lansing, MI.

Zabadal, T.J. 2002. *Growing Table Grapes in a Temperate Climate.* Michigan State University Extension Bulletin E-2774. E. Lansing, MI.

Zabadal, T.J. and J.A. Andresen. 1997. *Vineyard Establishment I—Preplant Decisions.* Michigan State University Extension Bulletin E-2644. E. Lansing, MI.

—5—

Pruning and Training

Authors
Andrew G. Reynolds, Brock University (Ontario, Canada)
Tony K. Wolf, Virginia Tech

Profitable grape growing requires that grapevines be managed so that they have a large area of healthy leaves exposed to sunlight. Such vines are likely to produce large crops of high-quality fruit on an annual basis. Grapevines must be appropriately trained and pruned to achieve this goal. The system by which grapevines are trained will generally dictate how they are pruned. Thus, pruning practices and training systems are discussed together in this chapter.

The expression "dormant pruning" refers to the annual removal of wood during the vine's dormant period, and dormant pruning is probably the single most important task annually performed in the vineyard. Grapevines are pruned primarily to regulate the crop and secondarily to maintain a vine conformation that is consistent with the desired training system. Dormant pruning has both short- and long-term effects on crop quantity and quality.

We manipulate the natural growth habit of grapevines to suit our cultivation practices and cropping goals. Consider the following points:

- Grapevines are botanically termed *lianas*; they derive their form based upon the structure or object they grow upon (for example, a trellis).
- Grapevines have indeterminate shoot growth and may, therefore, require summertime canopy management to maintain an optimal form when grown on a trellis.
- Grapevines can be extremely fruitful, but fruitfulness is governed by sunlight exposure of the developing buds. Optimizing the light environment of the canopy, particularly that region where the developing buds are found, is required for optimizing bud fruitfulness.

In a natural setting, a grapevine grows on trees or various man-made structures, and its height is determined only by the object it grows upon. If the vine becomes shaded, it grows towards the sun. In a commercial setting, a grape grower takes this unstructured entity and places it "in a box"—we decide upon its form by training; we curtail its indeterminate growth by hedging; and, consequently, we must constantly be vigilant about light levels in the canopy in order to maintain fruitfulness. This is the essence of viticulture.

Training positions the fruit-bearing wood and other vine parts on a trellis or other support. Except for renewal

This chapter was adapted from *Dormant Pruning and Training of Grapevines in Virginia* (Wolf 1988; Virginia Cooperative Extension publication number 423-011) and *The Mid-Atlantic Winegrape Grower's Guide* (Wolf and Poling 1995).

of damaged vine parts or system conversion, vine training is normally completed by the vine's third year in the vineyard. Training should uniformly distribute the fruit-bearing units (canes or spurs) in the vine's row space to facilitate perennial vine management, including pruning, and to promote high fruit yield and fruit quality.

GRAPEVINE TERMINOLOGY

Before discussing training systems, we need to be able to define the various parts of a grapevine. These will be the "horticultural parts" of the vine that will constitute the grower's working vocabulary.

Arm. A short, usually horizontal, woody perennial (two-year-old or older) portion of the vine. Arms usually arise from the "head" or other renewal region of the vine. Cordons are long arms.

Bearing units. Canes or spurs.

Bud. Normally refers to the compound structure from which shoots arise with spring growth. Dormant buds (often called "count" buds) are borne at nodes of the cane which are separated from each other by internodes. "Base" buds (sometimes called "non-count" buds) are those dormant buds that are borne at points other than nodes, often at the base of canes and spurs.

Cane. A ripened, woody shoot after leaf fall, following the shoot's growing season. A one-year-old bearing unit with four or more nodes is called a cane. Canes are retained for cropping.

Canopy. A collective term for the shoots, leaves, and fruit of a grapevine.

Canopy division. Training systems that divide a grapevine canopy into at least two distinct subcanopies—either vertically (for example, Scott Henry [page 121]) or horizontally (for example, Geneva Double Curtain [page 118]). In addition to the supporting hardware, canopy division is accomplished through shoot positioning.

Cordon. A long, usually horizontal, woody, perennial (two-year-old or older) portion of the vine from which canes or spurs arise. French, meaning "rope." Cordons are long arms.

Head. Upper portion of the vine comprising the top of the trunk(s) and junction of the arms. The top of a trunk is usually a renewal zone, particularly with cane-pruned training systems.

Internode. The portion of the stem between nodes.

Node. Conspicuous joints of shoots and canes. Count nodes have clearly defined internodes in both directions on the cane.

Periderm. The brown bark of canes, consisting of lignified, corky cells.

Renewal zone. The region within a vine from which next year's bearing units arise.

Shoot. Succulent growth arising from a bud—including stem, leaves, and fruit.

Spur. A one-year-old woody portion of the vine that is retained as a bearing unit and/or as a renewal for cane formation. A spur is usually one or two nodes, but may be up to 4 nodes under some conditions.

Trunk. A vertical, woody, perennial portion of the vine. The vine trunk is the vertical support structure that connects the root system and the fruit-bearing wood.

The current season's crop is borne as one to several clusters on shoots that develop from dormant buds (figures 5.1 and 5.2, page 100). Most buds are located at nodes, the conspicuous joints of shoots and canes. Buds are also present at the base of shoots and canes (figure 5.3, page 100). And buds can remain latent at the less conspicuous nodes of trunks and other perennial parts of the vine. Buds not borne at clearly defined nodes of canes are referred to as base buds (figures 5.3 and 5.4, page 100); and their shoots, which may or may not be fruitful, are termed base shoots. Shoots cease growth in late summer and become brown and woody during the cold acclimation, or "hardening" process. Lateral shoots often develop at the nodes of primary shoots. They too can become woody and persist after fall frosts. Buds are compound structures that consist of several growing points, or primordia (figure 5.1). The primary bud is the largest primordium; it is first to break bud (emerge) in the spring; and it usually bears more clusters than do shoots that develop from secondary or tertiary buds.

M. Goffinet, Cornell University

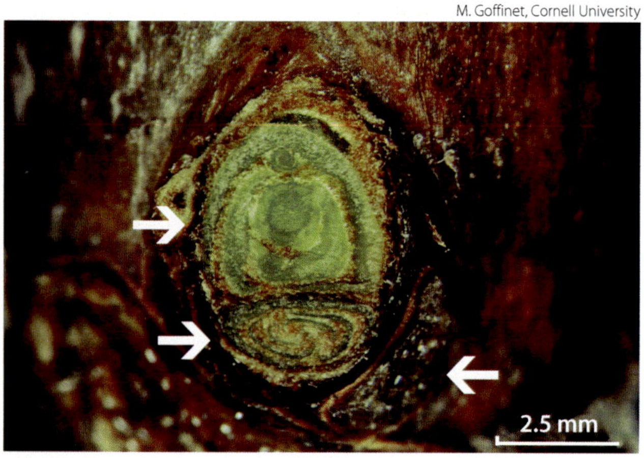

Figure 5.1 ▪ Dormant bud and node of a one-year-old cane.

The compound bud has been cut cross-sectionally to reveal the arrangement of the bud's internal structures. The primary bud is the large, central structure *(upper-left arrow)*; a secondary bud is below it *(lower-left arrow)*; and the summer lateral scar is visible. (*Note:* 2.5 mm is about 0.1 inch.)

M. Goffinet, Cornell University

Figure 5.3 ▪ Emerging shoots.

Shoots emerging from buds on a one-year-old spur borne on two-year-old wood. The two lower shoots originate from base buds; the two upper shoots originate from count buds or count nodes. (*Note:* 1 cm is about 0.4 inch.)

Figure 5.2 ▪ Primary shoots.

Recently emerged primary shoots at nodes of a one-year-old cane. Note flower bud clusters on shoots *(arrows)*.

Figure 5.4 ▪ Base buds.

Base buds developing from section of cordon where a spur has been removed.

Additional terms are used to describe grapevine parts in the context of a particular training system and the training system's integration with a trellis or other support. Trunks can have horizontal extensions of two-year-old or older wood, as with cordons (figure 5.5). These extensions might be short arms, as in the Umbrella Kniffin system (figure 5.6, page 102), or long cordons, as in the bilateral cordon system (figures 5.5a, 5.5c). Arms and cordons, in turn, usually bear spurs or canes. Trunks, arms, and cordons are generally retained for several to many years. The shoots of a vine and their leaves constitute the canopy of the vine. The renewal region is that region of the canopy where buds for the next season's crop develop. The renewal region is often, but not always, the fruiting region of the canopy.

DORMANT PRUNING

What is dormant pruning? Dormant pruning is defined as the deliberate removal of plant parts during plant dormancy to redirect or regulate growth, or to promote and control fruiting and flowering in the subsequent growing season.

a. Unpruned vine with two trunks, each of which is extended into a cordon.

b. Pruning of upper cane from last year's spur. The lower cane on the left will be shortened to a one- or two-node spur.

c. Same vine (from figure 5.5a) after pruning.

d. Early-season shoot development from spurs on cordon. Both a base bud and a count bud at a clearly defined node are shown in the spur close-up.

Figures 5.5 ▪ Cordon-training and spur-pruning.

Why do we prune grapevines? An unpruned grapevine has hundreds of nodes, most of which may bear fruitful buds and shoots. If left unpruned, the resulting crop would quickly unbalance our "grapevine in a box" to a point where growth and winter hardiness would be compromised. We therefore prune to regulate the crop. There are both immediate and long-term effects of overcropping grapevines. Immediate symptoms are observed in the current year and could include reduced rate of sugar accumulation in fruit, reduced pigmentation in berry skins, and decreased synthesis of flavor and aroma constituents. Rather than maturing into woody canes, the shoots of overcropped vines typically die back completely to older wood or they may mature only one or two basal nodes (nodes towards the base of the shoot). Poor wood maturation occurs because the carbohydrates necessary for wood maturation have been depleted by competitive fruit-maturation processes.

The long-term effect of overcropping is a reduction of vine vigor (rate of shoot growth) and vine size (cane pruning weight). Vine size reduction due to overcropping can occur without a noticeable degree of cane dieback. Although wood might appear to be mature, stored starch reserves in vines stressed by overcropping can be so low that the next year's vegetative growth and crop will be severely reduced.

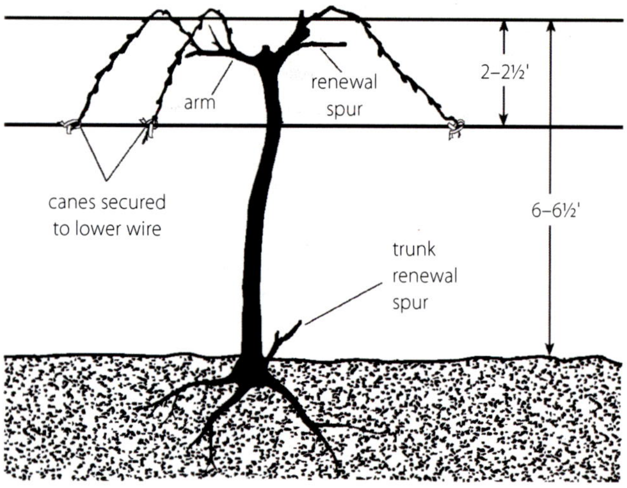

Figure 5.6 ▪ Umbrella Kniffin training system.

Although dormant pruning is the primary means of crop control, it will not provide adequate crop control in all situations. Additional crop control through flower or fruit cluster thinning is generally required with young vines (two years old or younger), with very fruitful varieties such as some of the interspecific hybrids, and in any case where the vine vigor and vine size are insufficient to fill the available trellis space.

Although the practice of dormant pruning is an established means of crop control in most grape production systems, it is not a universal requirement. "Minimal pruning" and no-pruning systems are used in certain situations, but these practices require a high degree of vineyard mechanization to regulate and ultimately harvest the crop. Minimal and nonpruned training systems are not common in wine production vineyards of the East.

The terms *crop yield*, *crop level*, and *crop load* are often used interchangeably in both the popular and scientific literature. The following definitions are used in this chapter:

Crop yield (or **crop size**). Yield per vine or per unit of land area (that is, tons per acre or pounds per vine).

Crop level. The number of clusters *retained* per shoot, per unit of cane pruning weight, or per unit of canopy length (feet or meters).

Crop load. The ratio of crop yield to cane pruning weights, with both variables measured in the same units (for example, pounds or kilograms per vine) and derived from the same growth season. A ratio of 10 to 1 through 12 to 1 is the presently accepted maximum for vinifera varieties (that is, 10 pounds of fruit produced on a vine that produced one pound of pruned canes).

Pruning and Cropping Theory According to Winkler

Pruning a woody plant has two important consequences. First, dormant pruning stimulates growth by reducing the number of growing points and, hence, concentrating growth into fewer shoots. Second, pruning during periods of active growth has a dwarfing effect by reducing photosynthetic surface area. There are two major take-home messages: a *balance* between vegetative vigor and crop is necessary to produce consistent yields of ripe fruit; overcropping will reduce shoot vigor and will ultimately reduce yield and fruit quality.

Thinning versus pruning for crop control. Dr. Albert Julius Winkler pioneered research that showed how pruning and cropping influenced yield and growth of the grapevine. Winkler was a viticulturist with the University of California at Davis from the early 1920s until his death in 1990. One of Winkler's many contributions to viticulture was to examine pruning of grapevines from a scientific viewpoint and to synthesize his research into key principles that he summarized in a comprehensive bulletin. Those fourteen principles are paraphrased below. Text in brackets was added to the original text by the authors for clarification.

1. The removal of . . . vegetative parts of the vine is depressing . . . pruning of any sort at any time decreases the capacity [vine growth and, hence, potential yield] of the vine. Pruning a vine involves removal of dry matter, which is important for maintaining vine growth. So pruning will reduce yield and growth potential compared to non-pruning. The yield and growth potential of the vine is termed capacity, whereas the more familiar term vigor, as we'll use below, refers to the rate of growth of shoots. Capacity can be measured by cane pruning weights (and, hence, has become synonymous with "vine size"), whereas vigor is measured by repeated measures of length or diameter of shoot growth.

2. The production of crop depresses the capacity of the vine. Just like pruning itself, crop removes dry matter from the vine during the maturation process; and the more crop that is retained, the greater the possibility of reducing vine capacity.

3. The vigor of individual shoots . . . varies inversely with the number of shoots that develop. If you severely prune a large vine and only five shoots develop, those five shoots could easily grow 20 feet long and be of undesirably large diameter. This is not desirable.

4. The capacity of a vine varies [directly] with the number of shoots that develop. Every shoot provides dry matter to the vine from photosynthesis in the leaves. It thus follows that increasing exposed leaf surface area per vine will lead to increased capacity.

5. The vigor of the shoots of a given vine varies inversely with the amount of crop it bears. In other words, as crop size is increased, shoot vigor is decreased.

6. Fruitfulness of the buds . . . varies inversely with the vigor of [the] shoots. Very vigorous shoots usually have large leaves and lots of lateral shoots with a high potential for creating shade. As canopy shade is increased, node fruitfulness is decreased. Node fruitfulness may also be directly affected by the rate of growth.

7. A . . . vine can properly nourish and ripen only a certain quantity of fruit; that is, its capacity is limited by its previous history and its environment. A vine that has been overcropped or water-stressed for several years will have greatly diminished capacity to produce and mature a crop. A vine with this type of history can mature less fruit than one that has been balanced for many seasons.

8. The fruitful buds of the vine occur most abundantly on one-year-old canes that arise from two-year-old wood. Certain hybrid varieties, however, have very fruitful base buds.

9. The more erect a shoot or cane . . . the more vigorously it will grow. A good example of this is to look at the top and the bottom canopies of a Scott Henry-trained vine (page 121). You will notice that the shoots that have been positioned downward stop growing, whereas those trained upward tend to be very vigorous. The same phenomenon occurs in a GDC-trained vine after the shoots are positioned downward. (GDC, or Geneva Double Curtain, is discussed on page 118).

10. The shoots starting farthest from the trunk . . . are the most vigorous. A caveat needs to be added here. Winkler was describing cane-pruned vines, and a physiological phenomenon known as apical dominance. Frequently, on vines pruned to long canes, the shoots that grow at the ends of the canes are the most vigorous. With cordon-trained vines, we often see the greatest vigor on shoots that are borne on spurs near the point of bending from vertical trunk to horizontal cordon.

11. Canes with internodes of medium length usually mature their wood best and have the most fruitful buds. Canes with medium-length internodes are usually those that have developed and matured in well-illuminated conditions and, hence, are most fruitful. "Shade" canes are frequently characterized as having poor periderm development, low fruitfulness, reduced winter hardiness, and long internodes.

12. A large cane or arm or vine is capable of greater production than a small one and, therefore, should carry more buds. This is basically just a description of balanced pruning. A large vine with high capacity should carry more buds than a weaker one, simply to achieve balance between crop size and vine size.

13. Well-matured canes have the best-developed buds. This is a reiteration of point #11; it says that if the periderm develops well, chances are that you have well-developed buds that will survive the winter. Cane maturity is achieved by optimizing light microclimate in the canopy.

14. By bending or twisting canes or shoots, their behavior may be modified for the purpose of regulating growth or fruiting. This is an interesting point. Our experience has been that bending a cane (into a training system, such as a pendelbogen [page 117] or Umbrella Kniffin [page 115]) will often reduce the apical dominance effect and create more uniform shoot growth along the cane than if the cane is positioned horizontally on the trellis wire. The looped-cane technique practiced in the Mosel region in Germany is perhaps the most extreme example of this technique.

How should Winkler's principles guide our pruning? We need to regulate both vigor and crop size by pruning to consistently and optimally mature fruit. Moreover, we need to strike a *balance* between vegetative growth and reproductive growth if we are to be successful winegrowers. Overcropping not only will reduce the vigor of individual shoots and, therefore, vine capacity but also will reduce yield and wine quality.

When to Prune

Vines can be pruned anytime between leaf fall and bud break the following spring. However, there is some evidence that fall-pruned vines are more susceptible to winter-injury than are vines pruned in late winter or early spring. Delaying pruning until late winter allows the grape grower to evaluate bud injury and to make compensations in the number of nodes retained where injury has occurred. Spring pruning does not harm vines, even where sap "bleeding" is observed; however, swollen buds and young shoots are extremely susceptible to breakage. Therefore, the removal of unwanted wood from the trellis should be completed before bud swell. Experienced pruners require thirty to forty hours to cane-prune an acre of vines. Somewhat less time is involved with spur-pruned vines. Cane pruning and spur pruning are described under "Grapevine Training" (page 109).

Double-pruning of vines is sometimes practiced in areas where spring frosts are common. Canes or spurs are retained with two to three times the desired number of nodes at the initial pruning in late-winter or early-spring. Buds nearest the pruning cut develop shoots as much as seven days earlier than the basal buds of the same cane or spur. To correct shoot density, a second pruning cut is made—after the threat of frost but before appreciable shoot growth has occurred—to shorten the spurs.

What to Retain

The selection and retention of suitable fruiting canes and/or spurs is extremely important. Select canes or nodes that show good wood maturation. This is far more important than selecting wood strictly on the basis of its location relative to a desired training system. Generally, medium- to dark-brown canes with short internodes (4 to 6 inches) are superior to lighter-colored canes that have internodes longer than 6 inches. Canes that have internode diameters of ⅜ to ½ inch are superior to canes outside this range. Well-matured lateral canes or spurs can be retained as fruiting wood if needed; however, medium-diameter canes that lack persistent laterals are superior to large canes that bear many persistent laterals. Canes associated with good bud-fruitfulness and cold-hardiness are located toward the exterior of the canopy where they received more sunlight than those that developed within the canopy.

Complications Due to Cold Injury

In many years, assessing and compensating for cold-injury are important aspects of pruning grapevines in this region. The retention of nodes is based on the assumption that buds of retained nodes are viable. If buds have been killed, such as by freezing, the number of retained nodes must be increased to compensate for this injury.

Bud injury is assessed before pruning by evaluating the viability of a representative sample of buds from a given variety. Dead buds are identified by a browning of their primordia, which occurs after the frozen buds are allowed to warm for a few days. To determine if a bud is dead, make several, consecutively deeper, cross-sectional cuts through the bud to expose the individual primordia (figure 5.7). A sharp, single-edged razor blade is the best tool for sectioning buds. The primary bud, located between the secondary and tertiary buds, is most susceptible to cold-injury. Dead buds will appear brown, whereas live buds will be a light-green color. If buds are sectioned too deeply, the primordia may be missed and the green tissue beneath the bud exposed. The novice

Figure 5.7 ▪ Cross section of dormant bud.
Central, primary bud killed by low temperature. Live secondary and tertiary buds shown on either side of the primary bud.

should gain some experience cutting live buds (such as those of a cold-hardy variety) to learn to recognize the individual primordia of a bud and to become familiar with the green appearance of live primordia.

Buds can be examined for viability on the vine, but it is generally more comfortable to collect 10 to 20 canes at random through a varietal block and examine the buds indoors. Collect only canes/nodes that might otherwise be retained at pruning. If there are large (30- to 40-foot) differences in elevation within a vineyard block, sample the regions separately because injury may vary spatially within the vineyard. Examine 100 to 200 buds of each variety and record the percentage of dead primary buds. If your bud assessment reveals 40% bud injury, then bud retention should be increased by roughly 40% in vinifera and native-American varieties. Many of the interspecific hybrid varieties, however, have fairly fruitful secondary and base shoots; and death of primary buds alone might not significantly reduce yields. The compensation for primary bud injury is, therefore, not as generous as with native-American and *V. vinifera* varieties.

Low temperature can also kill canes and trunks. Figure 5.8 shows in cross section a portion of a three-year-old grapevine trunk. Trunk tissues include (from exterior to interior): a corky periderm, or bark; the phloem, or food-conducting tissue; the vascular cambium; the xylem, or water-conducting tissue; and a central pith. The vascular cambium is a region of cell differentiation and division that produces new xylem and phloem cells on an annual basis. Canes have the same tissues as trunks but lack the annual rings of xylem. Cambium and phloem tissues are generally the most susceptible to cold injury. These tissues, like buds, will brown after being killed and subsequently warmed. Injury to the vascular cambium reduces or prevents the development of new xylem and phloem tissues. The old xylem tissue might sustain the initial water-conducting needs of the developing shoots in early spring; however, cold-injured vines often wilt and die in midsummer because the transpirational loss of water from leaves exceeds the ability of the impaired vascular system to transport water.

Cane and trunk cold injury is diagnosed by making shallow longitudinal cuts into the wood and examining the phloem and cambial regions for browning. These tissues form a thin cylinder immediately beneath the bark.

M. Goffinet, Cornell University

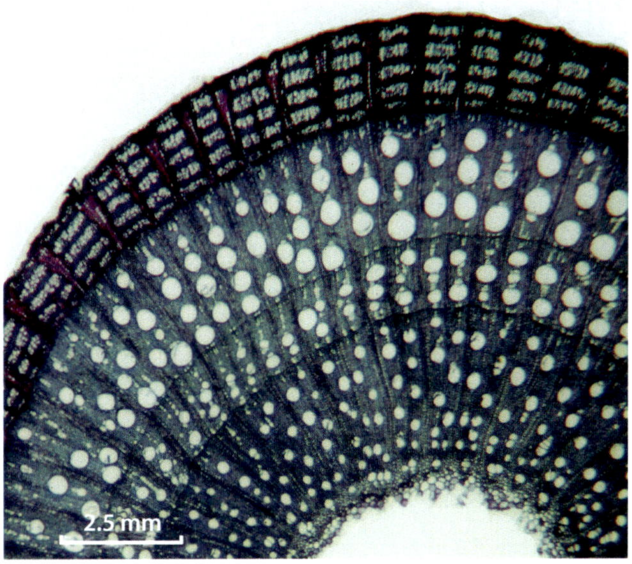

Figure 5.8 ▪ Cross section of trunk.
Cross-sectional view of a portion of a three-year-old grapevine trunk.

Browning or darkening of these tissues indicates injury. If wood injury is observed, retain extra canes at pruning. It's likely that injury will not be uniform and that some canes will be unaffected. Some of these extra canes can be removed or shortened after bud break if too many shoots are present. Trunk injury is also diagnosed by making shallow, longitudinal cuts into the wood. Injury may occur to any portion of the trunks and cordons, but it is often most readily apparent just above the snow line if the damaging cold event occurred when snow was present.

Cold-injured trunks frequently split or are affected by crown gall (page 234) one to two years after the cold-injury occurred. Vines will ultimately die if they must depend on a single, cold-injured trunk. Multiple-trunking is, therefore, highly recommended to assure the long-term survival of vines (see "Grapevine Training," page 109). Split, heavily crown-galled, or otherwise defective trunks should be sawn off and replaced with a cane that arises near ground level but above the graft union of grafted vines. "Renewal canes" that originate from renewal spurs left near the base of the vine (figure 5.6, page 102) facilitate trunk renewal, when needed.

The subject of avoiding, assessing, and compensating for winter cold injury is comprehensively covered in *Winter Injury to Grapevines and Methods of Protection.* (See References, page 123, Zabadal et al. 2007)

Complications Due to Bud Necrosis

In addition to cold-injury, a physiological disorder termed bud necrosis can also kill buds and complicate our efforts at arriving at a desired crop level. Bud necrosis (BN) results in the death of buds during the season of their initiation. If the buds are examined towards the end of summer, or into the fall, BN-affected buds appear darkened, desiccated, and possibly compressed by the greater development of secondary buds in the same compound bud (figure 5.9). The incidence of BN varies by variety, season, node location on the cane, and training system. Many varieties express up to 5% dead buds when buds are examined in late-summer—well before the advent of potentially damaging cold temperatures. Such low levels of BN do not measurably affect yield and are inconsequential. Other varieties—notably Viognier, Riesling, Syrah and Valdepeñas (Tempranillo)—express 50% or more BN in some seasons in the mid-Atlantic region, in which case the yields may be significantly affected. Early (around bloom time of the current season's flower clusters) destruction of primary buds may lead to increased development of secondary buds, including flower cluster formation, in which case the effects of BN on crop yield the following year may be less dramatic. In more extreme cases, BN can also affect secondary buds, and those nodes may fail to produce shoots. Examine buds prior to pruning, as described earlier for winter injury. Pruning can be adjusted if needed to compensate for primary bud loss, just as one can compensate somewhat for cold-injury to buds. In research conducted at Virginia Tech, the incidence of BN often increased with node position; so spur-pruning generally retained a greater proportion of good primary buds than did cane-pruning. Cane selection was also important, even if only a spur was retained. Canes with abundant laterals and those from shaded regions of the canopy had a higher proportion of necrotic buds than did moderate-sized, well-exposed canes. Follow the canopy-management guidelines (see chapter 6, page 124) to increase sunlight exposure of developing buds during the growing season to help reduce BN incidence.

Figure 5.9 ▪ Bud necrosis.

Cross-section of dormant bud revealing necrosis of primary bud, and further development of additional secondary buds.

How Many Nodes to Retain

Dormant pruning is the grape grower's primary means of regulating crop. Other factors not limiting, vines that are correctly pruned are likely to produce large crops of quality fruit. If vines are pruned incorrectly, vines and crop will ultimately suffer. It is imperative that the grape grower understand how many nodes to retain, as well as which nodes are associated with good cold hardiness and fruitfulness.

Eighty to ninety percent of the one-year-old wood is removed from vines at dormant pruning. Before pruning mature grapevines, the grape grower must decide how many nodes to retain. Overcropping and excessive canopy density will occur if too many nodes are retained. On the other hand, the crop will be needlessly reduced if too few nodes are retained. Furthermore, severely pruned vines are apt to produce excessively vigorous shoots because all of the stored energy in the trunks and roots is available to relatively few growing points. Excessive shoot vigor can reduce fruit set and delay shoot maturation in the fall.

Balanced Pruning

"Balanced pruning" was developed to help grape growers determine the appropriate number of nodes to retain at pruning. Balanced pruning is based on the concept that a vine's capacity for vegetative growth and fruit production is a function of the vine's size. The size of a vine is determined by the extent of growth of roots, shoots, and perennial wood. In its simplest terms, balanced pruning is based on the premise that a vine must be pruned in accordance with its capacity.

SIDEBAR

Pruning and Wine Quality—the Controversy over Yield

There is a controversial misconception that low yields are a prerequisite for high wine quality. This misconception refuses to go away despite much research worldwide that indicates the quality/yield relationship is not linear. Always remember that **crop size is regulated by pruning and cluster thinning.** One can both overprune (leave too few nodes) or underprune (leave too many), depending upon vine capacity. Many grape growers, sometimes at the behest of their winemakers, leave fewer nodes at pruning than necessary or may excessively thin clusters at various times during the season to reduce crop size and allegedly produce higher wine quality. Cluster thinning may certainly increase wine quality, but not universally; for example, a recent study in California showed that yield increases from 1.8 to 8.9 tons per acre actually *reduced* the intensity of vegetal aromas in Cabernet Sauvignon wines because the vines were more balanced at the higher crop sizes.

Because the growth of roots and other perennial wood cannot be conveniently measured, vine size is measured by weighing the one-year-old wood (canes) removed at pruning. Essentially, we balance the number of nodes retained against the weight of pruned canes: A large vine should retain more nodes than a small vine because the large vine has a greater capacity for both vegetative growth and crop production than does the smaller vine. The balanced pruning concept tells us that a large vine can carry more buds than a small one. Newton Partridge, a viticulturist in Michigan and a contemporary of Albert Winkler, noticed that many of the vineyards he examined contained a range of "vine size" (Winkler's "vine capacity"), from 0.1 to 6.9 pounds of cane prunings per vine. He found that by pruning large vines lightly (leaving more than 50 nodes per vine) and small vines severely (leaving less than 30 nodes), yields and vine size gradually became more consistent.

Nodes—specifically "count nodes" (colloquially termed "buds")—are the units counted in the pruning formulas. Count nodes are those that have clearly defined internodes in both directions on the cane; the first count node on a cane is colloquially defined as being more than a "thumb's width" from the base of the cane (figure 5.3, page 100). Having determined the appropriate pruning formula to use, the grape grower visually estimates the vine size, then calculates the number of nodes that should be retained on the pruned vine based on that estimate. This requires some experience, but 5- to 6-foot canes average about 0.1 pound. The vine is then pruned, leaving 10 to 15 extra nodes as a margin of estimation for error. The cane prunings are weighed with a handheld scale, and their weight is entered into the pruning formula to determine accurately the number of nodes to be retained. Nodes in excess of that number are then removed. Commercially, it is neither necessary nor practical to weigh cane prunings from every vine. In practice, most pruners acquire an ability to estimate closely the pruning weights and node retention. Thereafter, only an occasional vine is weighed to check one's estimates.

Occasionally, growers will retain an extra "kicker cane" that is normally used for insurance purposes in case a cane is damaged during tying or mechanical operations. Kicker canes are also used for vigor diversion in high-vigor vineyards; under these circumstances, the "kickers" are considered disposable and are usually removed sometime around 80% of canopy fill. Kicker canes can also increase yields in spur-pruned vineyards; if spur pruning does not result in satisfactory yields, sometimes kicker canes are retained to compensate for this.

Pruning formulas were developed for many varieties to calculate the number of nodes to be retained for a given pruning weight. A pruning formula of 20 + 20, for example, would require leaving 20 nodes for the first pound of canes removed plus an additional 20 nodes for each additional pound above the first. A 3.2-pound vine would, therefore, retain 64 nodes if the 20 + 20 formula was used at pruning. Weighing is done to the nearest 0.1 pound. There are both minimum and maximum numbers of nodes to be retained with all pruning formulas.

For example, a minimum of 10 nodes should be retained on vines that are 2 years old or older (at typical vine spacings of at least 4 feet in the row). Given 10 or more shoots, small vines will require some degree of cluster thinning to prevent overcropping, but the shoots and leaf area are needed to increase vine size. The maximum number of nodes to be retained on mature vines should be on the order of 4 to 6 nodes per linear foot of row or canopy (for example, 24 to 36 nodes for vines spaced 6 feet apart in the row). The lower number would be more appropriate for large-clustered varieties; the higher number would be acceptable for varieties with small- to medium-sized clusters.

The pruning formulas allow for additional shoots that will develop from non-count locations (base buds). Generally, the native-American and *V. vinifera* varieties do not produce many base shoots unless the vines have been pruned too severely. Many of the interspecific varieties, however, produce numerous, fruitful base shoots—even with moderate pruning. Balanced pruning of hybrid varieties, therefore, has limited utility. Crop control with some hybrid varieties, notably Seyval and Chambourcin, must be achieved through a combination of fairly severe pruning and shoot or fruit-cluster thinning. (See "Shoot Thinning," below, and "Cluster Thinning," page 140.)

Weight of Cane Prunings

A representative sample of vines from your vineyard should be pruned individually, the cane prunings separated from each, and weighed. Ideally, the weight of cane prunings should be in the range of 0.2 to 0.4 pound per foot of row or canopy. Many viticulturists refer to this synonymously as "vine size." Small vine size demands attention through reduced crop levels (by pruning or cluster thinning), soil fertility assessment and improvement, improving organic-matter levels, drainage, and tilling or subsoiling. Large vine size can be addressed through reduced fertilizer inputs, use of a cover crop, increased number of nodes, and canopy division.

Balanced Pruning and Shoot Thinning

French-American Hybrids

When vines of De Chaunac, Seyval, and other large-clustered varieties were first planted, it took some time before it was clear that their extremely fruitful base shoots provided them with a serious propensity to overcrop. Modifications in balanced-pruning formulas tended to have little effect on crop regulation due to the productivity of non-count nodes. In a New York study, severely pruned Seyval vines got smaller because of the extra crop load imposed by these fruitful base shoots. It was apparent that a strategy other than balanced pruning was required to maintain yield, vine size, and wine grape quality in large-clustered French-American hybrids. For another variety, De Chaunac, balanced pruning seemed totally useless; vines on which all "count" nodes were removed (leaving base buds and bare cordons only) yielded as much as conventionally pruned vines. Cluster thinning, it appeared, was to be the only effective mode of crop regulation in French-American hybrids.

Research in New York state first introduced the concepts of "shoot density" and "balanced thinning" for French-American hybrid grape production. A recommendation was made to retain 4 to 6 shoots per foot of row and to use balanced thinning (of flower clusters) to obtain consistent yields in Seyval. Under this concept, all vines are pruned identically to obtain a desired leaf area and canopy density; then vines are cluster-thinned according to vine size. That is, small vines are severely thinned (less than one cluster per shoot) and large vines are lightly thinned (approximately two clusters per shoot).

This concept provided a better approach to handling varieties with fruitful base buds than did balanced pruning. It specified a reasonable shoot density (4 to 6 shoots per foot of canopy, or 13 to 20 shoots per meter of canopy), and use of "balanced thinning" with large-clustered varieties. Hence, large and small vines carried the same number of shoots, but small vines were more severely cluster-thinned. In practice, all vines theoretically carry the same leaf area, but weaker vines are vegetatively stimulated by limiting the ratio of crop to leaf area (see chapter 6, page 124).

Other node retention schedules

There are other, more arbitrary means of determining the number of nodes to be retained at pruning. Node-retention figures are sometimes based on the linear row length or the square area that a vine occupies. For example, mature vines trained to conventional, nondivided-canopy training systems (for example, VSP) should generally retain 4 to 6 nodes per linear foot of row (see "Grapevine

Training System Examples," beginning on page 114, for details on VSP, or vertical-shoot-positioned, and non-divided-canopy systems). Expressing node retention on the basis of the linear measure of row or the square area of vineyard is convenient; however, it ignores individual variation in vine capacity and can lead to overcropping of small vines and/or undercropping of large vines. It is not as precise as balanced pruning or balanced cropping and is, therefore, not a recommended procedure, especially where variation in vine size is great.

Pruning Weights and Crop Loads

A mathematical basis for vine balance, that of the ratio of yield to cane pruning weight, was proposed in the early part of the twentieth century by Ravaz (the Ravaz Index) as a yardstick of vine balance. The Ravaz Index was redefined using the term crop load by Bravdo in Israel in the 1980s, who demonstrated that crop loads of 10 to 12 were the optimum, whereas 16 was the limit above which wine quality of Cabernet Sauvignon and Carignane might suffer. Work in western Canada with Seyval, Chancellor, and Riesling and in the USA with Chambourcin sets these limits much higher if proper canopy microclimate is provided and if sufficient heat units are available to mature the crop. On the other hand, optimal crop loads may be as low as 5 in regions where heat units during fruit maturation are sometimes insufficient. Hence, vine balance should not be envisioned as a simple, two-dimensional relationship between yield and vine size but as a triangular model that involves yield and vine size as the major components, with adjustments made where necessary for satisfactory fruit maturity.

Pruning Wounds as Sites for Fungal Infection

A final comment on pruning concerns the potential for vascular pathogens to invade pruning wounds and how our pruning practices might be adjusted to minimize this threat. Both *Eutypa lata* and *Botryosphaeria* species are slow-growing, fungal pathogens that can invade pruning wounds (see "Eutypa Dieback," page 232). With time, both fungi can form cankers and eventually kill affected cordons, trunks, or entire vines. Eutypa dieback and Botryosphaeria canker can start when the causal fungus invades large wounds—such as those caused by dormant pruning—during the dormant season. This is one reason why the traditional recommendations for Eutypa dieback avoidance include either double-pruning or a delay of making "large" pruning wounds until early spring or even after bud-break. Large wounds would include those made in two-year-old and older wood. Management of these diseases include delay of large pruning cuts until spring growth resumes, double-pruning, and regular cordon and trunk renewal to removed affected vine parts. The use of topical treatment of wounds with fungicides, detergent solutions, or boron-containing solutions is an active area of research, which may lead to US Environmental Protection Agency (EPA)-approved fungicidal management of wood-decaying fungi. Check with state or other local sources for current registrations and recommendations on this approach.

GRAPEVINE TRAINING

Grapevine training, like dormant pruning, is a prerequisite for high-quality grape production. Training may be defined as the "physical manipulation of a plant's form." Numerous training systems are in use today, and many are indigenous to the viticultural regions in which they are found. The training system used will depend upon the variety, the frequency of cold injury, the degree of vineyard mechanization, and the availability of skilled labor. Because these many objectives must be met, it is not surprising that training systems vary considerably throughout the world.

Factors Affecting Choice of Training System

The training system of choice is the one that adequately satisfies these objectives within the constraints of a particular site and variety.

- **Growth habit.** High or low renewal zone depends upon whether the variety has an erect or procumbent (drooping) growth habit.
- **Cold hardiness.** Cold-tender varieties are best cordon-trained and spur-pruned, provided that survival of the cordons is not an issue.
- **Fruitfulness of base buds ("non-count" buds) and proximal "count" buds (those closest to the base of the cane).** If fruitfulness is low (for example, Nebbiolo and Sauvignon blanc), cordon systems are not advised.

- **Adaptability of the training system to mechanization.** Cane pruning and unusual divided-canopy systems are not advised if mechanized pruning, shoot positioning, and harvesting are planned.
- **Facilitation of movement of equipment.** Promote efficient vineyard operation with respect to equipment traffic, fruit harvesting, pesticide application, and dormant pruning. That is, the bearing units (canes or spurs) must be distributed on the trellis to facilitate movement of equipment through the vineyard.
- **Cost-effectiveness.** The system of choice must be cost-effective.

An acceptable training system will:

- promote maximum exposure of leaf area to sunlight. Leaves must be well-exposed for photosynthesis to occur optimally.
- warm clusters for adequate sugar accumulation, acid degradation, and biosynthesis of flavor compounds in cool grape regions.
- create a desirable environment within the canopy (microclimate), particularly in the renewal region. Proper training can provide for a renewal zone to be formed, which ensures that the vine form is perpetuated and yield is maintained.
- arrange perennial wood and bearing units in such a way that shoot-crowding and leaf- and fruit-shading are avoided, thus optimizing wine quality, disease control, and yield.
- promote uniform bud break, especially with those varieties that exhibit pronounced apical dominance (described in the next section, "Initial Training of Grapevines").
- distribute vines and bearing units to avoid undue competition between vines.
- minimize the volume of perennial wood (for example, old trunks) in situations where the hazards of winter-injury outweigh the merits of perennial wood retention.

Initial Training of Grapevines

The growth potential of grapevines and the conditions under which vines are grown will not be uniform. Other factors being equal, vines grafted to pest-resistant rootstocks will generally develop faster and usually grow larger than will non-grafted vines. Variation in moisture and nutrient availability within a vineyard can cause differences in the extent of growth for a given variety. Training grapevines, therefore, requires discretion in evaluating the growth of individual vines during their establishment. Regardless of the intended training system, the initial training of grapevines has the following goals:

- Year 1: Develop large, healthy root systems.
- Year 2: Establish the initial components of the intended training system, including at least one semipermanent trunk.
- Year 3: Continue to develop or complete the training system, harvest a partial crop, and establish a second trunk, if not already accomplished.

There are several methods by which the above goals can be achieved. The following text and accompanying illustrations describe one means of establishing a low, bilateral cordon-trained vine using two semipermanent trunks. It is important to note that the method of training young vines discussed here is but one of several possible approaches.

Year 1: Bilateral Cordon-Trained Vine

It is recommended that the trellis posts and at least the lowest of the training wires be installed during the first growing season. This wire—and a slender stake set next to individual vines—will provide a support for shoot growth. Two to three shoots are allowed to develop on vines during the first year (figures 5.10). These shoots are trained vertically to the support stake and may eventually be tied loosely to the training wire if their growth warrants it. The objective in leaving several shoots on the first-year vine is to provide an abundant leaf area. Root growth is dependent on carbohydrates produced in the leaves. Thus, the greater the leaf area, the greater the root growth that occurs in late summer. Eliminating all but one shoot can also lead to excessive shoot vigor, especially if vines have large root systems when planted. Rapid and continued growth late into the fall can result in incomplete wood maturation, which increases the susceptibility to cold injury. In addition, the retention of several shoots, rather than one, provides some measure of compensation for possible wind damage, deer browsing, and other factors that can retard the development of young

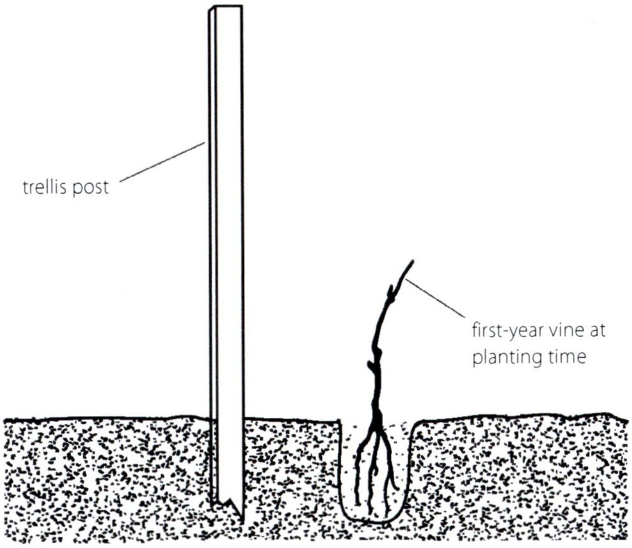

a. Year 1 — spring, at planting.

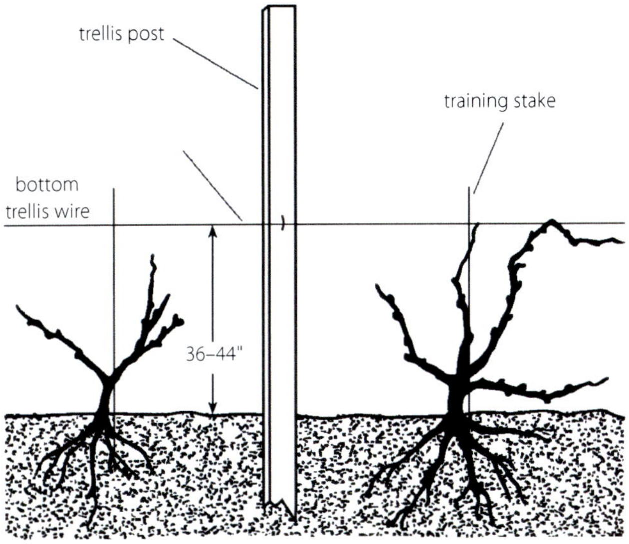

b. Year 1 — fall, at the end of the growing season.

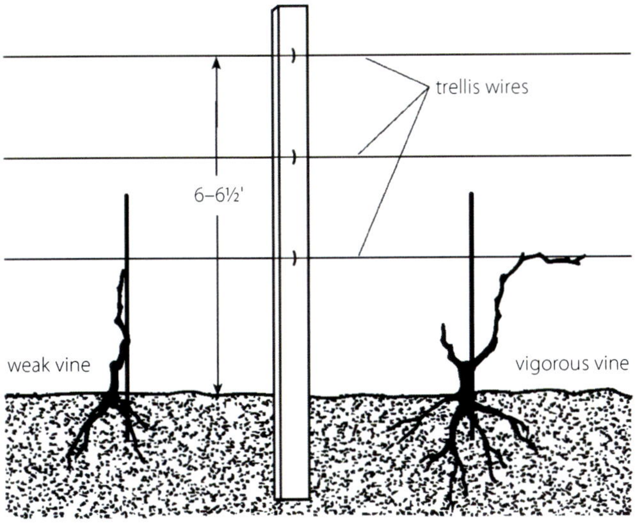

c. Year 2 — spring, after pruning.

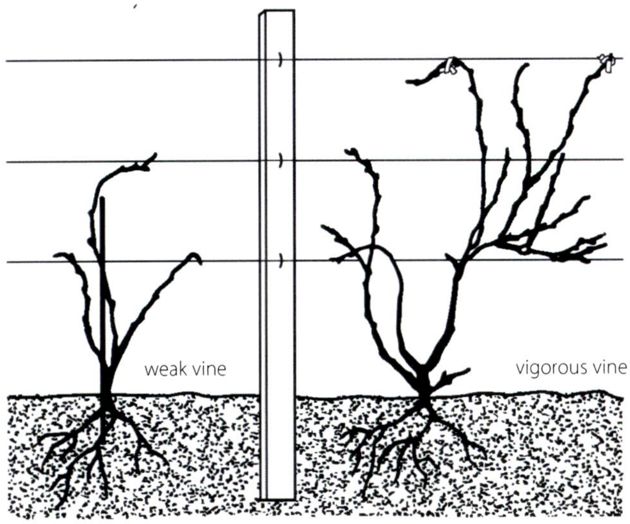

d. Year 2 — fall, at the end of the growing season.

Figure 5.10 ▪ Development of bilateral cordon-trained grapevines.

The vine on the left has demonstrated weak growth. The vine on the right grew vigorously and attained a greater size.

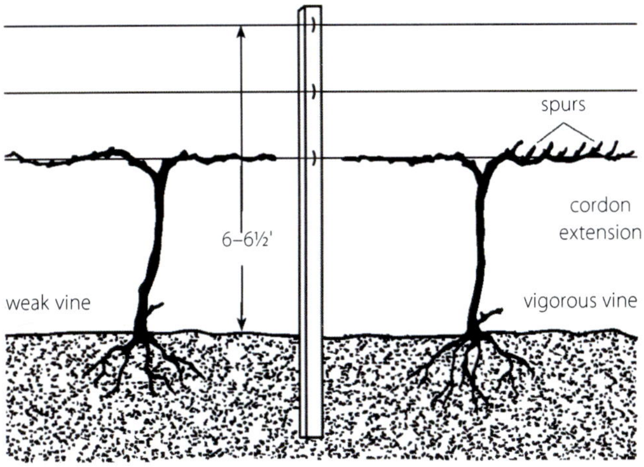

e. Year 3 — spring, after pruning.

vines. Some growers will remove lateral shoots from the primary shoots in order to promote primary shoot elongation. We do not generally recommend lateral removal for two reasons: it involves labor, and it removes valuable leaf area from the vine. Lateral shoot growth will be minimized if shoots are positioned upright and fastened to the support stake, rather than allowed to trail across the vineyard floor.

It is essential that young vines be protected from fungal diseases by appropriate fungicide applications. Powdery and downy mildews, particularly, can severely reduce the photosynthetic (food manufacturing) capabilities of leaves and retard the establishment of the training system. Deer, Japanese beetles, weeds, and other pests—as well as drought—will also have greater impacts on young vines than on older vines and must be diligently controlled. Young vines do not have the food storage reserves afforded by the large root systems and trunks of older vines.

Year 2: Bilateral Cordon-Trained Vine

The trellis should be completed before bud break of the second growing season. Training in the second year starts by evaluating the extent of growth achieved during the first year. If no canes reach the first wire, remove all but one cane. Prune this cane to three or four nodes and secure it to the training stake (figure 5.10c, page 111). Treat such a vine as a 1-year-old vine.

Vines that grew extensively in year 1 will likely have one or more canes suitable for retention as a trunk. If a cane is long enough to reach the lowest trellis wire and is of adequate diameter (approximately ½ inch), retain the cane as a trunk. The distal (towards the tip) portion of such canes can be trained horizontally along the training wire to serve as the basis for cordon establishment (figure 5.10c). If you elected to use a high training system, tie the cane vertically to the top wire of the trellis to form a trunk. In addition to the first trunk, retain a renewal spur of one or two nodes that originates near the soil line but above the graft union (figure 5.10c). If a second cane is large enough to serve as the second trunk, it too could be retained.

Cordon Establishment: Bilateral Cordon-Trained Vine

Cordon establishment can begin in the second season. It is best to establish cordons over a two-year period. Long canes (8 to 15 nodes) often exhibit poor shoot growth at mid-cane nodes. Shoots that develop near the terminal or distal end of a cane produce growth-regulating hormones that retard the development of mid-cane shoots. This so-called apical dominance of distal shoots is greatest when the cane is oriented vertically upright and is minimized when the cane is trained vertically down. To establish 3-foot-long cordons, use an 18-inch-long cane (or trunk extension) in year two (figures 5.10c and d) and complete the cordon in year 3 with another 18-inch-long cane that originates near the distal end of the short cordon (figure 5.10e). Canes that are used to establish cordons should be wrapped loosely around the trellis wire and securely tied at their terminal end with wire. The tying process will prevent the cordon from rotating or falling from the wire. If canes are wrapped too tightly around the cordon wire (greater than about two rotations in a four-foot length) they are apt to grow into the cordon wire within a few years. This does not impair vine performance, but it does prevent the cordon wire from being properly tensioned as it stretches with time.

During the second growing season, shoots of vigorous vines that originate below the lowest trellis wire are thinned to one or two near the graft union (figure 5.10d). Shoots that originate on the developing cordon are retained. Retain ten or more shoots, if possible, in year 2. Small or weak vines are handled as first-year vines during the second growing season (figure 5.10d). Remove all flower clusters from small vines. Where exceptional growth was achieved in year 1, it may be desirable to leave several fruit clusters per vine in the second growing season to slow vegetative growth. This token crop can be removed quickly in early summer if growth is less than expected. Shoots that develop in year 2 should be positioned and tied to the trellis wires to maximize sunlight exposure of their leaves. For cordon training, these shoots will form the spurs for shoot development the following year (figure 5.10e).

Year 3: Bilateral Cordon-Trained Vine

The basic elements of the training system should be completed in the third year. For low cordon-trained vines, canes that arise from the upper side of the cordon are pruned to one- to two-node spurs (figure 5.10e, page 111). For high cordon-trained vines (for example, figure 5.11), spurs would be retained on the lower half (lower 180°) of the cordon. Spurs should be spaced 4 to 6 inches apart. Develop a second trunk and cordon from a cane that originates near the graft union; follow the procedure outlined for the initial trunk. Retain a small crop (for example, one cluster per two shoots) on vines that had at least one pound of cane prunings from second-year

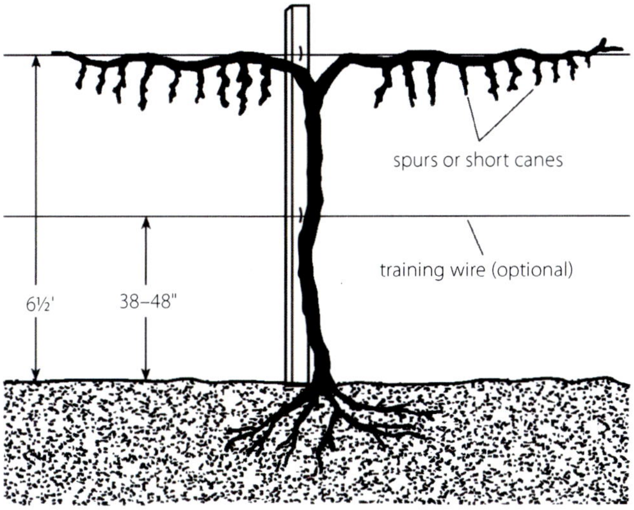

Figure 5.11 ▪ High-wire cordon-training system.
One example is the Hudson River Umbrella.

growth. Position and tie shoots to the upper trellis wires, as necessary, during the growing season. Treat weak vines as second-year vines and remove all crop.

Multiple-Trunking, Trunk Renewal, and Graft-Union Protection

Growing cold-tender grapevine varieties in eastern North America introduces problems not experienced in regions that have mild or more constant winter temperatures. Some degree of bud injury occurs regularly with cold-tender varieties but can generally be compensated for by retention of additional buds at dormant pruning. Cane and trunk injury occurs less frequently, and is much more difficult to compensate for. In some situations (cold-tender varieties planted on poor sites), complete vine loss has been experienced. Even in good to excellent sites, it is wise to anticipate cold injury to better compensate for its occurrence. Even in the absence of winter injury, other forms of injury can occur to vines, such as disease and mechanical damage by vineyard equipment.

One of the strongest recommendations for compensating for trunk injury is to use multiple-trunk training systems. This recommendation applies whether one is growing cold-tender vinifera or more hardy hybrids. All of the training systems illustrated in this chapter can be established using two or three trunks, as opposed to one. Cold injury or death of trunks is often not uniform. Frequently, only one trunk of a two- or three-trunk vine is killed. Similarly, the development of a wood-rotting disease such as Eutypa dieback may be observed initially on only one cordon or trunk (see "Eutypa Dieback," page 232). In either case, the removal of one of two trunks does not eliminate grape production from the affected vine. Second, it is usually easier to reestablish the lost unit from an existing trunk. The development of a multiple-trunk training system can be done starting in the first year as described above for a bilateral cordon system. Alternatively, a second or third trunk can be added in any year by training up a shoot originating near the graft union. Old trunks should not be replaced unless they are mechanically damaged, diseased, or cold-injured. To avoid Eutypa dieback infection, follow the precautions outlined in the "Eutypa dieback" section beginning on page 232 when removing injured trunks. The continual removal of shoots ("suckers" or watersprouts) from the base of the trunk will exhaust the latent buds that could be used to develop new trunks, and with such vines it is sometimes difficult to establish new trunks. Therefore, the maintenance of a one- or two-node renewal spur at the base of the vine (figure 5.6, page 102), while adding labor, does provide a continual supply of shoots and potential new trunks. A new shoot is trained up in advance of a planned trunk removal or at any time subsequent to an unpredicted trunk loss. Note: with grafted vines, suckers that develop from below ground level are usually arising from incompletely disbudded rootstock wood. These shoots can be recognized as rootstock variety by their distinctive leaf appearance and are obviously of no use in reestablishing the training system.

An additional strategy to compensate for winter injury with grafted vines is to protect the graft union and a portion (several inches) of the trunks with mounded soil in the fall. "Hilling up" of graft unions, which can be done mechanically with tractor-mounted implements, thermally protects a portion of the trunks. By providing a continuum with the relatively warm soil beneath the vine, the hilled soil insulates up to several inches of the trunk, including latent buds, above the graft union (figure 5.12, page 114). The insulating layer of soil is carefully removed ("dehilled") in early spring to prevent permanent scion rooting. In the event of very low winter temperatures, injury may occur to all exposed portions of the vine. This is a rare occurrence, but it has—and will—occur, especially in poor vineyard sites. If such damage occurs, the grape grower can reestab-

Figure 5.12 ▪ **Hilling up graft unions.**

Vine row after hilling up graft unions for winter protection with tractor-mounted plow. The soil berm is pulled down in the spring to avoid permanent scion rooting.

lish the training system by dehilling the vine and bringing up shoots that had been protected as buds by the soil. This would be faster and cheaper than replanting the vineyard. Hilling and dehilling represent an insurance practice, and their utility says much about the vineyard site. Hilling is definitely recommended if questions exist about a variety's hardiness or the suitability of a site. Hilling is not recommended, however, if the long-term experience (seven or more years) suggests that severe winter injury is unlikely. Why? Annual hilling and dehilling has resulted in considerable soil erosion in vineyards established on erosion-prone sites. Soil erosion—combined with the costs of the operations, including the potential mechanical damage to vines—has made the practice of dubious value in good to excellent vineyard sites. On the other extreme, if a grower finds that annual hilling and dehilling are mandatory for vine survival, the site/variety combination might not be suitable for profitable grape production.

Cordon Maintenance and Renewal

Long-term productivity of cordons can be a problem with varieties that are subject to winter cold-injury or with improper pruning of spurs. Cold injury or poor bud development can lead to areas of the cordon that lack spurs. Poor pruning can lead to a displacement of the one-year-old spurs away from the cordon on older wood. The latter problem can be minimized by retaining, where possible, buds that originate close to the cordon and by retaining base shoots that arise directly from the old wood of the cordon. Cordons that have poorly spaced spurs or wide gaps in spur positions should be renovated or replaced. If the cordon is free of disease, renovation may be all that is necessary to reestablish a uniform distribution of spurs along the cordon. Renovation entails removing all one-year-old spurs and spur extensions from the near-barren cordon. Leave a ¼- to ½-inch crown at the base of these extensions. The removal of this older wood stimulates a proliferation of base shoots from the retained crowns. The base shoots, which will be of low fruitfulness, can be trained and used to provide fruitful spurs for the following season. The severe pruning used in renovation is necessary to stimulate base-shoot development. Renovation temporarily reduces vine productivity, so it should be used only as needed and on a small proportion of vines in any one year. Replacement of cordons is advised if the cordon is diseased, cold-injured, or otherwise undesirable. It is extremely difficult to establish a new, parallel cordon while the original cordon is still alive and present. Therefore, cut out the old cordon at the time the new cane is laid down. The new cane can originate near the graft union, anywhere on the trunk, or anywhere proximal to the diseased or barren region of the cordon. Do not attempt to establish a cordon using a cane originating from the opposing cordon of a bilaterally cordon-trained vine.

GRAPEVINE TRAINING SYSTEM EXAMPLES

The range of variation in grapevine training systems used in a particular region might be bewildering to the novice. Nevertheless, most commonly used systems fall into four basic training/pruning combinations:

- **Head-trained/spur-pruned.** Basically a short trunk, terminating in a head, with two-node bearing units.
- **Head-trained/cane-pruned.** A short trunk with one or more longer bearing units.
- **Cordon-trained/spur-pruned.** Horizontal extension(s) of the trunk with several two-node spurs.
- **Cordon-trained/cane-pruned.** Similar to the previous combination (cordon-trained/spur-pruned), but with longer bearing units. Canes are usually tied to

trellis wires with head-trained systems but are usually free-hanging when used in conjunction with cordon training.

In addition to the above descriptions, training systems can be further characterized as having either divided or nondivided canopies.

The following training systems are appropriate for vineyards in the East. Trellis dimensions and the number of foliage catch wires used are provided as guidelines and might differ slightly from other references or as a function of grower preference. It is wise to visit many existing vineyards and formulate your own dimensions from a synthesis of those observations and discussions.

Nondivided-Canopy Training Systems

Nondivided-canopy training systems have a single curtain of foliage per unit length of row. They are more common than are the more elaborate and expensive divided-canopy systems described later.

Examples: Nondivided-Canopy Training Systems, Head-Trained Vines

Kniffin System

Three nondivided canopies that are still commonly used for interspecific hybrids and American (for example, *V. labruscana*) varieties are best discussed in a historical perspective. Two of these are head-trained, and one is a cordon-trained system. Back in 1852, William Kniffin began training his Concord grapevines by providing them with a high trunk and then running four individual canes out horizontally from the trunk and tying them to the wires. Anywhere from six- to ten-cane versions were tried. Over time, arms formed in the area where the canes originated. The regions where the canes originated each year are known as "renewal zones." This system, which is still in use, became known as the four-arm or six-arm Kniffin, depending on how many arms were used.

This system produced a vertically stratified canopy in which problems soon became apparent. The most obvious one was an apical dominance observed with the upper arms; the upper arms produced more vigorous shoots than did the lower arms. The result was shading of the lower portion of the canopy by the upper part. This shading reduced vigor, fruitfulness, and fruit "quality." Something needed to be done to deal with this renewal-zone shading.

Kniffin training is appropriate for small vines where the dominance factor tends not to be as apparent and where renewal-zone shading is generally low; however, Kniffin training is not recommended where vine size is expected to be moderate to high. The system is no longer in common use in the USA but is still widely used in Ontario on low-capacity sites.

Umbrella Kniffin

The obvious solution to the shading problem was to take out the shears and get rid of the lower arms. To maintain the same number of nodes, canes retained on the upper arms needed to be lengthened. Accommodating those long canes forced the grower to tie the canes in an arching fashion to the lower wire. Now, the renewal zone was at the top of the canopy and was not in deep shade since a Concord vine, for which this system is frequently used, tends to have a procumbent (drooping) growth habit. But there were still some shortcomings. There wasn't as much renewal-zone shade, but there was still some in large vines. Yet another solution was necessary for high-vigor vines.

Two- or three-wire trellises are used for Umbrella Kniffin training (figure 5.6, page 102). A trunk extends to a point 4 to 6 inches below the top wire. Short arms bear the fruiting canes, which are arched over the top wire and are tied to a lower wire of the trellis. Renewal spurs are retained in the head region to provide canes for the subsequent season. Large vines (for example, 3 pounds of cane prunings from vines that are spaced 7 feet apart in the row) might retain three to six canes. Smaller vines (for example, 1 pound of cane prunings) might retain only two canes to provide the appropriate number of nodes.

Advantages. A relatively simple, low-cost trellis is needed. Pruning decisions are easily learned. Apical dominance is reduced, and more buds are positioned in a unit length of row by bending the canes over the top wire. The high renewal region promotes good fruitfulness and good fruit quality with small- to moderate-sized vines.

Disadvantages. As with all cane-pruning systems, the mandatory tying of canes to trellis wires adds labor costs. There is little provision for shoot positioning; and shoot crowding can lead to shaded fruit with large, vigorous vines. The system is no longer in common use.

Keuka High-Renewal

Keuka high-renewal training system was developed in northern grape growing regions where frequent winter-injury aggravates the maintenance of large amounts of perennial wood and standardized training. The system permits flexibility in pruning and training, which is its chief asset. Keuka high-renewal resembles a Kniffin system except that multiple trunks position the renewal zone at a mid-trellis height (figure 5.13). The vines are pruned to short canes that originate from a dispersed head region. Canes are distributed and tied to trellis wires in a manner that promotes as uniform a shoot density as possible. Normally, four canes are tied horizontally on parallel wires, and two are tied upward at approximately 45° angles. Renewal-zone shading is high for procumbent varieties but not bad for upright-growing varieties.

Advantages. Keuka high-renewal training allows a flexible approach to winter-injury compensation. Short trunks minimize the maintenance of perennial wood. The system can be used for native varieties and hybrids, as well as *V. vinifera* varieties.

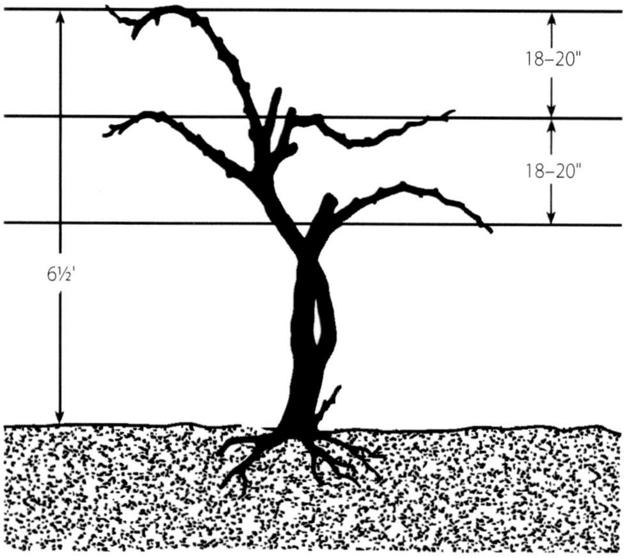

Figure 5.13 ▪ Keuka high-renewal training system.

Disadvantages. A considerable amount of time is expended with cane tying. Uniform canopy density is extremely difficult to achieve. The "flexibility" in training is difficult for inexperienced pruners to grasp. The dispersed fruit/renewal zone complicates other canopy management practices such as selective leaf removal from around fruit clusters.

Flat Cane (Synonyms: Guyot, Flachbogen)

This is perhaps the most common system used for *V. vinifera* in the northeast US. It consists of one or more trunks that are normally trained to about 30 inches in height. A head (renewal zone) is formed at the top of the trunk, from which two or more canes are retained. These horizontally positioned canes are tied to a fruiting wire at the 30-inch height. Two or more renewal spurs are retained in this zone at pruning to ensure that canes continue to be produced in the renewal zone. Multiple sets of catch wires are usually installed starting at one foot above the renewal zone to assist in forming the vertical-shoot-positioned (VSP) canopy. Shoots that emerge from the canes are positioned through the catch wires and trimmed at the 6-foot height.

Advantages. The flat cane system tends to use short canes and, therefore, reduces the possibility of apical dominance. Vines are, therefore, by necessity, positioned closely (3½ to 6 feet apart) in the row. Use of canes, as opposed to cordons and spurs, reduces the need for shoot thinning to achieve the desirable shoot density. The system is simple and, hence, easily convertible to vertically divided systems. As with all cane-pruned systems, the buds on canes are typically extremely fruitful relative to those on spurs. A uniformly positioned fruit zone facilitates canopy management practices such as selective leaf removal from around clusters.

Disadvantages. Cane tying increases labor. Apical dominance might be observed as weak shoot growth or blind buds in the mid-cane region if the vines are spaced more than 6 feet apart in the row. The remaining shoots that do grow may be excessively vigorous; and undercropping can, therefore, occur. Winter injury incidence, by many accounts, is increased in cane-pruned systems.

Arched Cane (Synonyms: Halbbogen, Pendelbogen)

This system utilizes a trellis similar to that used with the flat cane system. However, this system uses an extra fruiting wire at an 18-inch height, where the renewal zone is established. During tying, the two or more canes are brought up and over the fruiting wire at the 30-inch height, bent in an arch, and tied to the lower fruiting wire. The German term *halbbogen* normally refers to a system whereby the cane is bent into a broad arc; the term *pendelbogen* generally refers to a system whereby the canes are tied to form a very sharply defined arc. The halbbogen can accommodate more buds per linear measure of row than can flat cane systems; the pendelbogen can further increase this bud density.

Advantages. Bent canes reduce apical dominance and increase yield per linear measure of row compared to flat cane training. As with all cane-pruned systems, there is minimal shoot thinning of base buds. Arched-cane systems are simple to maintain and yet are easily convertible to vertically divided systems. Buds on the long canes are extremely fruitful.

Disadvantages. Cane tying increases labor. Also, despite the cane bending, apical dominance is still noticeable mid-cane in terms of weak shoot growth or blind buds. The dispersed fruit/renewal zone complicates other canopy management practices, such as selective leaf removal from around fruit clusters.

Examples: Nondivided-Canopy Training Systems, Cordon Systems

Hudson River Umbrella (High, Bilateral Cordon)

This next step along the evolution of training systems stretches the renewal zone out along the top wire at a height of about 6 feet through the use of a cordon, or horizontal extension of the trunk (figure 5.11, page 113). Cordons that extend in both directions from the trunk are referred to as bilateral cordons. The number of nodes can be accommodated by use of short (less than five nodes) bearing units. Use of combing (shoot positioning) allows the renewal zone to remain illuminated throughout the growing season; and as such, fruit composition is optimized, fruitfulness is high, and consistent yields are achieved. The problem that occasionally emerges is that the renewal zone of large vines can become shaded. If this happens, canopy division might be necessary.

Advantages. Inexpensive trellis installation. System lends itself to downward-growing (procumbent) species and varieties. Fruiting and renewal zone tend to receive optimal sunlight exposure. Pruning, shoot positioning, and harvest can be easily mechanized.

Disadvantages. High training may result in excessive fruit exposure, elevated phenolics, and increased sunburn potential, particularly in warm viticultural regions. Bird depredation and fruit rots can also be more severe, perhaps because of the greater exposure of fruit. Shoots of upright-growing varieties such as Cabernet Sauvignon may be difficult to train downward, increasing the risk of damage from passage of vineyard equipment.

Low, Bilateral Cordon

Low cordons are typically 36 to 42 inches above the ground and are spur-pruned; and shoots are trained upright in a thin, vertical canopy as the shoots develop from the spurs (figure 5.5a, page 101). At dormant pruning, one- to three-node spurs are retained at a uniform spacing along the upper side of the cordon (figure 5.5c, page 101). The vertically upright spurs encourage an upright growth habit to developing shoots. Cordons can extend either unilaterally or bilaterally from trunks; but in either case, cordons should ultimately span the distance between two adjacent vines in the row, leaving no gap between cordons of adjacent vines. Multiple sets of paired catch wires can be mounted on the trellis above the cordon to facilitate shoot positioning and to promote the development of a thin, vertical canopy. The first pair of catch wires should be no more than 10 inches above the cordon. This height reduces the likelihood that shoots will fall or be blown down before elongating through the catch wires, and thus the amount of shoot fastening to wires is greatly reduced. Bilateral cordon training, with vertically shoot-positioned canopies, is the most common training system for *V. vinifera* and hybrids in the mid-Atlantic and more southerly states of the eastern US.

Advantages. Spur pruning minimizes labor associated with cane tying and reduces the apical dominance effect previously mentioned with cane pruning. Fruit and renewal regions are at a uniform height, which facilitates canopy management, harvest, and pruning. In fact, cordon-trained systems can be mechanically pruned, shoot-positioned, and harvested without too much difficulty. If vines are manually pruned, the pruning is fast and easy to visualize for novice pruners. Because the buds at the base of canes mature first in the fall, they are usually the most cold-tolerant; and, hence, winter-injury may be reduced compared to cane-pruned systems.

Disadvantages. Basal nodes of a cane (those retained as a spur) are often not as fruitful as mid-cane nodes due to poor sunlight exposure during bud differentiation and development. Certain varieties (for example, Nebbiolo and Sauvignon blanc) have unfruitful base buds and proximal count buds. Shoot thinning is increased because of the increased proportion of base buds on cordon-trained vines. Cordons, like trunks, must be renewed in the event of winter-injury. There are also accounts of greater incidence of Phomopsis cane and leaf spot (page 224) and irregular shoot growth in cordon-trained, spur-pruned vines under certain conditions. Some growers have also reported greater problems with European red mites (page 249) with cordon-trained vines, a function perhaps of greater protection afforded overwintering eggs by the rough bark of cordons.

Divided-Canopy Training Systems

Canopy division provides one means of accommodating the large capacity of vines that are grown in high-vigor situations. Although rootstock, deficit irrigation, use of cover crops, and withholding fertilizer can help balance an overly vigorous vine, these cultural decisions cannot compare with canopy division's effectiveness in simultaneously increasing yield, improving wine grape quality, and optimizing vine balance.

Divided-canopy training systems consist of at least two curtains of foliage per unit length of row. Two systems, both having horizontally divided planes of foliage, are described here. Divided-canopy training systems are more elaborate—and thus more expensive to establish—than nondivided training systems. Canopy division can be used to take advantage of the large surface area of leaves that are produced by large vines. Conversion of nondivided-canopy vines to divided-canopy training has resulted in significant yield increases and, frequently, increased fruit and wine quality. Canopy division is not justified, however, when cane prunings average less than 0.4 pound per foot of row (for example, 2.8 pounds with vines spaced at 7 feet) in nondivided-canopy training systems.

Divided-canopy training, particularly horizontally divided systems, have inherently greater material costs of establishment compared to simpler, nondivided training systems. The decision to use divided-canopy systems, therefore, must balance the increased installation and maintenance costs against the generally accepted increased yield potential and the less consistent increases in fruit and wine quality that can accrue from superior canopy characteristics. To some extent, the crop-yield increases afforded by divided canopies can be approached at lower cost by establishing more closely spaced, nondivided-canopy rows. This was considered in chapter 1 (page 1), where economic returns of VSP-trained vines, spaced at 10-foot rows, were substantially improved by decreasing row width to 7 feet.

Finally, it should be noted that the added costs of divided-canopy training systems are wasted if the grower fails to shoot-position the canopies in order to create two separate, well-exposed canopies.

Examples: Divided-Canopy Training Systems, Horizontal Canopy Division

Horizontal canopy systems include Geneva Double Curtain (GDC) and various Lyre systems.

Geneva Double Curtain (GDC)

What if nondivided canopies don't work because of high canopy density? The Geneva Double Curtain (GDC) was the solution to large, high cordon-trained Concord vines, by basically creating two rows of canopy from one (figure 5.14). The key to GDC's success was shoot positioning. Shoots growing into the interior of the canopy need to be positioned outward and downward. The basis for the success of the GDC lies in creating exterior shoots from what would have been interior, shaded

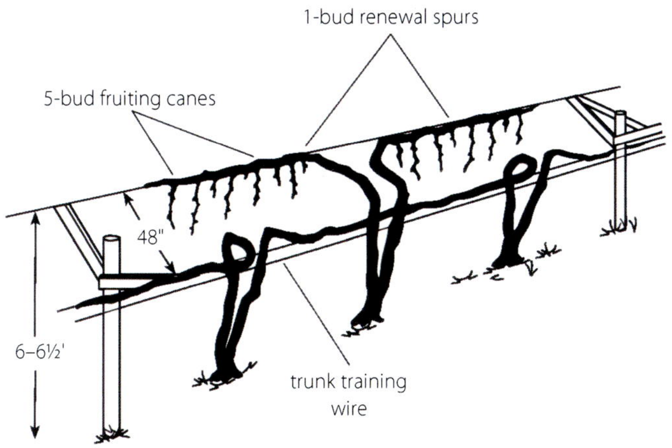

Figure 5.14 ▪ Geneva double curtain (GDC) training system.
Adapted from Jordan et al. (1981).

shoots. Interior shoots are characterized by low fruitfulness and yield, low winter hardiness, and poor fruit composition. Exterior shoots are well-illuminated and are, hence, high in fruitfulness, yield, and winter hardiness; and fruit composition is superior to that of shoots of canopy interiors. Since Dr. Shaulis first reported the effects of the GDC on Concord in New York state, others have reported on the beneficial effects of GDC on *V. labruscana* in other regions, on French-American hybrids, and on *V. vinifera*.

The top of the trellis is fitted with cross arms 4 feet wide (figure 5.14). This is accomplished using metal "arms" affixed to the tops of the posts, oriented upward at a 45° angle. This creates two walls of foliage about 4 feet apart. Compared to the Hudson River Umbrella, cordons were doubled in length, and vines were then tied alternatively to cordon wires on either side of the trellis. Cordon wires are supported on either end of the cross arms. Bilateral cordons extend from trunks that alternate, by vine, between one side of the trellis and the other. Cordons are pruned to spurs on their lower sides (lower 180° of the cordon). Short, 5-node canes are alternated with relatively short (1-node) renewal spurs (figure 5.14). The short canes will bear most of the current season's crop and will be removed or shortened during winter pruning. The renewal spurs will also produce fruitful shoots, but are primarily intended to give rise to shoots that will form the 5-node fruiting canes in the following year. The alternation of one-node and five-node bearing units was originally used with Concord grapes and may or may not be necessary with other varieties. Shoot positioning is required to maintain canopy separation and to promote sunlight penetration into the fruiting/renewal region (figure 5.15).

Advantages. The main advantages of GDC include large yield increases over conventional systems such as Hudson River Umbrella or VSP. This is normally accompanied by equal, if not higher, wine quality up to certain yield ceilings (varies with variety and site). The GDC is partially or fully mechanizable in terms of pruning, shoot positioning, and harvesting. Although the GDC system is well-adapted to varieties that have a trailing growth habit (such as those of native-American origin), it can also be adapted to many *V. vinifera* varieties as well.

Figure 5.15 ▪ Shoot positioning, Geneva Double Curtain (GDC)
Geneva double curtain (GDC) training system after leaf fall, with fruit retained to illustrate shoot positioning and position of fruiting/renewal region of two canopies.

Disadvantages. Initial capital cost is higher than with nondivided systems but not as expensive as the "U-shaped," divided-canopy system discussed next. Labor inputs are high (although labor per ton is lower than conventional systems). A large amount of perennial wood must be established, maintained, and potentially exposed to winter-injury. Considerable shoot positioning is required to achieve and maintain complete canopy division, especially with upright-growing varieties. Downward shoot positioning will "de-vigorate" shoots, and this may be viewed as a disadvantage in some circumstances. Mechanization of harvest can be difficult and costly. Finally, it can be easy to overcrop and go "over the top" on the yield/quality relationship (see sidebar, page 121).

Lyre, or "U-shaped," Training

The lyre design consists of a quadrilateral training system. Cordons are located 36 to 42 inches above ground (figures 5.16a, 5.16b). An elaborate trellis structure that consists of up to 14 catch wires is used to confine developing shoots to two independent and vertical curtains of foliage. The two curtains must be separated by at least 3, and preferably 4, feet at their bases (and slightly more at the top of the trellis). Shoots are trained to the independent curtains with the assistance of multiple catch wires. Shoot topping is performed when shoot tops elongate much beyond the top wires. Inner catch wires can be movable. It is imperative to maintain two independent curtains of foliage with this system.

This system originated in France and was devised for vigorous *V. vinifera* vines. Use of divided canopies for *V. vinifera* varieties may seem almost like sacrilege for the traditionalist. Nonetheless, they work, even in areas that experience cold winters. Where divided canopies have been evaluated for *V. vinifera*, yields and wine grape quality have been simultaneously increased by canopy division, while individual cane weight has decreased.

Advantages. The lyre system is better suited than GDC to varieties that exhibit a predominantly upright growth habit (for example, many *V. vinifera* varieties). Reestablishment of the training system following winter-injury is perhaps more rapid than with GDC. Greater yields can be achieved than with nondivided-canopy training systems of the same row width.

a. Shortly after budbreak.

b. Midsummer.

Figure 5.16 ▪ **Lyre, or U-shaped, training system.**
Cordons are separated from each other by 4 feet at the cordon height, slightly more at the top of the trellis. Multiple sets of adjustable catch wires are used to assist with canopy separation during the growing season.

SIDEBAR

High Crop Yields Compatible with High Wine Quality?

Are the high crop yields achieved with Geneva Double Curtain training compatible with high wine quality?

Two training experiments planted at Summerland, British Columbia in 1985 addressed the high-vigor problem of French-American hybrids through more innovative trellising and wide spacing. Yields for the GDC-trained vines were almost 16 tons per acre for Seyval, with very acceptable fruit composition. Incidence of bunch rot was substantially reduced by GDC training. Wines from GDC vines were also high in melon aroma and flavor and comparable to wines from much lower-yielding systems, while Kniffin-trained vines produced poor, acidic, vegetative wines. In the red wine variety Chancellor, yields from GDC training were even higher; and, again, wine grape quality was very acceptable. Furthermore, anthocyanins in the juice of GDC vines were higher than in all other systems. Wines from GDC vines were comparable to much lower-yielding systems. These results provide an example of the types of yield increases that may be expected, along with some other beneficial effects.

Disadvantages. Initial cost of trellis establishment is much greater ($3,000 or more per acre for hardware) than the initial cost of conventional trellis systems ($1,500 to $2,000). Shoot positioning and tying are still necessary to maintain complete canopy separation.

Examples: Divided-Canopy Training Systems, Vertical Canopy Division

Scott Henry

Growing wine grapes in his Oregon vineyard in the 1970s and early 1980s, Scott Henry was exasperated by his vertically shoot-positioned vineyard. The vigor was so great that the shoots grew up through the catch wires, over the top, and cascaded down each side of the vertical wall of foliage. Eventually, the shoots reached the ground and grew horizontally across the alleys. He hedged the shoots using a tractor and disc. Scott Henry was a NASA engineer, and he applied some of his design knowledge to come up with the now popular Scott Henry trellis (figure 5.17). In this system, one vertical canopy is oriented upward, and another directly below it has its shoots oriented down. The canopies are separated by a 6- to 8-inch "window" on the trellis. The shoot positioning operation is a bit tricky. First, one needs to move the shoots destined for the lower canopy away from those to be positioned upward. Once the upper canopy is positioned, a moveable rake wire can be moved downward from the central window in the trellis (between the two fruiting wires), hence positioning the lower canopy. The rake wire is kept in place by clips or J nails.

The Scott Henry system can be of many configurations. The original was four canes, much like a sideways letter *H*. Modifications have included:

Figure 5.17 ▪ Scott Henry training system.
The upper canopy originates from spurs borne on cordons, while the lower canopy originates from canes.

- Cordon/spur (upper) and canes (lower), as shown in figure 5.17 (page 121).
- Cordon/spur (upper and lower).
- Sideways S shape with one long cane each (upper and lower), which overlap reciprocal canes on the adjacent vines.
- Alternating upper and lower canopy vines with long canes, which overlap adjacent vines.

Advantages. Vertically divided canopies have several advantages over horizontally divided-canopy systems. They result in increased yields over nondivided systems without increasing shoot density per foot of canopy. Their improved fruit microclimate simultaneously improves fruit composition and wine quality. They are suitable for narrow rows with tight vine spacing and/or row spacing. In fact, vertically divided-canopy systems such as the Scott Henry and Smart-Dyson (discussed in the next section) are ideal when row spacing prevents horizontal canopy division.

Disadvantages. The Scott Henry system frequently uses cane pruning and carries with it all the disadvantages associated with cane pruning. Downward shoot positioning of the lower canopy can be difficult and expensive sometimes. The lower canopy may interfere with weed control and other soil-management practices. A "dominance" effect may be observed in the upper canopy over the lower canopy (Scott Henry gets around this by training "odd" vines "up" and "even" vines "down"). There are concerns that fruit of the lower canopy will ripen slower than fruit of the upper canopy will, because of direct or indirect effects of shoot orientation. The potential for asynchrony of ripening between canopies can be minimized by carrying a somewhat lower crop level on the lower canopy than on the upper (for example, two thirds of total crop on upper canopy and one third of crop on lower canopy).

Smart-Dyson and Smart-Dyson Ballerina

John Dyson, with some assistance from Richard Smart, found that the vertically divided Scott Henry canopies could be developed from a common cordon. In what became known as Smart-Dyson training (figure 5.18), bilateral cordon-trained vines are pruned with spurs oriented upward (to provide the upper canopy) as well as downward (to form the lower canopy). The cordon is placed slightly higher on the trellis (about 44 inches above ground) to accommodate the need for the lower canopy shoot development . Canopy separation is somewhat more laborious than with the Scott Henry system because there aren't already two distinct fruit zones and canopies to facilitate shoot positioning. A further modification to Smart-Dyson, the Smart-Dyson Ballerina, starts with a similar, common cordon, and creates a similar effect, but with less effort expended in trying to form a vertical lower canopy. Instead, shoots are allowed to arch outward and down on either side of the trellis at roughly 120° and 240° from the top of the trellis. It provides the advantage of not needing to worry about finding enough downward-oriented spurs; but the shoot positioning problems are the same as, if not worse than, the Smart-Dyson system's.

Figure 5.18 ▪ Smart-Dyson training system.
The upper and lower canopies originate from a common cordon, situated 42 to 44 inches above ground level. Canopy was photographed at partial leaf fall to better show cordon and orientation of canes on the two canopies.

Advantages. Generally the same as those described for Scott Henry. Smart-Dyson and its Ballerina version are very flexible systems to use so long as the cordon is initially placed high enough on the trellis to allow sufficient shoot length on the lower canopy. If vine vigor decreases with vineyard age, the system can be simply modified to a bilateral cordon with a single, VSP canopy.

Disadvantages. Generally the same as those described for Scott Henry. An additional disadvantage of the Smart-Dyson Ballerina is the need for somewhat wider row spacing in order to avoid significant damage to shoots due to vineyard machinery traffic.

Horizontal + Vertical Canopy Division

Experimental trellis designs exist that divide canopies in both directions—for example, the alternate double cross arm and the Ruakura Twin Two-Tier. Simultaneous vertical and horizontal canopy division is costlier both in terms of trellis materials and management labor. The alternate double cross arm trellis developed in British Columbia requires lots of perennial wood and obviously is expensive to establish. Results, however, were highly encouraging. Yields approached 13 tons per acre for Riesling, compared to about 5 tons per acre for a standard cordon-trained, spur-pruned vine. Standard harvest indices were equally encouraging, and these divided canopies also produced fruit with highest monoterpene flavor compounds.

CONCLUSION

A number of training systems are suitable for commercial grape production in eastern North America. Advantages and disadvantages can be cited for each. Growers should evaluate the growth potential of their vines, the growth habit of the intended variety, the availability of vineyard labor, and the potential for mechanization, as well as the hazards of winter-injury, before choosing a particular system. Conversion of inferior, existing systems to superior systems is a possibility. However, converting from a high-training system to a low-training system will be much more difficult than converting from a low to high system. Conversion of nondivided training systems to more elaborate divided-canopy training systems is also a possibility if rows are wide enough or if the trellis system is tall enough to accommodate a vertical division.

References

Bravdo, B., Y. Hepner, C. Loinger, S. Cohen, and H. Tabacman. 1985. Effect of crop level and crop load on growth, yield, must and wine composition, and quality of Cabernet Sauvignon. *American Journal of Enology and Viticulture* 36: 125–131.

Chapman, D.M., M.A. Matthews, and J.-X. Guinard. 2004. Sensory attributes of Cabernet Sauvignon wines made from vines with different crop yields. *American Journal of Enology and Viticulture.* 55:325–334.

Jordan, T. D., R.M. Pool, T.J. Zabadal, and J.P. Tomkins. 1981. Cultural practices for commercial vineyards. *Cornell University Cooperative Extension. Misc. Bulletin. No. 111.* 69 pp.

Reynolds, A. G., D. A. Wardle, M. A. Cliff, and M. King. 2004. Impact of training system and vine spacing on vine performance, berry composition, and wine sensory attributes of Seyval and Chancellor. *American Journal of Enology and Viticulture.* 55:84–95.

Shaulis, N.J., H. Amberg, and D. Crowe. 1966. Response of Concord grapes to light, exposure, and Geneva double curtain training. *Proceedings of the American Society for Horticultural Science.* 89:268–280.

Smart, R.E. 1994. Introducing the Smart-Dyson trellis. *Australian Grapegrower and Winemaker* 365:27–28.

Smart, R. E. and M. Robinson. 1991. Sunlight into wine: a handbook for winegrape canopy management. Winetitles, Adelaide, Australia. 88 p.

Winkler, A. J. 1934. Pruning vinifera grapevines. *California Agricultural Extension Service Circular 89.* University of California, Berkeley, CA. 68 pages.

Zabadal, T. J., M.L. Chien, I.E. Dami, M. C. Goffinet and T. M. Martinson. 2007. Winter injury to grapevines and methods of protection. *Michigan State University Extension Bulletin E2930.* 106 p.

—6—

Grapevine Canopy Management

Authors
Andrew G. Reynolds, Brock University (Ontario, Canada)
Tony K. Wolf, Virginia Tech

High-quality wines—those commanding premium prices—are produced only from high-quality grapes. Quality can be defined in different ways, but ripeness and freedom from bunch rots are two desirable grape features. Attaining ripe fruit with a minimum of rot and a maximum of varietal character is not an easy achievement in eastern North America. As described elsewhere in this book, the combination of climate, soils, and warm temperatures often leads to excessive vegetative vine growth. For reasons that will be reviewed, the luxurious vegetative growth can reduce vine fruitfulness, decrease fruit varietal character and other components of fruit quality, and increase fruit disease. These problems are exacerbated in cool, wet seasons that delay fruit ripening.

Canopy management is a broad expression used to describe both proactive and remedial measures that growers can use to favorably affect grapevine canopy characteristics. The grapevine canopy is defined by the shoot system of the vine, including stems, leaves, and fruit (figure 6.1). In the broadest sense, canopy management can entail decisions on row and vine spacing, choice of rootstock, training and pruning practices, irrigation, and fertilization, as well as the growing-season activities such as shoot hedging, shoot thinning, and selective leaf removal.

This chapter describes the common metrics used to define grapevine canopies, as well as the canopy-management principles and practices that have successfully been used to enhance fruit and wine quality in eastern North America. Several excellent references on this subject are cited at the end of the chapter (see page 134)—including the very informative text, *Sunlight into Wine: A Handbook for Winegrape Canopy Management*. Those seeking further insight into canopy management principles and practices should review those very worthwhile references.

THE BALANCED VINE

In a very basic sense, canopy management is part of a viticultural goal of achieving a balance between vegetative and reproductive (fruiting) growth. Balanced vines produce sufficient foliage to mature their crop, ripen next year's fruiting wood, and store sufficient carbohydrates to support the following season's initial growth. Foliage that is produced in excess of that basic requirement is unnecessary and is evidence that water and nutrient resources are excessive—a situation that is not easily remedied.

Figure 6.1 ▪ Grapevine canopy.

The grapevine canopy is defined by the shoot system of the vine —including stems, leaves, and fruit. Photo shows a Smart-Dyson-trained Traminette (page 122)—comprising both an upper canopy and a lower canopy, both originating from a common cordon at approximately the point where the person's hand is pointing.

Excessive Vigor

Excessive vine vigor is a common problem in eastern vineyards, especially when vines are grown on deep, fertile soils that have abundant water-holding capacity. The vines produce vigorous shoots with long internodes, large leaves, and strong lateral shoots. The fruit zone of such vines is often shaded, and the developing fruit is often inferior to that of more exposed canopies. Because the renewal zone (usually the current year's fruit zone) is shaded, the buds will produce shoots with fewer, smaller clusters or with reduced berry set; or the buds may fail to produce shoots the next season. Shaded clusters and leaves are also more susceptible to disease. Shaded clusters, in particular, are highly susceptible to powdery mildew (page 217) from a combination of low levels of sunlight, reduced fungicide coverage, and poor drying conditions in the canopy interior. The same conditions promote Botrytis bunch rot (page 227) as well as non-specific bunch rots (see chapter 11, page 216). In addition to increasing rot potential, the negative effects of shade on fruit composition can include elevated potassium, pH, and titratable acidity, and reduced phenols, pigmentation, varietal flavor, and soluble solids concentrations. Shade can also impart vegetative characters to fruit and wine. Collectively, this altered fruit composition can have significant negative effects on wine quality.

Insufficient Vigor

Drought, infertile soil, crop stress, and disease can all contribute to low vigor, sparse canopies, and sometimes too much fruit relative to the canopy or the vigor of the vine. Again, the vines are out of balance, but on the opposite side of the scale as the situation described above for excessive vigor. Because fruit competes with the shoots for needed carbohydrates, the shoots of overcropped vines may not adequately mature to withstand winter cold, and may die back completely to the older wood during winter. In the long term, if overcropping is severe, carbohydrate reserves will be inadequate for the next year's vegetative growth and vine size will decrease, and future crops will be severely reduced.

CANOPY MICROCLIMATE

Grapevine canopies create a localized microclimate that will vary with the variety, rootstock, training system, vine spacing, pruning and hedging strategies, and amount of cluster thinning. Soil hydrology and fertility will also affect the microclimate, as will our inputs of irrigation and fertilizer. Vineyard floor management may also affect canopy microclimate. The canopy microclimate, which is distinct from the climate immediately outside the canopy (table 6.1, page 126) is described or measured in familiar terms, such as sunlight (incident radiation), temperature, wind speed, and humidity.

Sunlight (Incident Radiation)

The amount and quality of fruit a vine produces are related to sunlight exposure, and profitable grape growing involves the efficient use of sunlight. Grape leaves utilize approximately 90% of the incident radiation (sunlight), transmitting only about 10% of this energy to subtending leaves. Leaves convert sunlight and carbon dioxide into carbohydrates, such as sugars, and other compounds by a process called photosynthesis. The interior of dense canopies can be shaded; shaded leaves are not as photosynthetically productive as are exterior leaves and can draw photosynthates away from the fruit. Shaded leaves

Table 6.1 ▪ Canopy microclimate differences as a function of canopy density

Microclimatic characteristic	Sparse canopy or exterior of dense canopy	Interior region of dense canopy
Sunlight	Most leaves and fruit receive some direct sunlight at some point during the day	Most leaves and fruit are in the shade
Temperature	Fruit and leaves are warmed by sunlight and may be cooler than air temperature at night	Most leaves and fruit are at approximately ambient temperature, both day and night
Humidity	Leaves and fruit experience ambient humidity values	Humidity can build up slightly within the canopy
Wind speed	Leaves and fruit are exposed to ambient wind values	Wind speeds are reduced within the canopy
Evaporative potential	Evaporation rates are similar to ambient values	Evaporation rates are reduced in the canopy

Adapted from Smart and Robinson (1991).

may also contribute excess potassium to the fruit, which under some conditions can contribute to elevated fruit pH and create problems for the wine maker. Shade can also reduce the fruitfulness of developing buds. Thus, crop yields from vines with dense canopies can be significantly less than yields from properly managed vines.

Temperature

Air temperature within the grapevine canopy does not vary greatly from air temperature immediately outside the canopy; however, fully exposed fruit can be 15° to 30°F warmer than surrounding air temperature because of radiant heating. That warming can be used to advantage in cool grape regions to reduce fruit acidity and to achieve a more optimal sugar/acid balance. On the other hand, excessive fruit exposure and radiant heating in warm to hot grape regions can be detrimental to temperature-sensitive flavor and aroma compounds. The degree of fruit exposure, from none to 100%, can, therefore, be modified based on your site, variety, and wine-making goals.

Wind Speed and Humidity

Dense canopies reduce canopy wind speed and ventilation. Even a slight breeze can help reduce fungal infections of fruit and leaves. Many of the fungi that attack grapevines in eastern North America require either free water or a period of high humidity for infection to occur. In addition to drying faster, the well-managed, well-ventilated canopy allows better pesticide penetration and coverage when vines are sprayed.

MEASURING CANOPY CHARACTERISTICS

Viticulturists quantify vine balance in various ways, including annual cane pruning weights; the average weight, diameter, and internode length of individual canes; the ratio of leaf area to crop; and the ratio of crop to cane pruning weight. Observations of lateral development, duration of shoot extension, and relative leaf size and color can also be used to judge canopy characteristics. Research conducted under a wide range of growing conditions and with many grape varieties has led to the development of an ideal or "ideotype" canopy—something to strive for to optimize fruit and wine quality. The ideotype canopy characteristics can be used as a "scorecard" during the period from veraison to harvest to evaluate average canopy characteristics in your own vineyard. The concept of canopy scoring is extremely helpful, even if all eight elements of the ideotype canopy (table 6.2) are not used. Some of the more commonly used measurements are cane pruning weights, shoot density, canopy transects, and shoot developmental features.

Cane Pruning Weights

The collection and use of cane pruning weight data were discussed in chapter 5 (page 98). Pruning weights in excess of 0.4 pounds per foot of canopy are symptomatic of excessive vine vigor and suggest that canopy division or other vine "de-vigoration" techniques are warranted.

Table 6.2 ▪ Characteristics of the ideal (or "ideotype") canopy

Canopy characteristic	Optimal value (between veraison and harvest, unless otherwise noted)	Justification
Shoot density	3–5 shoots per foot of canopy	Higher values promote canopy shade. Lower values cause excess shoot vigor and potential low crop yields.
Shoot length	15–20 nodes	Shoots less than 15 nodes (prior to trimming) are indicative of inadequate vigor. If they are trimmed to lengths less than 15 nodes, this may be excessive hedging. Untrimmed shoots longer than 20 nodes are an indicator of excess vigor.
Lateral shoot development	Ideally none	Excess lateral shoot growth may lead to shade. Presence of some laterals may provide carbohydrates to fruit and canes if leaves are net exporters of carbohydrates.
Growing shoot tip presence	Ideally none	Ideally, shoots should have stopped growing by veraison.
Individual cane weight	0.06–0.10 pound per cane (during dormancy)	Values under 0.06 pound per cane suggest inadequate vigor. Values above 0.10 pound are indicative of "bull wood" that is low in fruitfulness and susceptible to winter injury.
Cane pruning weights	0.2–0.4 pound per foot of canopy (during dormancy)	Values below 0.2 pound per foot of canopy are indicative of low vigor. Values in excess of 0.4 pound per foot of canopy are often indicative of canopy shade.
Ratio of leaf area to fruit weight	3–6 square feet per pound	Values less than 3 can be indicative of overcropping, although this value is somewhat variety-specific.
Ratio of crop weight to pruning weight	5–10	Values less than 5 are indicative of undercropping. Values greater than 10 are often indicative of overcropping, although this is very variety-specific.

Adapted from Smart and Robinson (1991).

Shoot Density

Shoot density is the number of shoots per unit length of canopy and usually relates well to overall canopy density: the greater the shoot density, the thicker or denser the canopy. Although shoot density can be assessed at any point in the growing season, it is usually done soon after bud break when corrective thinning can, if needed, be easily accomplished. Shoot densities should be in the range of 3 to 5 shoots per foot of row or canopy. For vines spaced 6 feet apart in a row and trained to a nondivided training system, this would translate to 18 to 30 shoots per vine. The lower number is more suitable for large-clustered, very fruitful varieties such as Seyval and Sangiovese. The higher limit is suitable for small-clustered varieties such as Norton or Riesling. As a starting point, most varieties will produce desirable yields of ripe fruit at a density of 4 shoots per foot of canopy.

Canopy Transects

Canopy transects, or point quadrat analyses, measure the thickness (leaf layers) of the canopy, its porosity (gaps in foliage), and the percentages of fruit and leaves that are exposed to sunlight. Such measurements are usually taken at or shortly after veraison. Canopy transects provide a considerable amount of information on canopy density. The fruiting region of the ideal canopy has the following characteristics: about 20% canopy gaps, 1 to 2 leaf layers, and 80% or more of clusters exposed to the canopy exterior for some portion of the day. Transects require at least two persons to complete. A thin metal probe is inserted horizontally at regular intervals (for example, every 6 inches) along the fruit zones of representative vines. A metal tape measure or ruled, wooden frame can serve as a guide for where to insert the probe (figure 6.2, page 128). The person inserting the probe

Figure 6.2 ▪ **Canopy transects (point quadrat analyses)**

A thin metal rod is inserted at even intervals (for example, every 6 inches) into the fruit zone of representative vine canopies. In this case, one person *(middle)* inserts the rod, one assistant *(right)* monitors the probe contact, and a second assistant *(left)* records the nature of the contact to generate data such as those of table 6.3.

should not watch the path of the probe through the canopy. An observer tracks the point of the probe and records the nature of all contacts as a leaf (L), a fruit cluster (F), or a gap (G). A gap is recorded when the probe fails to contact any leaves or fruit. No data is collected when the probe touches a shoot stem. Representative data should be collected from 5 to 10 panels or post lengths in a given vineyard block. Example data are given in table 6.3. Fifty probes were made, and the calculations of canopy density were as follows: 12% canopy gaps, 1.7 leaf layers, 80% exterior leaves, and 65% exterior clusters. (Calculations are shown at the end of table 6.3.)

Development of Shoots

Ideally, shoots should grow rapidly to 15 or 20 nodes and then cease elongating, with little subsequent lateral shoot development. In reality, shoots of vigorous vines often continue to grow well after veraison and can exceed more than 50 nodes in length if not hedged or topped. This is particularly true of upright-trained shoots, such as those of vertical shoot positioned canopies. Vigorous shoots often produce large lateral shoots, which further aggravate canopy shade with their additional foliage. Shoot vigor should be assessed periodically throughout the growing season; and the ends of shoots trimmed, if necessary. Lateral shoots can be removed if needed, as described below.

Table 6.3 ▪ **Canopy transect data**

Probe insertion number	Nature of probe contact [a]	Probe insertion number	Nature of probe contact [a]
1	LLFL	26	LLL
2	LLL	27	FLL
3	FLL	28	LL
4	LL	29	G
5	G	30	LL
6	FL	31	F
7	LF	32	LL
8	LL	33	FL
9	F	34	G
10	LL	35	LL
11	G	36	LFL
12	LL	37	LLL
13	FLLL	38	G
14	LL	39	LFLL
15	LFFL	40	LLLF
16	LLL	41	L
17	LL	42	G
18	LLL	43	LF
19	FL	44	LFL
20	LLL	45	LLL
21	LL	46	LL
22	LLF	47	F
23	LFLL	48	LF
24	F	49	LL
25	LL	50	LFL

L = leaf
F = fruit cluster
G = gap (or no contact)

Note: Representative data summarizes the nature of contacts made with fifty canopy insertions of a probe. All leaves (L) or fruit clusters (F) that are either the first or last contact are defined as "exterior." Pertinent data are summarized as follows:

- Percent canopy gaps = gaps/number of probes = (6/50) x 100 = 12%
- Leaf layers number = leaf contacts/number of probes = 85/50 = 1.7
- Percent exterior leaves/total leaf contacts = (68/85) x 100 = 80%
- Percent exterior clusters/total cluster contacts = (15/23) x 100 = 65%

[a] Contacts with shoot stems can be ignored.

AVOIDING UNBALANCED CANOPIES AND CORRECTING INFERIOR CANOPY CHARACTERISTICS

The canopy architecture is frequently far from ideal, and the question then becomes how to improve the balance between vegetative and reproductive (fruit) growth. Much could be written on canopy management, particularly when one considers the varied goals that arise from differences in varieties, training systems, climate, and other environmental factors. As previously mentioned, a useful, comprehensive manual on the subject of canopy management is *Sunlight into Wine: A Handbook for Winegrape Canopy Management* (Smart and Robinson, 1991). We will only summarize some of the details presented in that text.

Strategic canopy-management techniques are often prescribed when excessive vine vigor is anticipated. They may include a range of vineyard design options including canopy division, use of reduced fertilizer and irrigation inputs, vine size-limiting rootstocks and cover crops, and altered vine density. Once the vineyard is established, the options are more limited and include remedial or "Band-Aid" approaches such as shoot thinning, selective leaf and lateral removal, shoot hedging, and increased node number at dormant pruning. In high-vigor situations, canopy division might be possible and desirable if row spacing is adequate. We'll consider each of these techniques and where they could be used.

Reduced Fertilizer and Irrigation Inputs

Water, nutrients, and warm temperatures are the accelerator pedal of our grapevine "car," and eastern vineyards often run with the accelerator pedal pushed to the floor. It makes little sense to try pushing the brake pedal if our other foot is still on the accelerator. With this crude analogy in mind, one means of slowing the growth of imbalanced vines is to reduce the inputs of irrigation and fertilizers—to take our foot off the gas. Don't apply vine-stimulating nutrients unless they're needed (see chapter 8, page 141).

Rootstocks

The choice of rootstock is an important one where vigor may later become a problem. This topic has been discussed previously in this book (see chapter 3, page 37) for more information on rootstocks. It should be pointed out that the range of rootstock effects on vine size and vigor does not appear to be as dramatic in a water-surplus environment such as eastern North America as it does in more arid environments.

Vine Spacing

Vine spacing, or density, is a controversial topic. Our experience has shown that giving a grapevine more room in the row, within reason, will ultimately reduce vine size per linear measure of row or canopy, may display a slight reduction in soluble solids, but under some circumstances may produce fruit with higher monoterpene flavor compounds as a result of better fruit exposure. It is important to accommodate anticipated high vigor through wider in-row vine spacing; the concept of using root competition between grapevines to "de-vigorate" vines in the humid East is unproved at best and poor physiology at worst. It is generally accepted now that close in-row vine spacing will have no impact on vigor unless soil conditions are limiting to begin with. In-row vine spacing of 5 to 7 feet is typically recommended. Closer spacing (3 to 5 feet) is common for *Vitis vinifera* varieties on fine-textured soils in Ontario and New York. Elsewhere, growers will occasionally use close spacing, ostensibly for grape-quality reasons; but unless the close spacing results in more uniformity of grape ripening, the results may not merit the added establishment and operational costs inherent with closer spacing.

Cover Crops

Use of cover crops is a wise choice where high vigor is a consistent problem or where excess soil moisture and fertility cause persistent, late-season shoot growth. While perennial grass cover crops are more or less the "standard" row-middle treatment in mid-Atlantic vineyards, the use of under-trellis cover crops, such as dwarf fescue, has also been explored as a means of imposing additional water and nutrient stress on very high-vigor grapevines. The initial results of these grower trials have been encouraging; but we hasten to add that under-trellis vegetation can aggravate management of some arthropod pests, such as grape root borer (page 258), climbing cutworms (page 245), and European red mites (page

249). Furthermore, careful management of cover crops is necessary to prevent the tolerable stress from becoming intolerable, as can happen under extended drought conditions (see chapter 9, page 169).

Shoot Thinning

Shoot thinning is an effective technique to open and maintain a well-ventilated, light-porous canopy. Shoot thinning can also be used to control crop, especially for varieties that tend to overproduce crop. Shoots are thinned soon after bud break and preferably before they are more than 18 to 24 inches long. Removing longer shoots is more difficult, as the union of the shoot and older wood begins to lignify at about this stage. A convenient rule of thumb is to retain 3 to 5 evenly spaced shoots per foot of canopy. Nonfruitful shoots should be removed in preference to fruitful shoots, except where nonfruitful shoots might be needed to serve as spurs for the following year.

For cordon-trained, spur-pruned vines, remove distal shoots of spurs that bear more than one shoot in order to keep retained shoots closer to the cordon to serve as spurs in the following year. This strategy helps to keep the one-year-old, fruiting wood close to the cordon. Remove shoots in the upper 180° of the cordon of Geneva Double Curtain-trained vines (page 118), and leave those shoots that grow downward (lower 180° of cordon).

Shoot Positioning

Shoot positioning should be an integral part of vineyard management, regardless of the training system. The purpose of shoot positioning is to uniformly distribute leaf area and fruit to minimize mutual leaf shading and to improve fruit exposure and ventilation. The procedures differ somewhat as a function of training system, as follows.

Low-Cordon and Low Head Systems

For low-cordon systems and low head systems with cane pruning, such as used with vertical-shoot-positioned (VSP) systems, shoots are positioned vertically upright to form a thin plane of foliage. The shoots can be sandwiched between paired catch wires to prevent them from being blown free or sideways. The first set of catch wires is usually set 10 to 12 inches above the cordon, while a second and even third pair of catch wires may be used above the first. Commercial devices are available for tying or taping the shoots to trellis wires after they are positioned and for clipping or holding the paired catch wires together in the middle of the panel or post length.

Shoots should be positioned several times during the year. Positioning is best started when the shoots are about 18 inches long, as they begin to exceed the height of the first set of catch wires and before their tendrils have attached to wires or each other. At this point, the upward oriented shoots can be positioned between fixed catch wires or they can be positioned with the aid of movable catch wires. Shoots of some varieties are extremely susceptible to breakage at their point of attachment to older wood. Care and experience are needed to determine the optimum timing of the first round of shoot positioning. Depending upon the number of foliage catch wires used on the trellis, some repositioning of shoots may be necessary between bloom and veraison to maintain the desired canopy dimensions.

If outward-growing shoots are positioned upward too late, the canopy may develop a vase shape that is undesirably wide through the fruit zone and narrow through the catch wires. Some growers use movable catch wires to help position shoots. The wires are "parked" beneath the cordon or head region during the winter. During the growing season, the wires on either side of trellis posts are brought up to a fixed position above the cordon, and the shoots are brought up into a vertical plane in the process.

High-Cordon and High Head Systems

For high-cordon systems, such as Geneva Double Curtain, start shoot positioning a week or two before bloom and rake the shoots out and down. Wait a week if the shoots are too prone to breakage. The positioning process should be repeated about two weeks later. If the initial shoot positioning is delayed too long after bloom, the tendrils will intertwine with other shoots and trellis wires, and too many of the shoots will be broken in the positioning process.

The space between the two curtains of the Geneva Double Curtain training system needs to be kept open to allow sunlight to reach the clusters and the fruit zones of both canopies. Otherwise, it is not a "double curtain" and the benefits of canopy division will not be realized.

Mixed Systems

For Scott Henry (page 121), Smart-Dyson (page 122), and Smart-Dyson Ballerina (page 122), a two-stage process of shoot positioning is used. Start moving the shoots that will be trained downward two weeks or so before bloom. Move them to about 90° or 270° from vertical, and then reposition them just after bloom to further encourage them to grow in the full downward position. Movable catch wires are often used to facilitate and maintain a downward orientation of the lower canopies, and shoot tying or taping to these wires may be necessary to achieve the desired canopy architecture.

Selective Leaf and Lateral Shoot Removal

Selectively removing leaves and laterals from around fruit clusters is widely practiced with many varieties and with different goals in mind. Research in Virginia with Riesling and Chardonnay vines has illustrated that leaf and lateral removal can significantly reduce fruit rots by increasing ventilation and light penetration in the fruit zone. Removing a few leaves per shoot around the fruit cluster can also increase flavor compounds, improve color, reduce titratable acidity, reduce pH and potassium, increase bud fertility, allow better spray penetration for disease control, and reduce vegetative aromas in some cultivars. There are tradeoffs, however. Basal leaf removal can increase bird damage, slightly reduce soluble solids, sunburn the clusters of some varieties (such as Riesling) or produce hot, vegetative, or cooked flavors in wines made from some varieties (such as Gewürztraminer). Excessive leaf pulling can also lead to excessive reductions in acid and increased phenolic concentrations. Therefore, being selective is important. Do not pull leaves unless the canopy (and fruit) will benefit. If the leaf layer number is greater than 2, and most of the fruit clusters are hidden, there may be a compelling reason to pull some leaves. Moderation in leaf and lateral removal is important. At least 15 leaves per shoot are needed to adequately ripen the fruit and produce carbohydrate reserves for optimal winter hardiness. If leaf pulling is needed, the initial removal should be done soon after fruit set—preferably no later than three weeks after fruit set. Delaying beyond this point can lead to severe sunburning. Small lateral shoots in the fruit zone can also be broken out in this process. Occasionally, growers will also do a follow-up leaf pulling at or around veraison, sometimes in conjunction with crop-thinning procedures.

It is not necessary to remove all leaves in the fruit zone. Generally, pulling two or, at most, three leaves per shoot from around the fruit clusters is enough for a desired response. If more than three leaves need to be pulled, shoot density is excessive. Again, we are striving for an average of one to two leaf layers in the fruit zone after the leaves have been pulled, with an ultimate goal of about 50% to 80% of fruit exposed.

Leaf pulling is more efficient with training systems that have fruit placed uniformly along a cordon (such as vertical shoot positioning [VSP]) than with systems that have fruit scattered through the canopy (such as Geneva Double Curtain). With the latter systems, more basic aspects of canopy management such as training system conversion and shoot positioning might be more useful than resorting to leaf pulling.

Leaf pulling is often done by hand (figure 6.3a, page 132) in smaller vineyards, but tractor-mounted machines (figure 6.3b, page 132) are widely used and increase the speed with which the operation is done. A grapevine canopy that has been subjected to leaf pulling (figure 6.3C, page 132) shows a moderate degree of cluster exposure obtained with the technique.

Summer Pruning or Hedging

Summer pruning or "hedging" usually involves the removal of shoot tops from upright-trained canopies during the season. Only the leaves needed for adequate fruit and wood maturation are retained—that is, at least 15 leaves per shoot.

Hedging reduces the amount of fruit zone shading, reduces vine vigor, and ultimately produces more manageable vines. However, hedging can also reduce soluble solids, winter hardiness, and increase lateral shoot growth, which in turn leads to more fruit zone shading.

Hedging is most beneficial when applied to low- or midwire-trained vines that have vertically upright-positioned shoots. The shoots are cut (hedged) after they clear the top of the trellis and before they start to droop over and shade the fruit zone and original canopy (figure 6.4, page 133). The shoots of high-trained vines, such as those trained to Geneva Double Curtain, should be trained downward (combed) during the growing sea-

a. Manual leaf removal from lyre-trained Cabernet Sauvignon, about forty days after bloom.

b. Tractor-mounted, mechanical leaf remover.

c. Good but not excessive Chardonnay cluster exposure in mid-September, after selective leaf removal in early July.

Figures 6.3 ▪ Selective fruit zone leaf pulling.

son. These shoots can be subject to breakage if they are pinned to the ground by the tires of vineyard equipment. If this is a problem, the shoot tips can be tipped or "skirted" before the shoots reach the ground.

One school of thought suggests that hedging should be delayed as long as feasible to minimize lateral shoot development, preferably for 30 or more days after bloom. However, a grower must realize that hedging late in the season may reduce the leaf area upon which the vine depends for maturation and that, frequently, late-hedged vines (after bunch closure) have delayed maturity, especially if the hedging is severe. Also, late-season growth of lateral shoots may divert carbohydrates from ripening clusters. Our recommendation is early hedging (when berries are pea-sized), with a minimal amount of vegetation removed (6 to 12 inches). A second, light hedging can be done 2 weeks later to trim lateral shoots that may have grown as a result of the first hedging. More than two hedging operations per year is indicative of a high-vigor problem. A minimum of 15 (but preferably up to 20) primary leaves per shoot should be retained (not including laterals). It is not necessary to count leaves on every shoot; most shoots bear 15 to 20 leaves for every 4½ to 5 feet of length.

Heavy duty, scissor-type hedge shears are probably the most commonly used hedging tools. Gasoline or battery-operated cutter-bar-type hedgers have also been used. In either case, the process is less tiring if one works with arms at chest height by working from an elevated platform such as a trailer. Tractor-mounted canopy hedgers are commercially available and have become increasingly common as vineyard size increases (figure 6.5).

Increasing Shoot Density

For some varieties, increasing the shoot density can be a temporary measure to combat high vigor. The vines should respond with "de-vigorated" shoots, shorter lateral shoots, more fruit exposure, and acceptable fruit composition. Increasing the shoot density can, however, produce high crop loads that will not properly ripen, and the resulting wine may have an herbaceous or vegetative character. However, one can use increased shoot density as a tool for vigor diversion. Many shoots are retained on a vigorous vine, and these shoots are thinned after fruit set to an acceptable density (for example, four shoots per

Imed Dami, Ohio State University

a. Before hedging. Shoots have cleared the top of the trellis and have started to droop over.

Imed Dami, Ohio State University

b. After hedging. Shoot hedging of vertical-shoot-positioned (VSP) vines prevents shoot tips from shading or otherwise obstructing the basal portion of the canopy, including the fruit zone.

Figure 6.4 ▪ Hedging—removal of shoot tips from upright trained canopies during the season.

Fritz Westover, Texas AgriLife Cooperative Extension

Figure 6.5 ▪ Tractor-mounted canopy hedger.

foot of row). The later thinning of shoots is more laborious than early shoot thinning, but it results in a more significant vigor reduction.

CANOPY MANAGEMENT OF VINES THAT HAVE INSUFFICIENT CANOPY

The goal of canopy management during the first two years after a vine is planted is to develop a healthy root system, as well as the other permanent or semipermanent portions of the vine such as trunks and cordons. Well-exposed, healthy leaf area is needed for this development, and young-vine canopy management should strive to foster that goal. The steps in training young vines are reviewed in chapter 5, page 98, and should be reviewed in the context of young-vine canopy management.

Diseases, insect pests, and nutrient and water stress may also contribute to situations in which the vine size and leaf area are insufficient to fill available trellis space. Shoot hedging, leaf removal, and canopy division may be completely unwarranted under these conditions. The grape grower should instead investigate the constraints to inferior vine size and work to correct those deficiencies. Drought, nutrient deficiency, and overcropping may be relatively easily corrected; while poor root development, systemic diseases, or chronic cold-injury may be more difficult to deal with. In the latter case, the grape grower may need to make some fundamental changes in plant density, plant material, or vineyard site. Doing nothing is not a profitable option. (For in-depth discussions on nutrient and water stress, diseases, insect pests, and related topics, see chapter 8, page 141; chapter 9, page 169; chapter 11, page 216; and chapter 12, page 241.)

LONG-TERM SOLUTIONS TO VINE IMBALANCE

The canopy modifications described above are intended to improve the microclimate within the canopy and the balance between vegetative growth and crop production of existing vines. Some of these measures offer only short-term solutions, whereas others, such as canopy division, offer more lasting benefits.

Training systems that promote ventilation of fruit zones will have an advantage over those that tend to hide the fruit within shaded canopy interiors. The training system should promote maintenance of a thin canopy of foliage—no more than two leaf layers thick. For large, vigorous vines in established vineyards, conversion to divided-canopy training might be the most practical course of action to achieve the desired canopy density. Canopy division should be considered when the pruning weights of most vines are more than 0.3 to 0.4 pounds of prunings per foot of canopy in the absence of summer pruning.

There are several approaches to canopy division, and specific guidelines should be sought before pursuing this course. The establishment of multiple trunks prior to the year of conversion makes the process much easier and avoids loss of crop in the conversion year.

If a VSP cordon is placed at least 42 inches above the ground, the VSP canopy can be later converted to a Smart-Dyson or Smart-Dyson Ballerina training system if vine size and vigor warrant such a conversion. The number of shoots left on the cordon is increased, and the shoots emerging from the lower half of the cordon are trained downward. Cordons closer than 42 inches to the ground might not develop enough leaves to ripen the fruit, or the fruit will not ripen at the same time as the fruit on the corresponding upper canopy.

VINE VARIABILITY IN AGING VINEYARDS

As vineyards age, they tend to become increasingly non-uniform with respect to individual vine canopy characteristics. With that in mind, the grape grower should strive to treat vines individually, as appropriate, for their canopy management needs. Moderately vigorous vines may require little canopy modification. Large vines in the same row may require more intensive management, while small vines may require substantial fruit thinning and other vigor stimulation. This graduated response is often difficult to convey to unskilled workers but is necessary to achieve uniform fruit-ripening conditions within the vineyard.

References

Dami, I., B. Bordelon, D.D. Ferree, M. Brown, M.A. Ellis, R.N. Williams, and D. Doohan. 2005. *Midwest Grape Production Guide*. The Ohio State University Extension Bulletin 919. 155 p.

Reynolds, A.G., C.G. Edwards, D.A. Wardle, D.M. Webster, and M. Dever. 1994a. Shoot density affects 'Riesling' grapevines. I. Vine performance. *Journal of the American Society for Horticultural Science*. 119: 874–880.

Reynolds, A.G., C.G. Edwards, D.A. Wardle, D.M. Webster, and M. Dever. 1994b. Shoot density affects 'Riesling' grapevines. II. Wine composition and sensory response. *Journal of the American Society for Horticultural Science*. 119: 881–892.

Reynolds, A.G., R.M. Pool, and L.R. Mattick. 1986. Effect of shoot density and crop control on growth, yield, fruit composition, and wine quality of 'Seyval blanc'. *Journal of the American Society for Horticultural Science*. 111: 55–63.

Smart, R. E. and M. Robinson. 1991. Sunlight into wine: a handbook for winegrape canopy management. *Winetitles*, Adelaide, Australia. 88 p.

—7—

Crop Yield Estimation and Crop Management

Author
Tony K. Wolf, Virginia Tech

Crop prediction, or estimation, is a process in which we attempt to quantify the amount of crop that a vineyard or vineyard block is carrying. The most obvious reason for acquiring this information is to estimate the amount of crop that will be harvested. A second, and no less important, reason is to know early in the season whether the vines are over-cropped; bearing more crop than can be adequately ripened. Knowing this may enable the vineyardist to thin or remove some portion of the crop to bring the vines into better balance. This chapter reviews the basic procedures for crop estimation and explains the consequences of overcropping and the rationale for crop thinning when crop yields are excessive.

CROP ESTIMATION

Crop estimation requires time for sampling the vineyard block in an accurate and representative way. The experienced grower may be tempted to omit this procedure and rely instead on data from past vineyard performance. However, relying on past yield performance may result in errors in estimation, particularly in regions that are subject to spring frosts, winter injury, or variable weather at bloom, as these conditions can greatly affect year-to-year vineyard productivity.

Crop yield estimation relies on several pieces of critical information: (a) a good historical (multiyear) record of average cluster weights at harvest; (b) an accurate count of current bearing vines per acre or block; and (c) an accurate determination of the average number of clusters per vine at the time of the crop estimate. The component most subject to year-to-year variation is average cluster weight. The principles of crop estimation are based on the above three requirements along with an understanding of the components of grape yield. Those components are listed diagrammatically in figure 7.1 (page 136).

As figure 7.1 illustrates, we can differentiate between yield components that contribute to fruit clusters per block and those yield components that determine the average cluster weight. Variability in yield per acre can be traced back to variation in one or more of the many components that, collectively, determine yield. Looking specifically at cluster weights (pounds per cluster in figure 7.1), it is common to see yearly variation in the percentage of flowers that set fruit. Reductions in set may occur due to poor weather during or immediately after bloom, poor vine nutrient condition, and possibly other factors such as pesticide phytotoxicity. Berry weight also will vary from year to year due to variable moisture conditions of the growing season and variations in seed

Pounds per cluster X **Clusters per block** = $\frac{\textbf{Tons}}{\textbf{Block}}$

Sources of variation

- Flowers/cluster
- Berries/flower (% set)
- Seeds/berry
- Weight of fruit/seed

Sources of variation

- Vines/block
- Nodes/vine
- Shoots/node
- Clusters/shoot

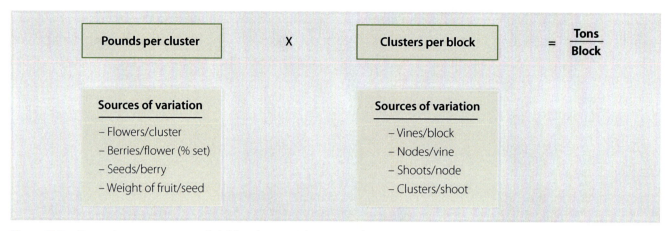

Figure 7.1 ▪ Grapevine components of yield and potential sources of variation.

number. Regardless of the cause, several years' of average cluster weight data will be more meaningful than data obtained from a single year. The number of clusters per block is also subject to year-to-year variation. The number of (bearing) vines per block tends to decline through attrition as a vineyard ages unless the vineyardist is conscientious about vine replacement. The number of nodes per vine is a function of how many nodes are retained at dormant pruning. The number of shoots per node will vary with variety, vine vigor, and the use of shoot thinning as a canopy management practice. The number of clusters per shoot is affected by variety and clone, the extent of bud injury, and the growing conditions of the vine during the previous season. Compared to well exposed shoots, those that develop in dense shade are more likely to have nodes with less fruitful shoots the following year. While the above discussion provides some understanding of crop variation, it is not essential to consider each component of yield to estimate crop. In practice, a simple equation can be used (equation 7.1).

As previously stated, the key components in this equation are: (a) bearing vines per block or per acre; (b) average clusters per vine; and (c) average harvest cluster weight. The "1/2000 lbs" converts pounds (as used in average cluster weight) to tons. There are more sophisticated procedures for estimating crop, but equation 7.1 provides a reasonably accurate prediction. The following discussion provides some specific recommendations for each of the three critical components of the equation.

Bearing Vines per Block or per Acre

The maximum number of vines per acre is determined by row and vine spacing when the vineyard is established. A fully planted acre of vines spaced at 9 (row) by 6 (vine) feet will have about 807 vines. However, the actual number of bearing vines in most vineyards is somewhat less than the theoretical maximum. In poorly maintained vineyards, the actual number of vines may comprise less than 70% of maximum vine space. Yield estimates can error significantly if the estimates do not account for missing vines. For example, an estimate based on 807 bearing vines per acre might predict 4.4 tons of crop per acre. Using the same average cluster weight (0.6 pounds) and clusters per vine (18), the actual yield would be only 3.9 tons per acre if 10% of the vines were missing or were nonbearing. Unfortunately, it is not uncommon for a vineyard to have 10% of its vines missing. Therefore, it is important to ensure that crop estimates are based only on bearing vines. A vineyard can and should be segregated into separate blocks if there is variation in vine size or trellis fill or if other factors are present that would

$$\text{Estimated Yield (tons/block)} = \frac{1}{2{,}000\text{ lbs}} \times \frac{\text{Vines}}{\text{block}} \times \frac{\text{Clusters}}{\text{vine}} \times \text{Average cluster weight (lb)}$$

Equation 7.1

contribute variability to the crop estimate. Obviously, blocks that differ by variety, age, vine density, or training should be sampled separately for crop estimation.

In some vineyards, the trellis spaces created by missing vines are filled in by extending cordons from adjacent vines. This practice might be helpful in maintaining vineyard productivity, but it complicates accurately counting vines per acre. An alternative is to count the number of panels (the distance between two consecutive posts in a row) per acre and to make counts of clusters per panel rather than clusters per vine.

Average Clusters per Vine

The average number of fruit clusters per vine is determined by counting clusters on representative vines in the vineyard and deriving an average figure from those counts. Crop can be estimated any time after all the flower clusters are exposed on developing shoots. One advantage in waiting until after fruit set, however, is that the percentage of berry set can be gauged as well. The selection of vines on which to count clusters should be methodical. One suggestion is to sample on a grid; for example, every twentieth vine in every third row. Avoid edge rows and vines at the end of the rows. The number of vines on which clusters are counted will be determined by vineyard size and the uniformity of vines within the vineyard. In a small 1- to 2-acre vineyard, with vines of a uniform age, size, and training system, one might sample only 10 to 15 vines. The vine sampling in larger, non-uniform vineyards should be stratified to account for variation among distinct areas of the vineyard. Bear in mind that the sampling is attempting to determine the average clusters per vine for the entire vineyard; the larger the sample size, the greater the likelihood that the sample average will be closer to the actual vineyard average with a normally distributed (bell-shaped) population of clusters per vine.

Average Harvest Cluster Weight

Average cluster weights for each variety from a vineyard should be obtained annually at harvest. These data are then averaged over years and used in crop estimates. Clusters can be collected from picking bins after harvest, but the tendency in such a sampling will be to select clusters that are larger than average. It is better to pick all the fruit from a sample of representative vines, such as those used in cluster counts, into picking lugs. For each vine, record the total number of clusters picked, collectively weigh the clusters, and divide the total weight by the number of clusters to obtain the average cluster weight. Subtract the weight of the empty picking bins from the total fruit weight. Picking all clusters from representative vines will ensure that you take into consideration the extremes in cluster size found on a vine. Again, 10 to 15 vines may be a sufficient sampling size for a small, uniform vineyard. You may use data from other sources as a starting point; however, keep in mind that clonal variation and other factors may affect results. Having determined bearing vines per block (or acre), the average number of clusters per vine, and an average cluster weight to work with, you can then predict the crop yield per block or acre (equation 7.1).

Unfortunately, the above discussion may over-simplify the crop prediction process. Even with thorough sampling, accurate vine counts, and cluster weight data collected over the course of many years, the actual crop tonnage at harvest can vary significantly from the predicted tonnage. Many experienced producers are satisfied if the difference between predicted and actual yields is less than 15%, but the variance is often far greater than this. The most variable component of the crop prediction equation (equation 7.1) is the average cluster weight at harvest. This variation will arise from variation in the cluster weight components listed in figure 7.1. Furthermore, disease, pests, and the environment may affect cluster weights. A dry summer, for example, will tend to reduce berry size and thus decrease average cluster weights. For example, a 1/10-pound change in average cluster weight can translate to a yield difference of nearly three-quarters of a ton per acre. Note, too, that the predicted yield does not account for fruit lost to bunch rots, birds, deer, or other unpredictable factors.

The crop prediction model can be refined to provide a more accurate estimation of actual crop if a grower is willing to invest the extra time. One means of refinement is to use repeated measures of cluster weight during the growing season. These measures can then be used to adjust the average harvest cluster weight predicted at harvest. One can assume that a cluster weight that is above average at mid-season will also be above average at harvest. The harvest cluster weight can be estimated by fitting the seasonal cluster weight data to a regression model and then using that model to predict the final cluster weight. Regression analysis is a tool used to describe how a unit change in one variable, for example days after bloom, affects a dependent variable such as average cluster weight. However, to derive a meaningful model (one in which the regression model accounts for a significant proportion of variation in cluster weight), the vineyardist must sample cluster weight on a number of days during the growing season. This is both time-consuming and destructive to the crop.

A more streamlined approach involves determining the average cluster weight at the "lag phase" of cluster development and using that single measure to adjust the harvest average cluster weight. This approach is based on the observation that seasons that result in cluster weights that are smaller than average at lag phase will also result in cluster weights that are smaller than average at harvest. Similarly, mid-season clusters that are larger than average will increase the average harvest cluster weight. To use this method, the grower must develop an historical average lag phase cluster weight for the vines. The lag phase of cluster growth corresponds to the lag phase of berry expansion that occurs with seed coat hardening (see chapter 16, page 282). The end of the lag phase can be observed as the point at which a sharp knife will no longer cleanly slice through berries; instead, the knife will tend to drag against the hardening seed coats. If berry weight or volume were measured every several days, the lag phase would be measured as a transient slowing of the otherwise linear increase in cluster weight throughout the season (figure 16.4, page 286). There is, however, a method of simplifying the determination of lag phase. Depending on variety, the lag phase occurs 30 to 40 days after full bloom, or after the accumulation of approximately 1100 to 1200 seasonal heat units (50°F base). Much, but not all, of the current season variation in harvest cluster weight will be determined at this stage. A note on timing: it is more important to use the same method or benchmark each year (for example, the seasonal accumulation of 1200 heat units) than to pinpoint a specific event such as termination of lag phase.

To determine the average lag phase cluster weight, collect about 300 clusters, weigh the clusters, and derive an average by using the same method you would use for calculating the average harvest cluster weight. You can estimate the onset of lag phase with the knife approach or by summing heat units (1100 to 1200 growing degree units); however, the latter approach is not valid for all varieties. Again, it is preferable to completely harvest half of 10 to 15 vines per block rather than trying to randomly select 300 clusters per block.

The crop prediction model is then modified to use both an historical average lag phase cluster weight and the average lag phase cluster weight for the current season to adjust the average harvest cluster weight as shown in equation 7.2.

Fitting some hypothetical numbers into this refined model illustrates how a small change in the cluster weight during the lag phase will correspond to a change in the average harvest cluster weight. Timing the lag phase of berry development is a potential source of variation with this technique, but if the grower is willing to gather this information on an annual basis, the accuracy of the overall crop estimation will be greatly improved when compared to the results obtained when no mid-season cluster weight modifier is used. Even using lag phase cluster weights, growers must be aware of seasonal changes in water surpluses or deficits, as these can measurably affect cluster weights very close to harvest. Don't be discouraged if first attempts at crop estimation are inaccurate; the more experience and data acquired, the more accurate the estimates will become.

$$\text{Estimated yield per block} = \frac{\text{Vines}}{\text{block}} \times \frac{\text{Clusters}}{\text{vine}} \times \left(\frac{S}{A} \times H\right)$$

where: S = lag phase cluster weight for current season
A = historical average lag phase cluster weight (several years' data)
H = average harvest cluster weight (several years' data)

Equation 7.2

In conclusion:

- Good average cluster weight data are essential to accurate crop prediction. Do not rely on average cluster weight data from other vineyards. Long-term data will be more meaningful than data from a single year.
- Cluster-to-cluster variability is thought to be greater than vine-to-vine variability. Sample entire vines to determine the average cluster weights.
- Nonuniform vineyard blocks (for example, those that vary with soil, topography, vine age, vine training, or other factors) should be divided into uniform sub-blocks.
- Accuracy of yield estimates depends on representative sampling.
- Sampler variation can be significant. Use the same person each year to estimate crop.

CROP MANAGEMENT

Crop management refers to the intentional quantification and subsequent manipulation of crop level. Crop management may be used to target a specific crop level or to bring individual vines or entire vineyard blocks into optimal balance and uniformity of ripening. Thinning of crop can be done either before or after bloom, but the results will vary depending on the timing.

Crop management may serve the following purposes:

1. **Eliminating a potential overcrop situation.** Fertility of the buds varies from year to year. If weather conditions were good during cluster initiation and development in a given year, and if the vineyard was well managed, the following season's crop can be extremely large, even if the vines were correctly pruned in the dormant period.

2. **Maintaining consistent yields and wine grape quality.** Cordon trained, spur-pruned vines may have very fruitful base shoots as well as secondary and tertiary buds. The clusters can become crowded along the cordon resulting in poor ventilation and increased bunch rots. Shoot thinning may alleviate some of this problem, but cluster thinning may also be necessary with large-clustered varieties.

3. **Avoiding overcropping in large-clustered varieties.** Some large-clustered varieties, such as Seyval Blanc, Chambourcin, and Tannat, tend to be out of balance even if the correct dormant pruning formulas are used. Crop thinning may be necessary above and beyond shoot thinning to regulate crop on these varieties.

4. **Ripening the crop.** Additional clusters may need to be removed during cool, wet years to promote fruit ripening. This decision can be delayed until mid-summer or even up until veraison in order to gauge how the season is progressing.

Measuring Crop Load

Crop load is the ratio of the weight of crop produced by a given vine in a given year to the weight of pruned canes produced during that same season. This ratio is one measure of whether vines are balanced. To determine crop load, weigh the fruit at harvest from 10 to 20 representative vines per vineyard block. Weigh the canes removed at pruning from those same vines at dormant pruning. You will need to identify the location of the vines in the vineyard in order to use the same vines for the two measures. For each vine, divide the crop weight by the cane pruning weight to determine the crop load for that vine. Use the average crop load of the representative vines to assess the crop load of the vineyard block. If possible, use the same vines every year to develop a long-term database on the vineyard block's performance. The same vines can be used for other canopy measures and for the crop estimates previously described.

Crop load ratios for most varieties should have a range of 5 to 10. For vines with cane pruning weights of 2.5 pounds, crop yields should be in the range of 12.5 to 25 pounds per vine. Vines with crop loads outside this range of 5 to 10 could be out of balance and should be evaluated. Crop loads less than 5 may indicate excessive vegetation, although this would be normal for young, lightly bearing vines. Also, some producers prefer to significantly restrict crop levels in order to produce more concentrated flavors, particularly with some red-fruited varieties, and this might result in crop load ratios of 3 to 4. Crop loads greater than 10 might indicate too much fruit produced for the vine—a condition that has the potential to impair wine quality.

Cluster Thinning

Shoot thinning and dormant pruning are frequently inadequate for bringing into balance large-clustered vines with fruitful basal buds. Reducing the number of shoots may be counterproductive if the leaf area-to-crop ratio is adversely affected. Cluster thinning is the intentional removal of some portion of clusters per vine to target a desired level. It is usually done only after crop estimation, as previously discussed. Once the existing crop level is known (for example, 7.5 tons per acre), a reduced crop level (for example, 4.5 tons per acre) can be achieved, if desired, by cluster thinning. In this example, our crop estimate of 7.5 tons per acre is based on average harvest cluster weights of 0.5 pounds and vine density of 807 vines per acre. Vines are bearing an average of 37 clusters per vine prior to thinning. To achieve our 4.5 tons per acre target crop level, we would need to remove about 15 clusters per vine. Depending on when it is done, cluster thinning can improve berry chemistry by increasing the soluble solids, anthocyanins (color components), monoterpenes, other flavor and aroma compounds, and, ultimately, the varietal character of the wine.

When and What to Thin

If flower clusters are removed before or during bloom, the vines may produce larger clusters, chiefly as a function of increased set and increased berry size. For this reason, thinning to regulate crop level is usually performed after fruit set. Thinning shortly after fruit set can also increase cluster size and cluster compactness due to a compensatory increase in berry size of the remaining crop. The increased berry size and cluster compactness are generally undesirable features due to decreased wine potential quality and increased disease pressure that results with compact clusters. To minimize the compensation, crop thinning is often delayed until 4 to 6 weeks after bloom; however, the longer one waits to thin surplus crop, the less dramatic will be the improvement in remaining crop quality.

In making the determination of what to thin, note that it is generally advisable to thin the crop uniformly off the vine, rather than removing all the clusters from just several shoots or a section of cordon. Clusters borne on short, stunted shoots are good candidates for removal, as are the more distal clusters borne on average-sized shoots that bear two or three clusters. It might also be desirable to remove a cluster where two or more clusters are intertwined. Tangled clusters often harbor insect pests or foster bunch rots as the fruit ripens.

Crop thinning at veraison ("green thinning") is occasionally used as a means of rouging clusters that are retarded in ripeness, thereby increasing the uniformity of ripeness in the remaining crop. In practice, one waits until most of the clusters have turned blue-black and then a small percentage (for example, 10 percent) of clusters that have failed to turn color, or that are apparently retarded in this process, are removed. The same process can be used in white varieties, although somewhat greater visual judgment is required to differentiate degrees of ripeness.

—8—

Nutrient Management

Authors
Terence R. Bates, Cornell University
Tony K. Wolf, Virginia Tech

In the most basic sense, grapevines require air (carbon dioxide), sunlight, water, and sixteen essential nutrients for normal growth and development. Photosynthesis and canopy management have been studied to optimize carbon assimilation and sunlight interception for a given production system. Ground-cover management and irrigation aim to optimize vine-water relations. Vineyard nutrition integrates the natural processes of soil biology and plant physiology with the grower's goals of economically sustainable production and environmental stewardship.

Vineyard nutrient management is one component of a comprehensive vineyard management program. The issues here concern adequacy of vine nutrition; cost of fertilizer and fertilizer application; effects of nutrients on crop yield, fruit quality, and other aspects of vine performance; and the potential for pollution of groundwater and surface bodies of water with misapplication of fertilizers.

Ensuring adequate vine nutrition begins in the preplant phase of vineyard establishment (see chapter 2, page 14). Soil samples should be collected to determine if lime or other fertilizers are needed. Soil depth, texture, and internal drainage should be evaluated before vineyard establishment because they have a strong influence on rooting volume, nutrient uptake, and resultant vine performance.

THE ESSENTIAL MINERAL NUTRIENTS

Grapevines require sixteen essential nutrients for normal growth and development (table 8.1, page 142). Carbon (C), hydrogen (H), and oxygen (O) are obtained from water by roots or from air by leaves. The remaining thirteen nutrients are termed essential mineral nutrients—or mineral nutrients—and are primarily obtained by root uptake from the soil. The essential mineral nutrients all meet three criteria: each one is needed to complete the plant life cycle, the function of the element cannot be replaced by another mineral element, and each element must be directly involved in plant metabolism.

The macronutrients—nitrogen (N), phosphorus (P), potassium (K), calcium (Ca), magnesium (Mg), and sulfur (S)—are used in relatively large quantities by vines and are often supplemented as applied fertilizers in agricultural systems because release of these elements from natural sources is often insufficient during periods of high plant demand. The micronutrients—iron (Fe), manganese (Mn), copper (Cu), zinc (Zn), boron

Table 8.1 ▪ Sixteen essential nutrients for normal grapevine growth and development

Element	Chemical symbol	Form used in plant uptake	Available in the soil through . . .
Macronutrients			
Hydrogen	H	H_2O, HCO_3^-	Water and air
Carbon	C	CO_2	Water and air
Oxygen	O	CO_2, H_2O, oxyanions	Water and air
Nitrogen	N	NH_4^+ (ammonium), NO_3^- (nitrate)	Microbial decay of organic matter, mineral weathering, cation exchange
Potassium	K	K^+	Mineral weathering, cation exchange, mineral dissolution
Calcium	Ca	Ca^{2+}	Mineral weathering, cation exchange, mineral dissolution
Magnesium	Mg	Mg^{2+}	Mineral weathering, cation exchange, mineral dissolution
Phosphorus	P	$H_2PO_4^-$ (dihydrogen phosphate), HPO_4^{2-} (hydrogen phosphase)	Microbial decay, ligand exchange, minerals
Sulfur	S	SO_4^{2-} (sulfate)	Microbial decay, oxidation, minerals, acid rain
Micronutrients			
Chlorine	Cl	Cl^-	Dissociation from salts
Boron	B	H_3BO_3 (boric acid)	Desorption
Iron	Fe	Fe cations or chelates	Mineral dissolution, desorption, dissociation, cation exchange
Manganese	Mn	Mn^{2+} or chelates	Mineral dissolution, desorption, dissociation, cation exchange
Zinc	Zn	Zn^{2+} or chelates	Mineral dissolution, desorption, dissociation, cation exchange
Copper	Cu	Cu^{2+} or chelates	Mineral dissolution, desorption, dissociation, cation exchange
Molybdenum	Mo	MoO_4^{2-} (molybdates)	Desorption

(B), molybdenum (Mo), and chlorine (Cl)—although no less essential to a grapevine's life cycle, are needed in relatively smaller quantities. Adverse physical or chemical soil properties, as well as restricted plant root growth, can limit nutrient uptake and lead to deficiency symptoms. When one or more of these elements are deficient, vines may exhibit foliar deficiency symptoms, may have reductions in growth and crop yields, and may be predisposed to winter injury or death. Conversely, an oversupply can have equally serious consequences, such as excessive plant growth or stunting and plant death due to toxicity. Maintaining adequate but not excessive nutrient availability is, therefore, critical for optimum vine performance and profitable grape production.

UNDERSTANDING THE SUPPLY OF ESSENTIAL MINERAL NUTRIENTS

Mineral nutrients are supplied to the soil solution through mineral weathering, ion exchange, organic matter decomposition, or fertilizer applications. Living organisms—such as insects, fungi, and bacteria—play an essential role in the release of mineral nutrients and in nutrient cycling. Grapevine roots themselves alter nutrient availability through rhizosphere modification and associations with microorganisms. Ions in the soil solution are available for plant uptake by roots. Once inside the plant, the mineral nutrients are incorporated and used in plant structure or metabolic function. Vineyard nutrient management is more than the application of inorganic fertilizers to the vineyard floor. It consists of understanding the natural nutrient supply in a particular vineyard soil, the nutritional demands of the grapevine variety, and the manipulation of that supply-and-demand relationship to achieve desired production goals.

Soil is the medium in which roots grow and nutrients are supplied. Soil consists of a solid phase of minerals and organic matter; a liquid phase of water and dissolved ions; and a gaseous fraction containing the atmospheric gases of oxygen, carbon dioxide, and nitrogen. The balance and interaction of the three soil phases are important in plant nutrition and vineyard productivity. Minerals and organic matter in the solid soil phase are a reservoir of mineral nutrients that are exchanged with the liquid soil phase. Plants take up mineral nutrients mostly in the liquid soil phase; therefore, understanding the solid-liquid exchange of nutrients in a given vineyard soil is important for nutrient management.

In most agricultural soils, the total amount of essential mineral nutrients in the solid soil phase greatly exceeds the amount needed by grapevines; however, very little of these nutrients are in a plant-available form. These reserve sources exchange nutrients with the soil solution in the form of ions, organic complexes, or uncharged molecules. Commercial fertilizers supply nutrient ions when dissolved in the soil solution, but even these typically do not remain in solution for long periods. The ions provided through inorganic fertilizers are taken up by the plants, leached through the soil profile, or immobilized in the solid soil phase.

Understanding Cations: Cation Exchange Capacity (CEC), Base Saturation, Soil pH

Ion exchange plays an important role in the transfer of nutrients between the solid soil surface and the soil solution. Clay minerals in soil are composed of silica (Si) and alumina (Al) sheets. Si^{4+} and Al^{3+} in the sheets can be replaced by other positively charged ions (cations) of a lesser charge in a process called isomorphous substitution, giving a net negative charge to soil particle surfaces. To stay electrically neutral, negative clay surface charges attract cations from the soil solution. The absorption and desorption of cations between soil surfaces and the soil solution is called cation exchange. Cation exchange is an important process in supplying grapevines with potassium (K^+), calcium (Ca^{2+}), magnesium (Mg^{2+}), ammonium (NH_4^+), and non-essential sodium (Na^+) cations, although sodium is rarely of importance in high-rainfall regions. Other sources of soil cation exchange capacity are the broken edges of clay particles and organic matter (humus) when the soil pH is favorable.

Cation Exchange Capacity (CEC)

Cation exchange capacity (CEC) is the total amount of negative charge a soil has to attract cations and is defined on soil tests as the sum of the exchangeable bases plus total soil acidity. That is, all the cations held by the soil that can be exchanged with other cations plus the residual acidity that cannot be exchanged with another cation in an unbuffered situation. Cation exchange capacity is largely determined by the amount and type of clay, the amount of organic matter, and the soil pH. Soils with greater clay content, particularly of clays with greater isomorphous replacement, and humified organic matter will have higher CEC. A soil with higher CEC has greater plant mineral nutrient-holding potential. However, other factors (such as soil pH) determine how much of that potential is achieved.

The strength at which soils hold cations is not equal and is dependent on the charge intensity and hydrated size of the cation. Attraction strength is as follows:

Strong Weak

$$Al^{3+} > Ca^{2+} > Mg^{2+} > K^+ = NH_4^+ > Na^+$$

Attraction strength and ion concentration in the soil solution both have an influence on plant nutrient availability. For example, in limestone-based soils, the abundance and attraction strength of calcium and magnesium out-compete potassium in exchange sites. However, increasing the concentration of potassium in the soil solution with potash fertilizer will temporarily displace the more strongly attracted ions of calcium, magnesium, and aluminum. In a strongly acidic soil, aluminum is highly soluble; the combination of high solubility and high binding strength of aluminum causes the displacement of plant-essential cations and results in multiple nutrient-deficient symptoms in the plant.

Base Saturation

Base saturation is the fraction of CEC that is satisfied by the basic cations of Ca, Mg, K, and Na. Sodium is generally a minor component of CEC in our soils, given its weak ionic attraction to soil colloids and the high rainfall normally experienced in the East. In rare situations, as in some coastal areas, Na levels higher than 15% on the exchange site could result in soil dispersion, poor water infiltration, and possible sodium toxicity to vines. The remaining cation fraction of CEC in typical acidic soils includes Al, Mn, and H. For a particular soil, there is a measurable positive correlation between base saturation and soil pH. As the soil pH decreases from 5 to 3.5, aluminum solubility increases; and it is the free and exchangeable aluminum ions, as well as hydrogen ions, that fill the cation exchange sites in the soil displacing the desirable Ca, Mg, and K cations, as well as Na, if present. Therefore, as the soil pH decreases, so does the base saturation value. In contrast, as the soil pH increases, aluminum precipitates out of the soil solution, hydrogen ion concentration decreases, and more cation exchange sites are occupied by Ca, Mg, K, and Na (figure 8.1). Therefore, as soil pH increases from a pH of 5.0 to 6.0 the percent base saturation (the percent of the cation exchange sites occupied by Ca, Mg, K, and Na) may increase from 20% to 60%.

Soil pH

Soil pH is a measure of the acidity or alkalinity (the concentration of H in the soil solution). The pH scale is numbered from 0 to 14. A value of 7 is neutral. pH values less than 7 reflect acidity, whereas numbers above 7 are indicative of alkaline conditions. The pH scale is logarithmic (pH = –log H concentration); a pH of 5.0 is ten times more acidic than a pH of 6.0 and one hundred times more acidic than a pH of 7.0. Soil pH, which can range from 3 to 10, is affected by many factors, including the parent material, the amount of organic matter, the degree of soil leaching by precipitation, and additions of lime or acidifying fertilizers. While soil pH can be precisely quantified, we often use qualitative terms to describe soil pH based on grapevine performance. "Strongly acidic" soil, from a viticultural standpoint, will generally be a pH of 5.5 or lower and will most likely exhibit mineral-nutrient imbalances. "Slightly acidic"

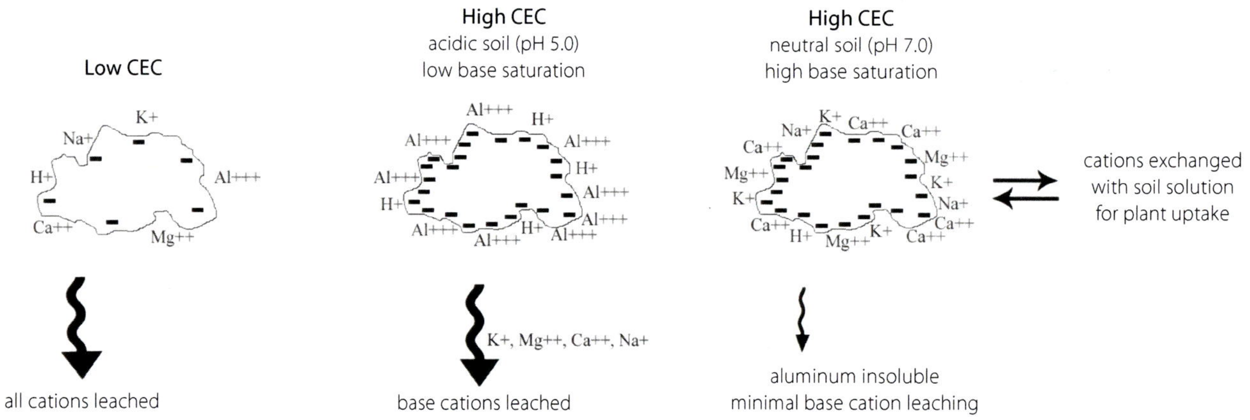

Figure 8.1 ▪ Three soil situations influencing grapevine mineral nutrition

Left: Soils with low cation exchange capacity (CEC), such as a sandy soil with little clay content and low organic matter, have little negative surface charge and little nutrient-holding capacity.

Middle: Soils with high CEC have more negative surface charges because of higher clay and organic matter content. Acidic soils can have high CEC; however, the exchange sites tend to be dominated by aluminum ions, which displace plant-essential cations.

Right: Soils with high CEC and with the exchange sites dominated by calcium (Ca), magnesium (Mg), and potassium (K)—high base saturation—can desorb nutrients into the soil solution for plant uptake.

(pH 5.6 to 6.9) and "neutral" (pH 7.0) vineyard soils generally have better nutrient balance for plant growth. Calcareous soils—those that contain free calcium carbonate—may be "slightly to strongly alkaline" with soil pH greater than 7.0 or greater than 8.5, respectively. Vineyard soils with a pH greater than 7.5 are rare in eastern North America but will typically exhibit nutrient imbalances where they do exist.

Soil pH, CEC, and base saturation values are used together to determine a lime recommendation (the amount of calcium carbonate equivalent needed to raise the soil pH by a desired amount). First, CEC is directly related to soil's buffer capacity. It takes more lime to raise a soil pH from 4.5 to 6.0 in a soil with high CEC (that is, a clayey soil with a high percentage of organic matter) than it does in a soil with low CEC (that is, a sandy soil with a low percentage of organic matter). Second, there is a positive correlation between soil pH and base saturation, but the relationship is not linear. The chemistry of aluminum and carbonate buffer the soil at low and high pH, respectively. Therefore, it takes more lime to change the soil pH from 4.0 to 5.0 (because of aluminum) or from 6.0 to 7.0 (because of carbonate) than it does to change the soil pH from 5.0 to 6.0 (where aluminum and carbonate are not factors).

Aside from improving the percent base saturation, there are other good reasons to maintain soil pH at optimal levels. In strongly acidic soils, high amounts of free aluminum and iron precipitate phosphorus (P) out of the soil solution, making P unavailable to the plant. Aluminum toxicity can also affect root growth by inhibiting cell division in the root apical meristem. In addition, soil microorganisms, which improve soil fertility through the breakdown of organic matter and metabolism of nitrogen, can be inhibited in acidic soils (figure 8.2).

The availability of micronutrients (for example, Zn, Fe, Mn, and Cu) also changes with soil pH. In general, availability is high in acidic soils and low in alkaline soils. High availability at low soil pH can cause direct toxicity symptoms on the vine or cause indirect deficiency of another element. For example, high availability of zinc and iron may limit phosphorus availability for root uptake. In contrast, low availability of zinc and iron at high soil pH can lead to zinc or iron deficiency. Zinc deficiency is common in California vineyards with sandy, high-pH soils. Some grape varieties are susceptible to

Figure 8.2 ▪ **Acidic soil sickness.**

Healthy leaves *(top row)* and leaf symptoms of acidic soil sickness *(bottom row)*. Concord, Traminette, Noiret, White Riesling, and Cabernet Sauvignon leaves are shown *(left to right in each row)*. Aluminum toxicity in acidic soils leads to poor root growth, low phosphorus availability, and decreased cation uptake.

iron deficiency in vineyards where the soil is alkaline (a pH greater than 7.0), cool, and wet.

Cation Balance

As described earlier ("Base Saturation," page 144; figure 8.1, page 144), Al, Mn, and H are the dominant cations which tend to exchange Ca, Mg, and K in strongly acidic soils. The addition of lime to increase the soil pH improves soil fertility by allowing Ca, Mg, K, and Na to become the dominant cations on soil exchange sites. After target soil pH and percent base saturation are achieved, the next factor to watch is the relative balance of beneficial cations. Just as aluminum competes with the other cations under acidic soil conditions, the beneficial cations also compete for exchange sites at optimum soil pH for plants. The adjustment of soil pH with limestone has a large effect on the balance between K and Mg. At low soil pH, magnesium availability can be depressed and the addition of potassium fertilizer to a low pH soil can further reduce magnesium availability. In higher pH soils—either naturally high limestone soils or soils adjusted with lime—magnesium availability can be high and will compete with potassium in the soil and in root uptake. It is important to note that soil pH itself has a significant effect on magnesium availability, irrespective of how the soil pH increase was achieved. The choice to adjust soil pH with dolomitic (high Mg) limestone instead of calcitic limestone has an additional, but smaller, effect on magnesium availability. When applying lime to a vineyard, it is wise to monitor vine potassium and magnesium concentrations and apply potash if necessary. The response of soil calcium availability to soil pH is similar to magnesium; therefore, calcium competes with potassium in soil CEC. However, magnesium is relatively more competitive with potassium than calcium in vine uptake.

Nutrient Anions and Organic Matter

Because they lack sufficient positive surface charge, most soils have little electrostatic anion exchange capacity (the attraction of negatively charged ions to positive soil surface charges). However, soils will absorb specific anions such as phosphate (hydrogen phosphate [HPO_4^{2}], dihydrogen phosphate [$H_2PO_4^-$]), and sulfate (SO_4^{2-}) through different chemical reactions. Phosphate anions bind to the surface of minerals by replacing OH groups bound to iron or aluminum. This association can be relatively strong and can limit the mobility of phosphorus in most soils with the exception of very sandy soils. In alkaline soils with high calcium levels, phosphorus can precipitate with calcium, limiting phosphorus availability.

While relatively more mobile nutrient elements translocate to roots through the bulk flow of soil water or through diffusion in the soil solution, the uptake of immobile elements, like phosphorus, relies more heavily on the absorptive surface area of roots. Plants with greater root surface area—due to increased root branching and/or root hairs—are more efficient at phosphorus uptake. Grapevines, in general, have low rooting density and little root hair development. Grapevines do, however, increase their effective root absorptive surface area though mycorrhizal associations, which presumably improve the acquisition of immobile nutrients, such as phosphorus.

Organic Matter Supply

Organic matter comprises a large reserve of slow-release nitrogen, sulfur, and phosphorus, all of which become available to plant uptake with organic matter decomposition. More nitrogen is needed by grapevines cropped at average, sustainble yields than any other mineral nutrient. However, relatively little nitrogen is provided to the soil solution for plant uptake through ion exchange. Most plant-available nitrogen in non-legume crops, such as grapes, comes from the decomposition of soil organic matter.

Organic matter consists of both living organisms and dead organic material. There are three stages of organic matter decomposition. Macrofauna and microfauna first break up raw organic substrates. The most abundant natural substrate is above-ground plant material that falls to the soil, as well as dead root tissue and root cells sloughed off during normal plant growth. Large organic compounds, indigestible by general microbes, are broken down by specialized bacteria and fungi into smaller organic compounds. General microbes may then assimilate and metabolize these compounds into carbon dioxide, ammonium ions, and other inorganic ions that can enter the soil solution for plant uptake.

Humus is the final solid organic by-product that resists microbial decomposition. Phenolic macromol-

SIDEBAR

Cation Management: What is the Best Situation for Eastern Vineyards?

Given the concepts of soil pH, CEC, base saturation, and cation balance, what is the best situation for eastern vineyards? This is a simple question with a complex answer based on soil type, grape variety, and production goals. A vineyard soil with low CEC or severe cation imbalances may lead to visual nutrient deficiencies, low vine size, and poor production. On the other hand, an overly fertile vineyard soil with high CEC and nutrient-holding capacity may lead to undesirably excessive vegetative growth.

Because soil pH has such a large effect on cation availability and soil pH is best adjusted during vineyard preparation, it is an obvious starting point. Aluminum toxicity is a concern with all cultivated grape varieties when the soil pH drops below 5.2. Theoretically, low soil pH (mild aluminum toxicity) could be used as a tool to control excess vine vigor through reduced root growth. However, since soil pH is a logarithmic function (a soil pH of 4 has a ten-fold increase in hydrogen concentration compared to a soil pH of 5), small changes in soil pH can have a large effect on nutrient availability and vine growth—making soil pH adjustment a risky tool for vigor control. In general, targeting soil pH between 5.5 and 6.5 decreases aluminum solubility and increases the availability of potassium, calcium, and magnesium.

After soil pH adjustment, the nutritional focus moves to the supply and balance of the other base cations. In the Concord juice industry, where production goals strive for large vines and large yields, soils with high CEC are desirable. Furthermore, the large demand for potassium by heavy yielding vines requires a cation ratio that favors potassium. In the wine grape industry, where production goals strive for vigor and yield control, soils with more moderate CEC may be helpful in controlling canopy growth and maintaining vine balance.

While there is interaction between cations as described above, the suggestion that there is some strict ratio among them in order for the grapes to grow well is not true and is discredited by most soil scientists. For example, the practice of arbitrarily lifting the calcium base saturation level to above 70% or more by liming to achieve a specific ratio of Ca, Mg, and K is not only wasteful but also may adjust the pH to a level that can cause micronutrient deficiencies.

ecules make up a large portion of humus and give humus its characteristic dark brown color and "earthy" odor. Carboxyl groups on the phenolic molecules of humus are weakly acidic and will absorb or desorb H^+ depending on soil pH. This allows humus to buffer soil pH, have a pH-dependent surface charge (cation exchange capacity), and chelate certain mineral nutrients.

In production viticulture, the addition or loss of surface organic matter is quite variable. Mechanical cultivation in vineyards with sloping topography and high rainfall leads to the breaking apart of organic materials and the erosion of soil with organic matter. In contrast, yearly incorporation of green manures or surface mulching can continually add to the supply of organic matter substrates for decomposition.

Any organic material added to the vineyard floor—whether it is chopped-up cane prunings, herbaceous plant tissues, pomace, or earthworm castings—will eventually decompose given the right biological and environmental conditions. The rate of decomposition is dependent on the composition and physical state of the substrate material, temperature, water, nutrients, and oxygen.

Different types of organic matter and the stage of decomposition can be characterized by the material's carbon-to-nitrogen (C:N) ratio (table 8.2, page 150). This is important in vineyard nutrition because decomposition of organic matter will either release or immobilize nitrogen based on the C:N ratio.

When a decomposing material has a low C:N ratio or when nitrogen is in excess, microbes release the excess nitrogen in a plant-available form into the soil solution. When a decomposing material has a high C:N ratio, microbes will immobilize nitrogen through assimilation until the decomposition process lowers the C:N ratio.

SIDEBAR

Mycorrhizae Increase and Phylloxera Decrease Nutrient-Absorbing Root Surface Area

Mycorrhiza (figure 8.3) is the beneficial association between plant roots and mycorrhizal soil fungi. Several studies on *V. vinifera* roots document the beneficial association of vesicular-arbuscular mycorrhizae (VAM) in acquiring immobile soil nutrients such as phosphorus. In this mutual relationship, hyphae from the fungus extend into the soil, increase the nutrient-absorbing surface area, and transfer the acquired nutrients back to the host plant. In return, the fungus receives photosynthates from the plant for growth and reproduction.

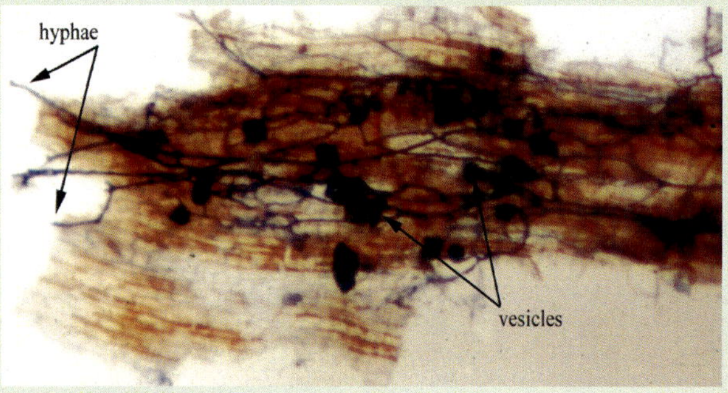

Figure 8.3 ▪ Mycorrhizae.
Concord root squash with mycorrhizae hyphae and vesicles.

Fortunately, in unfumigated vineyard soils in the eastern North America, mycorrhizal fungi are abundant and readily form associations with grape roots. In most cases where nursery vines are grown under natural field conditions, this beneficial association has already been established and is transferred to the new vineyard on the nursery stock. There is no evidence at this time indicating that adding additional spores under normal, unfumigated field conditions aids in vineyard establishment (because natural mycorrhizae are already abundant).

In contrast to mycorrhizae, grape phylloxera (figure 8.4) is a major insect pest of grapes and has had a profound impact on viticultural practices worldwide. *V. vinifera* are particularly susceptible to the root form of phylloxera, theoretically because *V. vinifera* did not coevolve with the pest. Because phylloxera infestation can be lethal to *V. vinifera,* it is generally necessary to grow *vinifera* varieties on a tolerant rootstock where phylloxera has become established. The situation is different for Concord *(V. Labruscana)* production in the eastern North America, where the majority of Concord vineyards are own-rooted and non-irrigated. Concord has phylloxera-resistant leaves and phylloxera-tolerant roots. Although phylloxera infection does not kill Concord grapevines, the roots do support large populations of phylloxera in eastern North America. Phylloxera nodosities weaken the root system and potentially decrease the water- and nutrient-absorbing capacity of Concord, as well as many interspecific hybrids. The same effects weaken and usually kill non-grafted vinifera.

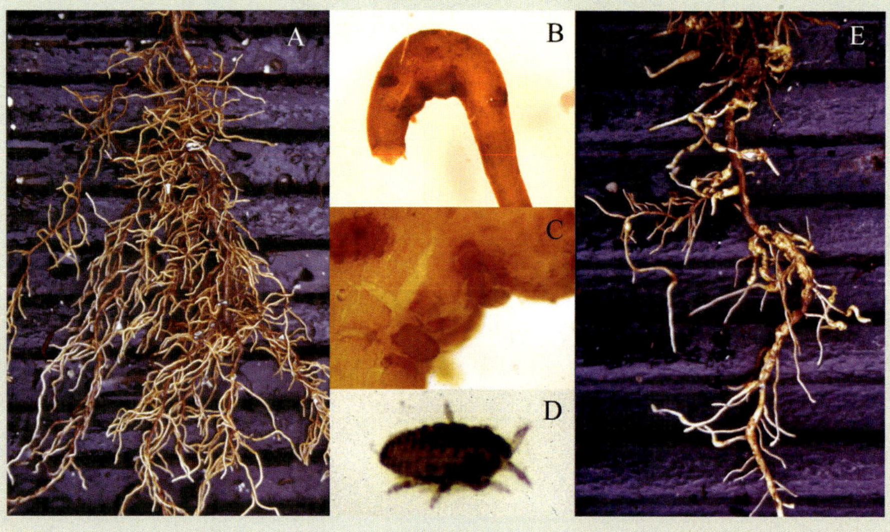

Figure 8.4 ▪ Phylloxera.
Phylloxera on Concord roots. Fine root sub-samples of Concord roots without *(A)* and with *(E)* phylloxera nodosities, illustrating the impact on fine root surface area. Characteristic root swelling and hook of phylloxera infection *(B)*. Phylloxera crawlers inside the root hook *(C)*. A single crawler *(D)*.

Immobilization of nitrogen for the decomposition of organic matter with a high C:N ratio can lead to vine nitrogen deficiency. The rule of thumb is that if the C:N ratio is lower than 20 or the material's nitrogen content is above 2.5%, nitrogen will be released. If the C:N ratio is above 20, nitrogen will be immobilized until sufficient decomposition has taken place.

There are several methods to avoid nitrogen deficiency when adding organic matter with a high C:N ratio to vineyards. Supplemental nitrogen fertilizer or a high-organic-nitrogen source (for example, bloodmeal) can be added to make up for the immobilized nitrogen. Organic matter can be left on the surface of the vineyard floor and not incorporated, resulting in a thin decomposition zone at the soil surface, which is slow because of dry conditions and few microbes. A third option is to lower C:N ratio of the organic matter through composting prior to applying it to the vineyard floor.

Although living cover crops and green manures can be used to raise or maintain vineyard organic matter over time, attention to vine water and nutrient status should be monitored, especially in non-irrigated vineyards. Living plant material on the vineyard floor will compete with the vines for water and nutrients. In vineyards with excessive vegetative grapevine growth, a permanent sod cover can be used to help check vine growth. In low-fertility or drought-susceptible vineyards cover crops may further depress vine growth through competition for water and nutrients. Mulch, herbicides, and properly timed green manures can be used to conserve vineyard soil moisture and improve mineral nutrient availability.

AVOIDANCE AND CORRECTION OF NUTRIENT DEFICIENCIES

Physical soil features are evaluated in the site-selection process (see chapter 2, page 14). The soil must meet minimum standards of depth and internal water drainage. Soil survey maps should be consulted to determine the agricultural suitability of any proposed site. The history of crop production at the site or in nearby vineyards can provide some indication of grape production potential. As described in chapter 2, the ideal vineyard soil is neither impoverished nor overly fertile.

As vines mature and crops are harvested, many vineyards will require periodic application of one or more nutrients as well as pH adjustments with lime. Vineyards are sometimes fertilized on the basis of habit, salesmanship, or wishful thinking. At the other extreme, some growers neglect to fertilize vines even when the symptoms of deficiency are readily apparent. In other cases, entire vineyard blocks might be fertilized when, in actuality, only specific areas of the block require fertilizer. Aside from the expense, inappropriate vineyard fertilization can result in inadequate or excessive vine vigor, poor fruit set, impaired leaf photosynthetic ability, and reduced fruit quality. In some cases, such as with boron, excess availability can cause vine injury. It is, therefore, important that growers have a sound basis for determining the fertilizer needs of their vines.

There is no single method for accurately assessing the nutrient needs of established vines. Instead, a combination of soil analysis, plant tissue analysis, and visual observation is used.

Soil Analysis

Detailed soil analyses must be made prior to vineyard establishment to determine pH, primarily, and also soil fertility. Soil test kits are available from some cooperative extension offices or from commercial laboratories. Soil samples can be collected either with a shovel or a cylindrical soil probe. In either case, samples must be representative of the area to be planted. Sites greater than two or three acres should be subdivided and sampled separately if differences in topography or soil classification exist. Collect samples when the soil is moist and not frozen; fall is an excellent time. Each sample should consist of ten to twenty subsamples that are thoroughly mixed. Exclude surface litter, sod, large pebbles, and stones; and retain about a pound of the mixed soil for testing. The top few inches of soil are usually quite different from deeper soil with respect to pH, organic matter, and nutrient availability. For this reason, it is best to divide each soil probe into two samples: one at an 8-inch depth or less, and a second at an 8- to 16-inch depth. Grape roots can grow much deeper than 16 inches in loose, well-aerated soil. But because of limited ability to significantly alter soil characteristics below this depth, there is little point in collecting deeper samples.

Soil test results will indicate whether adjustments to pH and mineral nutrients are necessary. Soil test data are not customarily used to assess the need for nitro-

Table 8.2 ▪ Nitrogen content and carbon-to-nitrogen (C:N) ratios for organic matter

Organic matter source	Value	Nitrogen (N) content (%)	Carbon-to-nitrogen (C:N) ratio	Affect on nitrogen (N)
Woody plant tissue	Range	0.1-0.5	80-400	N immobilizing
Paper (domestic refuse)	Typical	0.2-0.25	127-178	N immobilizing
Straw-general	Range	0.3-1.1	48-150	N immobilizing
Corn cobs	Range	0.4-0.8	56-123	N immobilizing
Tree leaves	Range	0.5-1.3	40-80	N immobilizing
Corn stalks	Typical	0.6-0.8	60-73 [a]	N immobilizing
Shrub trimmings	Typical	1.0	53	N immobilizing
Horse manure [b]	Range	1.4-2.3	22-50	Variable
Fruit wastes	Range	0.9-2.6	20-49	Variable
Apple pomace	Typical	1.1	48	N immobilizing
Corn silage	Typical	1.2-1.4	38-43 [a]	Variable
Hay-general	Range	0.7-3.6	15-32	N immobilizing
Cattle manure [b]	Range	1.5-4.2	11-30	Variable
Herbaceous plant tissue	Range	1.0-5.0	10-30	Variable
Seaweed	Range	1.2-3.0	5-27	Variable
Grass clippings	Range	2.0-6.0	9-25	Variable
Sheep manure	Range	1.3-3.9	13-20	N releasing
Vegetable produce	Typical	2.7	19	N releasing
Swine manure [b]	Range	1.9-4.3	9-19	N releasing
Tree trimmings	Typical	3.1	16	N releasing
Turkey litter	Average	2.6	16 [a]	N releasing
Garbage (food waste)	Typical	1.9-2.9	14-16	N releasing
Broiler litter	Range	1.6-3.9	12-15 [a]	N releasing
Animals and microbes	Range	1-12	5-10	N releasing
Cottonseed meal	Typical	7.7	7	N releasing
Soybean meal	Typical	7.2-7.6	4-6	N releasing
Fish-processing sludge	Typical	6.8	5.2	N releasing
Poultry carcasses	Typical	2.4 [c]	5	N releasing
Blood wastes (slaughterhouse)	Typical	13-14	3-3.5	N releasing

Source: Ranges and averages are representative values derived from On-Farm Composting Handbook*, NRAES-54. Reprinted with permission.*
Note: More C:N ratios can be found at this web site: <HTTP://COMPOST.CSS.CORNELL.EDU/ONFARMHANDBOOK/APA.TABA1.HTML>.

[a] Estimated from ash or volatile solids data.
[b] Variability in manure C:N is highly affected by storage length and any composting done with the material.
[c] Mostly organic nitrogen.

gen, although most labs can determine and report the percentage of organic matter in soil samples. Depending upon the lab used, specific recommendations can be provided in the soil test report to increase the soil level of nutrients that are below optimal levels. Standards and methods used to calculate fertilizer application rates are presented below, under specific nutrients.

Adjusting Soil pH

Soil pH adjustments in eastern North American vineyards, with few exceptions, are made to increase, rather than decrease, pH. The pH of acidic soils can be raised by applying lime. That simple statement unfortunately oversimplifies the complexity of soil acidity problems—particularly in established vineyards. It is extremely difficult to increase the pH in all but the top few inches of soil once vines are planted. This is particularly true once a permanent cover crop has been planted and cultivation is no longer desirable. For that reason, it is extremely important to determine soil pH—and raise it if necessary—before the vineyard is established. The applied lime should be incorporated as thoroughly and as deeply as possible. Common agricultural-grade liming materials (for example, ground limestone) have very low solubilities and will move very little, if at all, beyond the first few inches when applied to the soil surface. Even if cultivation is used, lime incorporation beyond about 12 inches is unlikely with conventional tillage equipment. Subsoil pH can be raised somewhat by applying lime and deeply cultivating (12 to 18 inches) with a chisel plow or subsoiler. Some research has been conducted with lime injectors, but that technology will not be readily available to most grape growers.

Most vineyard soils tend to become acidic even if they are limed to pH 6.5 at the time of establishment. Acidification occurs due to leaching of basic ions from the soil profile, microbial activity, and the addition of acidifying fertilizers such as ammonium sulfate or ammonium nitrate. Fungicidal sulfur applications will also reduce soil pH over time. Soil tests should, therefore, be conducted every two to three years after vineyard establishment to monitor pH.

The materials commonly used for agricultural liming are the oxides, hydroxides, and carbonates of calcium or mixtures of calcium and magnesium. Commercial bulk application of lime will typically involve spreading of ground limestone, which contains calcium carbonate or mixtures of calcium and magnesium carbonate. Limestone that contains a high proportion of magnesium carbonate is termed dolomitic limestone and can be useful in situations where available magnesium is low. The oxides and hydroxides (hydrated lime = calcium hydroxide) are more reactive and have a greater neutralizing value than the carbonates. These materials are, however, unpleasant to handle. They absorb moisture and can cake, and they can irritate skin and injure tissues of the eyes, nose, and mouth. Oxides and hydroxides are also more expensive than carbonates. In addition to dry materials, liquid lime formulations are available from some distributors. The choice of what liming material to use will be determined in part by what is locally available. Much of the cost of liming is due to transportation and spreading.

The effectiveness of a liming material for correcting soil acidity depends on two factors—chemical purity and physical fineness. These two factors, along with water content, must be considered in determining how much lime to apply per acre.

Purity refers to the acid-neutralizing capacity of the liming material. It is measured and reported as "percent calcium carbonate equivalence" (CCE). Pure calcium carbonate would test 100%. Purity can be reduced by impurities such as shale or chert, soft rock that has no neutralizing value. A number of other substances also have an acid-neutralizing capacity. Pure magnesium carbonate has a CCE of 119%; dolomite, 108%; and calcium hydroxide (hydrated lime), 120% to 150%.

Fineness refers to the particle size of the liming material. Many ground limestones contain a range of particle sizes from dust to the size of coarse gravel. When crushed limestone is applied to soil, the fine particles dissolve and chemically react with the soil acid neutralizing them. The coarse particles react very slowly and have little value.

In general, the amount of lime specified in a soil test report is based on a purity of 85%, and a fineness of 60% (that is, 60% of the material will pass a 100-mesh screen—an industry standard measure of fineness, about the consistency of flour). More lime would be needed if the purity is less than 85%. Ask vendors to provide the lime specifications before purchasing it.

As introduced above, the amount of lime needed for a particular acidity problem is affected by a number of factors including: pH, texture, and amount of soil organic matter; grape species to be planted; and type and particle size of lime used. Table 8.3 provides guidance for liming based on initial pH and cation exchange capacity (CEC). Both soil pH and soil CEC can be found on a standard soil test. Again, because of the soil mobility issues with lime, soil pH adjustment with limestone is best done with deep incorporation of the material before vineyard planting. In established vineyards, surface lime application should not exceed 2 tons per acre per year. For new vineyards, greater amounts of lime can be applied (see table 8.3) provided the lime is incorporated and mixed well with the vineyard soil.

While it is preferable to correct soil pH before planting—not after—the natural tendency of many eastern North American soils will be to decrease in soil pH over time. Again, that process is hastened with the addition of fungicidal sulfur and acidifying nutrients. If the soil pH falls below approximately 5.5, one can expect to see some evidence of impaired nutrient uptake. Corrections of acid pH are difficult to achieve in the absence of soil/lime mixing, as by deep cultivation. Deep cultivation damages both vine roots and established cover crops. The twin dilemmas for established vineyards are how to avoid or at least slow the reductions in soil pH, and how to correct low soil pH in existing vineyards without injuring the vines. Reductions in soil pH can be slowed by routine soil testing and addition of high-quality lime. Shallow cultivation (2- to 3-inch depth) can increase the effectiveness of surface-applied lime. Most commercial lime application equipment is too large for narrow vineyard rows. Options are to ask fertilizer distributors or grower organizations about the availability of small application equipment or to use pelletized lime that will feed through a 3-point hitch-mounted broadcast spreader. Once the vineyard soil pH has dropped below 5.5, more intensive correction measures are warranted. One approach is to broadcast the recommended rate of neutralizing lime and cultivate the lime into the soil with a chisel plow to a depth of 8 to 10 inches. Cultivate alternate rows in alternate years to minimize the potential effects of vine root damage. Such damage is more likely to occur with shallow-rooted vines than with vines with roots that extend some feet into the soil. Consider the timing of this approach from the standpoint of potential vine root damage and reestablishment of a row middle sod, if a sod is used. If the cultivation will destroy an existing cover crop, make the lime application in late winter and cultivate prior to bud burst. Drill or broadcast a fast-growing cereal on the disturbed row middles, followed by mowing and light cultivation in late summer, after which a permanent grass cover crop can be sown or drilled. Deep cultivation might be risky in late summer, particularly in dry years, because of the damage to vine roots and potential for drought injury to vines that are carrying a crop.

Gypsum and Aluminum Toxicity. Deep liming is an effective method of raising soil pH and neutralizing exchangeable aluminum, which can inhibit root growth and decrease nutrient and water uptake. However, deep liming and soil mixing increase cost and are typically restricted to the pre-plant period. Subsequent surface lime applications do not readily move down through the soil profile. The use of gypsum (calcium sulfate) to combat subsoil acidity has received recent attention in perennial crops because surface-applied gypsum can more readily move down through the soil profile. Gypsum is not known as a pH-altering material. Although gypsum can slightly increase soil solution pH by exchanging SO_4^{2-} with OH^-, this pH change is typically small and is masked by natural soil-pH variation under field conditions. Alternatively, gypsum acts directly on the concentration of exchangeable aluminum by forming $Al\text{-}SO_4^+$ ion pairs, which are less phytotoxic than free aluminum ions in the soil solution. In field crops, the formation of $Al\text{-}SO_4^+$ ion pairs, along with an increase in exchangeable calcium with gypsum application, has led to increased root growth and crop yield. The use of gypsum to ameliorate acidic eastern North American vineyard soils and improve grapevine productivity has not yet been substantiated, is not a common recommendation, and has not replaced the use of limestone as the first line of defense against acidic soil conditions. If one wished to try this approach, applying 2 to 3 tons of gypsum per acre would be a reasonable starting point.

Plant Tissue Analysis

Plant tissue analysis provides an objective means of determining the nutrient status of grapevines. Tissue analysis reveals the concentration of essential nutrients or elements absorbed by or within vine tissues. To be meaningful, tissue analysis must entail: a standardized tissue-sampling

Table 8.3 ▪ Lime needed to adjust soil pH

Initial soil pH	Initial soil CEC (cation exchange capacity)(meq/100g)[a]															
	2	3	4	5	6	7	8	9	10	11	12	13	14	15	16	17
	Lime (tons/acre) needed to adjust soil pH to 6.0															
5.9	0.22	0.33	0.45	0.56	0.67	0.78	0.89	1.00	1.11	1.23	1.34	1.45	1.56	1.67	1.78	1.90
5.8	0.25	0.37	0.49	0.62	0.74	0.87	0.99	1.11	1.24	1.36	1.48	1.61	1.73	1.86	1.98	2.10
5.7	0.27	0.41	0.55	0.68	0.82	0.96	1.10	1.23	1.37	1.51	1.64	1.78	1.92	2.05	2.19	2.33
5.6	0.30	0.45	0.61	0.76	0.91	1.06	1.21	1.36	1.52	1.67	1.82	1.97	2.12	2.27	2.42	2.58
5.5	0.33	0.50	0.67	0.84	1.00	1.17	1.34	1.51	1.67	1.84	2.01	2.18	2.34	2.51	2.68	2.85
5.4	0.37	0.55	0.74	0.92	1.10	1.29	1.47	1.66	1.84	2.02	2.21	2.39	2.58	2.76	2.94	3.13
5.3	0.40	0.61	0.81	1.01	1.21	1.41	1.61	1.82	2.02	2.22	2.42	2.62	2.82	3.03	3.23	3.43
5.2	0.44	0.67	0.89	1.11	1.33	1.55	1.77	2.00	2.22	2.44	2.66	2.88	3.11	3.33	3.55	3.77
5.1	0.49	0.73	0.98	1.22	1.47	1.71	1.95	2.20	2.44	2.69	2.93	3.18	3.42	3.66	3.91	4.15
5.0	0.53	0.80	1.06	1.33	1.60	1.86	2.13	2.39	2.66	2.92	3.19	3.46	3.72	3.99	4.25	4.52
4.9	0.57	0.86	1.14	1.43	1.71	2.00	2.28	2.57	2.85	3.14	3.43	3.71	4.00	4.28	4.57	4.85
4.8	0.61	0.91	1.21	1.51	1.82	2.12	2.42	2.73	3.03	3.33	3.63	3.94	4.24	4.54	4.84	5.15
4.7	0.65	0.98	1.30	1.63	1.95	2.28	2.60	2.93	3.25	3.58	3.90	4.23	4.55	4.88	5.20	5.53
4.6	0.71	1.06	1.41	1.77	2.12	2.48	2.83	3.18	3.54	3.89	4.24	4.60	4.95	5.30	5.66	6.01
4.5	0.75	1.13	1.51	1.89	2.26	2.64	3.02	3.40	3.77	4.15	4.53	4.91	5.28	5.66	6.04	6.42
4.4	0.81	1.22	1.62	2.03	2.43	2.84	3.24	3.65	4.05	4.46	4.86	5.27	5.67	6.08	6.48	6.89
4.3	0.86	1.29	1.72	2.15	2.58	3.01	3.44	3.87	4.30	4.73	5.16	5.59	6.02	6.44	6.87	7.30
4.2	0.91	1.37	1.82	2.28	2.73	3.19	3.64	4.10	4.55	5.01	5.46	5.92	6.38	6.83	7.29	7.74
4.1	0.96	1.43	1.91	2.39	2.87	3.35	3.83	4.30	4.78	5.26	5.74	6.22	6.70	7.17	7.65	8.13
4.0	0.99	1.49	1.99	2.48	2.98	3.48	3.97	4.47	4.97	5.46	5.96	6.46	6.95	7.45	7.95	8.44
Initial soil pH	Initial soil CEC (cation exchange capacity)(meq/100g)[a]															
	2	3	4	5	6	7	8	9	10	11	12	13	14	15	16	17
	Lime (tons/acre) needed to adjust soil pH to 6.5															
6.4	0.13	0.20	0.27	0.33	0.40	0.47	0.54	0.60	0.67	0.74	0.80	0.87	0.94	1.00	1.07	1.14
6.3	0.16	0.24	0.32	0.40	0.47	0.55	0.63	0.71	0.79	0.87	0.95	1.03	1.11	1.19	1.27	1.35
6.2	0.18	0.27	0.36	0.45	0.55	0.64	0.73	0.82	0.91	1.00	1.09	1.18	1.27	1.36	1.46	1.55
6.1	0.21	0.31	0.41	0.51	0.62	0.72	0.82	0.93	1.03	1.13	1.23	1.34	1.44	1.54	1.65	1.75
6.0	0.23	0.35	0.46	0.58	0.69	0.81	0.92	1.04	1.15	1.27	1.39	1.50	1.62	1.73	1.85	1.96
5.9	0.26	0.39	0.51	0.64	0.77	0.90	1.03	1.16	1.29	1.41	1.54	1.67	1.80	1.93	2.06	2.19
5.8	0.29	0.43	0.57	0.71	0.86	1.00	1.14	1.28	1.43	1.57	1.71	1.86	2.00	2.14	2.28	2.43
5.7	0.32	0.47	0.63	0.79	0.95	1.11	1.26	1.42	1.58	1.74	1.90	2.05	2.21	2.37	2.53	2.69
5.6	0.35	0.52	0.70	0.87	1.05	1.22	1.40	1.57	1.75	1.92	2.10	2.27	2.45	2.62	2.80	2.97
5.5	0.39	0.58	0.77	0.97	1.16	1.35	1.55	1.74	1.93	2.13	2.32	2.51	2.70	2.90	3.09	3.28
5.4	0.42	0.64	0.85	1.06	1.27	1.49	1.70	1.91	2.12	2.34	2.55	2.76	2.97	3.19	3.40	3.61
5.3	0.47	0.70	0.93	1.16	1.40	1.63	1.86	2.09	2.33	2.56	2.79	3.03	3.26	3.49	3.72	3.96
5.2	0.51	0.77	1.02	1.28	1.54	1.79	2.05	2.30	2.56	2.82	3.07	3.33	3.58	3.84	4.09	4.35
5.1	0.56	0.85	1.13	1.41	1.69	1.97	2.25	2.54	2.82	3.10	3.38	3.66	3.95	4.23	4.51	4.79
5.0	0.61	0.92	1.23	1.53	1.84	2.15	2.45	2.76	3.07	3.37	3.68	3.99	4.29	4.60	4.91	5.21
4.9	0.66	0.99	1.32	1.65	1.98	2.31	2.63	2.96	3.29	3.62	3.95	4.28	4.61	4.94	5.27	5.60
4.8	0.70	1.05	1.40	1.75	2.10	2.45	2.79	3.14	3.49	3.84	4.19	4.54	4.89	5.24	5.59	5.94
4.7	0.75	1.13	1.50	1.88	2.25	2.63	3.00	3.38	3.75	4.13	4.50	4.88	5.25	5.63	6.00	6.38
4.6	0.82	1.22	1.63	2.04	2.45	2.86	3.26	3.67	4.08	4.49	4.90	5.30	5.71	6.12	6.53	6.94
4.5	0.87	1.31	1.74	2.18	2.61	3.05	3.48	3.92	4.36	4.79	5.23	5.66	6.10	6.53	6.97	7.40
4.4	0.94	1.40	1.87	2.34	2.81	3.27	3.74	4.21	4.68	5.14	5.61	6.08	6.55	7.01	7.48	7.95
4.3	0.99	1.49	1.98	2.48	2.97	3.47	3.97	4.46	4.96	5.45	5.95	6.44	6.94	7.44	7.93	8.43
4.2	1.05	1.58	2.10	2.63	3.15	3.68	4.20	4.73	5.25	5.78	6.31	6.83	7.36	7.88	8.41	8.93
4.1	1.10	1.66	2.21	2.76	3.31	3.86	4.41	4.97	5.52	6.07	6.62	7.17	7.73	8.28	8.83	9.38
4.0	1.15	1.72	2.29	2.87	3.44	4.01	4.58	5.16	5.73	6.30	6.88	7.45	8.02	8.60	9.17	9.74

Source: A&L Eastern Labs, Richmond, Virginia.

Note: Use the soil pH and CEC values from your soil test. Locate the pH in the first column and the soil CEC in the top row (of the desired soil pH sub-table). The intersection of the column and row is the amount of lime needed to bring the soil pH to a desired value.

[a] meq/100g—milliequivalents of charge per 100 grams of dry soil.

procedure, accurate and precise analytical methods for determining the elemental concentrations of tissue samples, standard references with which to compare diagnostic sample values, and a means of interpreting diagnostic data and making fertilizer recommendations. In practice, a grower will collect the tissue sample and submit it to a lab for analysis. The lab follows standardized procedures for determining the mineral nutrient concentration of the tissue. Elemental concentrations of the diagnostic sample are compared with standard grapevine tissue references from healthy vines. Based on those standards, elements or nutrients in the diagnostic sample are classified as being either adequate, high, or low/deficient. Fertilizer recommendations to increase nutrients that are low or deficient can be made either by the lab or by a grape specialist. Growers should contact university or commercial labs for further information on submission procedures.

The specific recommendations for tissue sample collection will depend on the grower's objectives. There are basically two different reasons to conduct plant tissue analyses. One reason is for routine nutrient status evaluation. The other is to diagnose an observed visual disorder for which a nutrient deficiency is suspected.

Routine Nutrient Status Evaluation

The general nutrient status of vines should be evaluated on an annual or two-year basis to gauge the vineyard's need for, or response to, applied fertilizer. Such tests will usually detect deficiencies before visual symptoms are manifested. This strategy is intended to avoid having to make corrective fertilizer applications in favor of maintenance applications.

Time to Sample

The concentration of most essential nutrients varies in the plant throughout the growing season. For example, the concentration of nitrogen in grape leaves is higher at bloom time than at veraison (onset of rapid fruit maturation) or near fruit harvest. For other nutrients, such as potassium, research has shown that foliar concentrations at late summer (70 to 100 days after bloom) are better correlated with vine performance than are concentrations diagnosed at bloom time. One might ideally sample vines at different times of the season to evaluate different nutrients; however, that is both inconvenient and expensive. A compromise is to choose a well-defined stage of vine development that provides useful information for the majority of plant-essential nutrients. For these and other reasons, it is recommended that samples be collected at full bloom, which is considered to exist when about two-thirds of the flower caps have been shed. Because the tissue concentrations of many of the essential elements are rapidly changing in the early part of the growing season, it is important to sample as close to full bloom as possible.

Tissue to Collect

Sample each variety separately because nutrient concentrations may vary somewhat among varieties and because varieties may differ in time of bloom. Collect 100 petioles from leaves that are located opposite the first or second (from bottom) flower cluster of the shoot. Petioles are the slender stems that attach the leaf blade to the shoot (figure 8.5). Collect petioles systematically throughout the vineyard block to ensure that the total block is represented. If different portions of the vineyard (hills versus low areas, for example) produce differences in vine growth, separate samples should be collected from each of those areas. Collect no more than one or two petioles per vine. Choose leaves from shoots that are well exposed to sunlight and that are free of physical injury or disease. Immediately separate the petioles from leaf blades and place the petioles in a small, labeled paper bag or envelope. Oven-dry the samples at 200°F for 30 minutes prior to shipping to avoid decomposition or molding. Ship the petioles in a paper envelope or cardboard

Figure 8.5 ▪ Petioles.

SIDEBAR

Soil Testing versus Tissue Testing: Which Is Better?

The debate on the superiority of soil or tissue nutrient testing is common enough to warrant a brief discussion. In research on vineyard water relations, scientists refer to the soil-plant-atmosphere continuum. The supply-and-demand relationship for water in a vineyard is determined by soil factors (soil type, soil depth, water-holding capacity); root characteristics (rootstock, rooting volume, phylloxera damage); canopy growth (amount of exposed leaf area, stomata function); and environment factors (relative humidity, wind). All of the factors work together to give the total picture of vineyard water supply and demand. Vineyard nutrient relations work in a similar way. Soil tests give information on the relative amount and availability of nutrients. Tissue tests give information on how much of the available nutrients in the soil are being taken up by the vine. In a healthy and well-balanced vineyard with adequate root function, soil and tissue tests often go hand in hand. However, analogous to water relations, the supply and demand of vineyard nutrients can be influenced by the soil's physical and chemical properties (soil pH, compaction, anoxia); root function (root distribution, phylloxera); crop size; and environmental factors (excessive rainfall or drought). Often, because of the interaction of these factors, there is disconnection between the information obtained on a soil test and a tissue test. Testing just soil or just tissue can easily lead to a poor recommendation. In contrast, comparing soil and tissue tests can strengthen each test by identifying a nutrient supply-and-demand problem, leading to further investigation and a correct nutrient recommendation.

box. Samples put in plastic bags will rot, and decomposition may alter test results.

Commercial and some university labs will provide an interpretation of tissue analysis results if the grower requests that information. Sufficiency ranges for nutrient concentrations from vines sampled at bloom or late summer (70 to 100 days after bloom) are presented in table 8.4, page 156, along with target values for soil samples. Concentrations that exceed the sufficiency range do not necessarily indicate a problem. For example, some fungicides contain manganese, copper, or iron, which—if the materials were recently applied to the vines—will elevate the test results for those elements. On the other hand, elements that are lower than sufficient may need to be applied as fertilizers.

Certain elements, notably potassium, are best evaluated in late summer when their concentrations become more stable. Where bloom-time samples indicate questionable nutrient levels, particularly with potassium, growers should make a second collection 70 to 100 days after bloom. These late-summer samples should consist of 100 petioles collected from the youngest fully expanded leaves of well-exposed shoots. The youngest fully expanded leaves will usually be located from 5 to 7 leaves back from the shoot tip. If the shoots have been hedged (summer-pruned)(see page 131), collect primary leaves near the point of hedging (not lateral re-growth). Separate the petioles from leaf blades and submit only the petioles as described above.

For "troubleshooting" suspected nutrient deficiencies, sample at any time during the season that symptoms become apparent. Collect 100 petioles from symptomatic leaves regardless of their shoot position. In addition, collect an equal number of petioles from nonsymptomatic or healthy leaves of the same relative shoot position from which affected leaves were collected. Label and submit the two independent samples so that their elemental concentrations can be compared. Leaf blades (lamina) can also be collected for tissue analysis when troubleshooting a presumed nutrient problem between healthy and symptomatic vines. Petioles and lamina will have different nutrient concentration values when taken at the same time. It has been suggested that lamina may be superior to petioles for troubleshooting nutrient deficiencies because they reflect a cumulative storage of nutrients. However, there is currently a much stronger

Table 8.4 ▪ Sufficiency ranges for nutrient concentrations

Nutrient	Chemical symbol	Target values		
		Soil	Bloom petiole	Late-summer petiole (70–100 days after bloom)
Total Nitrogen	N	— [a]	1.2%–2.2%	0.8%–1.2%
Phosphorus	P	20–50 ppm	0.17%–0.30%	0.14%–0.30%
Potassium	K	75–100 ppm	1.5%–2.5%	1.2%–2.0%
Calcium	Ca	500–2,000 ppm [b]	1.0%–3.0%	1.0%–2.0%
Magnesium	Mg	100–250 ppm	0.3%–0.5%	0.35%–0.75%
Boron	B	0.3–2.0 ppm	25–50 ppm	25–50 ppm
Iron	Fe	20 ppm	30–100 ppm	30–100 ppm
Manganese	Mn	20 ppm	25–1,000 ppm	100–1,500 ppm
Copper	Cu	0.5 ppm	5-15 ppm	5-15 ppm
Zinc	Zn	2 ppm	30-60 ppm	30-60 ppm
Molybdenum	Mo	— [c]	0.5 ppm	0.5 ppm
Aluminum	Al	< 100 ppm [b]		
Organic matter		3–5%		
pH		5.5 (*V. labrusca*) 6.0 (hybrids) 6.5 (*V. vinifera*)		

Note: ppm is parts per million.

[a] Soil nitrogen is not normally evaluated for vineyards in eastern North America.

[b] Calcium level is normally adequate when pH is in the proper range for the grape variety. The same is true for aluminum.

[c] Adequacy of soil molybdenum for grapevines is uncertain.

database for bloom and fall petiole standard values than there is for lamina (satisfying one criteria for tissue nutrient analysis—standard references with which to compare diagnostic sample values). Therefore, if collecting lamina samples, it is critical to collect tissue from symptomatic and healthy leaves for an internal comparison.

Visual Observations

Foliar symptoms of nutrient deficiencies and observations of vine vigor and crop size are important clues as to whether vines are suffering nutrient stress. It is, however, possible to be misled by foliar disorders because some are of non-nutritional origin. For example, some herbicide toxicity symptoms bear similarities to certain nutrient deficiencies. And, to the inexperienced, European red mite feeding injury (page 249) may be misinterpreted as a nutrient deficiency. The correct interpretation of foliar disorders requires a certain amount of experience and understanding of pattern expression. In general, there are three different patterns of symptoms to examine: patterns within the vineyard, patterns on a given vine, and patterns on a particular leaf.

Variation in symptoms within the vineyard can provide useful clues as to whether a nutrient deficiency is the cause of observed symptoms. With undulating or hilly topography, nutrient-deficiency symptoms are usually first observed on the higher sites, especially where soil erosion has occurred. In particular, nitrogen, potassium, magnesium, and boron deficiencies may be anticipated to occur first at higher sites because of thinner

SIDEBAR

Percent, Parts per Million (PPM), Pounds per Acre

Soil and petiole nutrient values are typically reported on lab tests as concentrations by weight. Meaning, there are *x* grams of the nutrient in *y* grams of the soil or tissue analyzed. Confusion starts when different units are used to describe the same concentration.

Using metric units, there are 1,000,000 milligrams (mg) in 1 kilogram (kg). Likewise, there are 1,000,000 micrograms (μg) in 1 gram (g). If a soil test measures 1 microgram of boron in a 1 gram sample of soil, then the boron soil concentration is reported as 1 μg/g, 1 mg/kg, or 1 part per million (ppm).

$$mg/kg = \mu g/g = ppm$$

This also means that there is one pound of boron in 1,000,000 pounds of vineyard soil. One acre-furrow slice in agriculture is defined as 2,000,000 pounds of soil, not 1,000,000. Therefore, multiply ppm by 2 to calculate pounds/acre. One ppm boron in our example could also be reported as 2 pounds/acre.

$$ppm \times 2 = pounds/acre$$
$$pounds/acre \div 2 = ppm$$

Macronutrient tissue concentrations can be listed in percent, which is simply parts per hundred instead of parts per million. If a tissue test measures 0.01 grams of potassium in a 1 gram tissue sample, then the potassium tissue concentration is reported as 1%.

top-soil and reduced moisture availability. Soil moisture aids movement of nutrients to the root/soil interface (rhizosphere); and under droughty conditions, nutrient deficiencies can develop.

Vine-to-vine variation in symptoms can also provide meaningful clues. Generally, a nutrient deficiency will affect sizable portions of a vineyard and rarely only one or two vines at random. Peculiar symptoms that appear on only a few vines throughout the vineyard, or where healthy vines alternate with symptom vines, suggest a biological pest. Leafroll virus (page 238), for example, will produce distinct foliar symptoms on red-fruited varieties, and affected vines may be directly adjacent to healthy vines.

The position or age of symptomatic leaves on a given vine also provides information about which nutrient might be causing the deficiency symptoms. Generally, deficiencies of the mobile elements such as nitrogen, potassium, and magnesium will appear on older or mid-shoot leaves. Deficiency symptoms of some of the less-mobile trace elements, notably iron and zinc, first appear on the youngest leaves of the shoot.

Aside from foliar symptoms, observations on vine vigor and fruit set/yield can be used to further diagnose a suspected nutrient deficiency. Uniformly weak vine growth, for example, may point to a need for added nitrogen; however, first consider that water stress, overcropping, or disease will also constrain growth. Poor fruit set, straggly clusters, and uneven berry size and shape could suggest a boron deficiency; again, however, similar symptoms might point to tomato ringspot virus infection (page 239). It should be obvious, then, that the diagnosis of a nutrient deficiency depends on experience and should be confirmed with a combination of visual observations and laboratory tests.

SPECIFIC NUTRIENT DEFICIENCIES AND THEIR CORRECTION

Fortunately, relatively few of the sixteen essential elements required by grapevines are commonly deficient in the eastern North American vineyards. We occasionally find nitrogen, phosphorus, potassium, magnesium, and boron at low to deficient levels. This section provides an overview of the role of these nutrients, the symptoms of deficiencies, and options for correcting the deficiencies.

Nitrogen

Role of Nitrogen

Vines use nitrogen to build many compounds essential for growth and development. These compounds include amino acids; nucleic acids; proteins (including all enzymes); and pigments, including the green chlorophyll of leaves and the darkly colored anthocyanins of fruit.

Symptoms and Effects of Nitrogen Deficiency

Nitrogen deficiency is not as easily recognized as are deficiencies of certain other elements such as magnesium or potassium. The classic symptom is a uniform light green color of leaves, as compared to the dark green of vines that receive adequate nitrogen (figures 8.6, 8.7, and 8.8). Nitrogen deficiency is considered severe if leaves show this uniform light green color. Other clues that point to nitrogen deficiency are slow shoot growth, short internodal length, small leaves, and an unusually red color of leaf petioles and shoot stems. Insufficient nitrogen can also reduce crop yield through a reduction in clusters, berries, or berry set. Thus, nitrogen deficiency might be noted as a reduction in yields over several years. It is important to remember, however, that other factors such as drought, insect and mite pests, and overcropping can also cause symptoms.

Excessive Nitrogen

Nitrogen stimulates vegetative growth; and if both excess nitrogen and water are available to vines, excessive vine growth may occur. Shoots of such vines can grow late into the fall, attain a length of 8 to 10 feet, and produce abundant lateral shoots. Conventional trellis and training systems do not accommodate such extensive growth, and some form of summer pruning might be needed to create an acceptable canopy microclimate for fruit and wood maturation. The percentage of shoot nodes that mature (become woody) can also be decreased when excessive nitrogen causes growth to continue late in the season and such growth is killed by early frost.

Yields can also suffer from excessive nitrogen uptake. Yield reductions can result from reduced bud fruitfulness caused by shading of buds in the previous year. Yields can also be reduced because of inadequate fruit

Figure 8.6 ▪ Nitrogen deficiency on Vidal.
A clue to nitrogen deficiency is the presence of prematurely yellowing leaves despite good soil moisture and sunlight exposure of leaves, as in the canopy shown.

Figure 8.7 ▪ Nitrogen deficiency on Vidal.
The leaf on the left is showing early signs of nitrogen deficiency as evidenced by light green color, relative to healthy, dark green leaf on right.

Figure 8.8 ▪ Nitrogen deficiency on Concord.
A nitrogen-deficient leaf *(left)* and a healthy leaf *(right)*.

set in the current year. In the latter situation, vigorous shoot tips can provide a stronger "sink" than the flower clusters for carbohydrates, nitrogenous compounds, and hormones that are necessary for good fruit set.

Some growers believe that any added nitrogen will reduce the cold hardiness of vines. This is an unfortunate misconception because nitrogen stress increases—not decreases—winter hardiness. If vines exhibit poor vigor and are not producing good crops as a result of nitrogen deficiency, the addition of moderate amounts of nitrogen (30 to 50 pounds of actual nitrogen per acre) will not reduce their cold hardiness and will undoubtedly improve their overall performance.

Causes of Nitrogen Deficiency

Nitrogen is the essential element used in greatest amounts by vines and, in older vineyards, is the nutrient most commonly applied on a regular basis. Once absorbed by the vine, nitrogen can be lost through fruit harvest and the annual pruning of vegetation. Since grape berries contain approximately 0.18% nitrogen, a five-ton crop will remove approximately 18 pounds of nitrogen per acre from the vineyard. The depletion of nitrogen will be even greater if cane prunings (about 0.25% nitrogen) are removed from the vineyard. Given an annual export (crop) of nitrogen with no input (fertilizer), most mineral soils will eventually be depleted of readily available nitrogen. Nitrogen depletion will occur most rapidly with soils that have a low organic-matter content. As described earlier (page 146), much of the nitrogen in soils is associated with organic matter. Through a series of reactions involving soil organisms, the pool of organic nitrogen is converted to forms (ammonium- and nitrate-nitrogen) capable of being absorbed by vines and other plants. When soil nitrogen reserves are exhausted, nitrogen must be applied to satisfy the vines' needs.

Vines grafted to pest-resistant rootstocks (for example, *Vitis vinifera* varieties) are often more vigorous than non-grafted vines, and their requirements for nitrogen fertilizer may be substantially less than that for own-rooted vines. However, grafted grapevines are not immune to nitrogen deficiency. The robust root system of grafted vines is capable of exploring a large volume of soil. Even so, continued cropping or soil mismanagement will eventually exhaust available soil nitrogen.

Assessing the Need for Nitrogen Fertilizer

There is no single index that serves well as a guide in assessing the vine's need for nitrogen fertilizer. Instead, a number of observations made over several consecutive years can be used to determine the vine's nitrogen status. Vines can be grouped into three general categories with respect to their nitrogen status: deficient, adequate, and excessive.

Nitrogen-Deficient. Nitrogen deficient vines commonly exhibit combinations of the following symptoms:

- Vines consistently fail to fill available trellis with foliage by August 1.
- Crop yield is chronically low.
- Cane pruning weights are consistently less than 0.25 pound per foot of row or per foot of canopy for divided-canopy training systems (for example, less than 1.75 pounds for vines spaced 7 feet apart in the row).
- Mature leaves are uniformly small and light green or yellow; leaf reddening may occur with red-fruited varieties, and leaf petioles may be unusually red.
- Shoots grow slowly and have short internodes.
- Shoot elongation ceases in midsummer.
- Fruit quality may be poor, including poor pigmentation with red-fruited varieties.
- Bloom-time petiole nitrogen concentration is less than 1.0%.

Adequate Nitrogen. Vines typically exhibit the following characteristics when nitrogen status is adequate:

- Vines fill trellis with foliage by August 1.
- Yields are acceptable.
- Cane pruning weights average 0.3 to 0.4 pound per linear foot of canopy.
- Mature leaves are of a size characteristic for the variety and are uniformly green.
- Shoots grow rapidly and have internodes 4 to 6 inches long.
- Shoots cease growth in late summer or early fall.
- Fruit quality and time to maturation are normal for the variety.
- Bloom-time petiole nitrogen concentration is between 1.2% and 2.2%.

Excess Nitrogen. Vines that have excess nitrogen availability and uptake may present the following symptoms:

- Shoots fill trellis with an excess of foliage: shoots 8 to 10 feet long by mid-July.
- Fruit yields are low because there are few clusters, poor fruit set, or both.
- Cane pruning weights consistently exceed 0.4 pounds per foot of row (for example, 2.8 or more pounds of cane prunings for vines spaced 7 feet apart in the row).
- Mature leaves are exceptionally large and very deep green.
- Shoot growth is rapid, and internodes are long (6 or more inches) and possibly flattened.
- Shoot growth does not cease until very late in the fall.
- Fruit maturation is delayed.
- Bloom-time petiole nitrogen is greater than 2.5%.

Again, the occurrence of symptoms listed as typical of nitrogen-deficient vines does not prove that a lack of nitrogen is limiting growth. Drought, in particular, can cause similar symptoms. Nitrogen fertilizer will not overcome problems that arise from the lack of water or other growth-limiting factors.

Correcting Nitrogen Deficiency

It is preferable to avoid nitrogen deficiency rather than wait until correction of a deficiency is necessary. Maintaining an appropriate nitrogen status is based on experience, observations of vine performance, and supplemental use of bloom-time petiole analysis of nitrogen concentration. Once nitrogen deficiency symptoms are visually detected, yield or quality losses have already been sustained, and the deficiency will require time to correct.

If nitrogen fertilizer is warranted, a prudent starting point is to apply at a rate of 30 to 50 pounds of actual nitrogen per acre. Do not be surprised if an initial application of nitrogen has no pronounced effect in the year of application. It occasionally takes a full year for added nitrogen to have an impact on vine performance because much of a vine's early-season nitrogen needs are met by nitrogen stored in the vine from the previous growing season. Thus, nitrogen applied to vines in the current year may have its greatest benefit in the following season.

Several forms of nitrogen fertilizer are commercially available. As described earlier (page 160), all will satisfy the vines' needs, and the relevant issues are cost per unit of nitrogen and availability (table 8.5). Urea will likely be the most economical form. Ammonium-based fertilizers such as urea and ammonium nitrate should be incorporated into the soil to minimize volatilization (and, hence, loss) of ammonia. Rain within one or two days of application is a convenient but unpredictable means of incorporation. Alternatively, soil cultivation, as by dehilling of grafted vines, is acceptable. Recommendations for application of actual nitrogen must be translated into rates based on commercial formulations. A recommendation for 40 pounds of actual nitrogen per acre, for example, would require 87 pounds of urea per acre, 190 pounds of ammonium sulfate per acre, or 250 pounds of calcium nitrate per acre.

Nitrogen fertilizer should be applied only during periods of active uptake to minimize loss through soil leaching. These times include the period from bud break to veraison, and the period immediately after fruit harvest. Historically, routine maintenance applications of nitrogen were made at or immediately after bud break. That timing coincides with normal precipitation patterns that increase the likelihood of soil incorporation. Unfortunately, much of this applied nitrogen can be leached out of the root zone before it is absorbed by the vine. Contemporary research has illustrated that greater nitrogen uptake efficiency occurs when nitrogen is applied between bloom and six weeks after bloom, at least in cool climate viticulture regions. Similarly, where applications of more than 75 pounds of actual nitrogen per acre are required, a split application should be used with 50% of the total nitrogen applied at bloom and the balance applied six weeks after bloom. This method extends the absorption across the most efficient phase of nutrient uptake; the disadvantage is the extra labor involved.

Apply nitrogen where the vine roots are concentrated. Grapevine roots are concentrated in the weed-free strip under the trellis in vineyards that that have sodded row middles; fertilizers are typically banded under the trellis in these situations. By contrast, vine roots may extend well into the row centers in vineyards that have clean cultivated or herbicided row middles; fertilizers should be broadcast over the vineyard floor in this case. Under-trellis application can be done either by placing the fer-

Table 8.5 ▪ Common nitrogen-containing fertilizers

Nitrogen source	Percentage of actual nitrogen	Price per 50-pound bag [a]	Cost per pound of actual nitrogen [b]
Urea	46%	$29.06	$1.26
Ammonium sulfate	21%	$13.60	$1.30
Calcium nitrate	16%	$13.59	$1.70
Diammonium phosphate	18%	$35.01	$3.89
Monoammonium phosphate	11%	$36.00	$6.55
10-10-10	10%	$15.13	$3.03

Liquid nitrogen, anhydrous ammonia, and other "complete" fertilizers such as 10-10-10 could also be considered. However, specialized equipment for application, or greater cost per unit of nitrogen should be considered with those forms.

[a] Prices quoted in northern Virginia in 2008. Prices are significantly cheaper if product is purchased in bulk. However, the quantities of nitrogen that would be applied in small (5- to 20-acre) vineyards do not warrant the inconvenience of bulk handling.

[b] Calculated by taking price per bag (in column three) and dividing by (50 [pounds per bag] x percentage of actual nitrogen [in column two]). For example, for urea, $29.06 ÷ (50 x 0.46) = $1.26.

tilizer in a ring around individual vines or by banding it with a tractor-mounted fertilizer spreader with directed output. The quantities of nitrogen used are so small that ringing individual vines (at 12 to 18 inches from trunks) is a practical alternative for small vineyards. Regardless of the method used, apply nitrogen only where it is needed. Poor vigor is more apt to be observed in vineyard regions of soil export, or erosion, rather than in regions of soil import; fertilize accordingly.

Nitrogen can be applied to foliage for short-term correction of N deficiency. Many commercial, water-soluble fertilizers supply N. Some contain other nutrients that may or may not be needed yet add to the cost. Aside from the cost of foliar N fertilizers, and their short-term benefits (as opposed to soil-applied N), two precautions warrant mentioning. First, when tank-mixed with certain pesticides—especially at temperatures above 90°F and/or at high relative humidity (greater than 90%)—the cocktail can cause significant fruit and foliage burn (phytotoxicity). Therefore, if one chooses to use a foliar N fertilizer, it should be applied independently of other chemicals. Second, low-grade urea may contain contaminants that will cause phytotoxicity if applied to the foliage. If urea is used, the buyer should ensure that the urea is of sufficient quality for foliar application. Excessive rates of even high-quality urea can cause phytotoxicity (figure 8.9).

Figure 8.9 ▪ **Nitrogen burn and nitrogen deficiency.** Nitrogen burn from foliar applied urea *(left)* and nitrogen deficiency on American type grapevine *(right)*.

Appendix 1 (page 296) provides a logical method to assess N status based on tissue analysis and visual symptoms and, based on those criteria, to formulate a measured fertilizer response. This approach is provided for all essential grapevine nutrients, starting with N.

Phosphorus

Role of Phosphorus

Phosphorus is needed for plant metabolism and reproduction. It's the P of adenosine triphosphate (ATP)—one of the fundamental currencies of energy used by organisms. Phosphate is also a part of nucleic acids, coenzymes, and phospholipids. As with other essential nutrients, phosphorus is needed for normal grapevine growth and fruit production, and shortfalls in supply can impair performance.

Symptoms and Effects of Phosphorus Deficiency

Phosphorous is mobile in the plant and moves to the growing points. Leaf deficiency symptoms, like those in figure 8.10, will, therefore, start on older leaves and move up the shoot as deficiency intensifies. In red-fruited varieties, the leaves acquire interveinal ("between the veins") reddening. Merlot is very expressive of these symptoms. Generally, white-fruited varieties exhibit chlorosis (yellowing) around the leaf margins. Severe phosphorus deficiencies can stunt the growth of grapevines, and shoots might have shortened internodes. Phosphorus deficiency can be confused with leaf roll virus (page 238) or mite injury (page 249), particularly with red-fruited varieties. Chronic deficiency will reduce crop yield. There may be blind buds, or poor or no fruit set, loose clusters, and smaller clusters and berries than usual.

Phosphorus adequacy will vary somewhat with scion and rootstock variety. In general, 20 to 50 ppm in the soil, 0.17% to 0.30% for petioles at full bloom, and 0.14% to 0.30% for petioles collected in late summer would be considered adequate (appendix 1, page 296). By way of illustration, the Merlot leaves in figure 8.10 were of vines that revealed only 0.06% phosphorus in petioles collected 5 weeks post-bloom with corresponding soil phosphorus of only 7 ppm in the top 8 inches, and 2 ppm in the 10- to 18-inch stratum. Merlot, apparently, is also more likely to express phosphorus deficiency symptoms than certain other varieties; Cabernet Sauvignon vines in a block adjacent to the symptomatic Merlot were relatively free of symptoms.

Figure 8.10 ▪ Phosphorus deficiency

Advanced phosphorus deficiency symptoms on Merlot leaf.

Acidic soil contributes to phosphorus deficiencies. Phosphorus tends to become immobilized as soil pH decreases below 5.5. Furthermore, phosphorus can react with iron (Fe) and aluminum (Al) at low pH, or with calcium (Ca) at high pH, to form insoluble compounds that are unavailable to the plant. The soil pH in the above-referenced Merlot case ranged from 5.4 to 5.6—marginally acceptable. Both the pH and the soil phosphorus levels need to be examined, as both may require adjustment.

Correcting Phosphorus Deficiency

Phosphorus, as phosphates, can be applied anytime, including the dormant season because phosphorus is relatively immobile in the soil. The fertilizer should be applied in a band under the trellis. The band creates concentrations that are high enough to overcome the rapid phosphate fixing (adsorption) that occurs with soils, leaving some of the phosphorus available to the plant. Light cultivation after leaf-fall in the fall can assist with moving the phosphorus into closer association with vine roots. Elevation of soil pH (if less than 5.5) would be of benefit for a range of reasons that go beyond simply increasing phosphorus availability.

Sources

Monoammonium phosphate (MAP) is 48% phosphate (P_2O_5) and 11% nitrogen, and has an acidic soil reaction. Diammonium phosphate (DAP), which is 46% phosphate and 18% nitrogen, also has an acidic soil reaction. Two sources of phosphate that do not have nitrogen are concentrated superphosphate (also called "triple superphosphate") and superphosphate. Concentrated superphosphate has about 46% phosphate, and about 87% is water-soluble and readily available to the plant. Superphosphate has 18% to 20% phosphate, and about 85% of the phosphate is water-soluble. If nitrogen and potassium are also needed, standard fertilizer mixtures such as 10-10-10 and 20-20-20 can be used or the proportions custom-blended to meet specific needs. Fertilizer mixtures labeled 10-10-10 contain 10% N, 10% P_2O_5, and 10% K_2O; those labeled 20-20-20 contain 20% of each of the three nutrients. See appendix 1 (page 296) for a summary analysis of P status and a graduated response to deficiency.

Potassium

Role of Potassium

Potassium functions in a number of regulatory roles in plant biochemical processes, including carbohydrate production, protein synthesis, solute transport, and maintenance of plant water status. Although potassium can account for up to 5% of tissue dry weight, it is not normally a component of structural compounds.

Symptoms and Effects of Potassium Deficiency

Foliar symptoms of potassium deficiency become apparent in midsummer to late summer as a chlorosis (or fading) of the leaf's green color. This yellowing commences at the leaf margin and advances towards the center of the leaf. Red-fruited varieties are apt to express red pigments in the leaves rather than leaf yellowing. Leaf tissue adjacent to the main veins remains darker green, at least under mild potassium deficiency. Basal to mid-shoot leaves are the first to express these symptoms.

With advanced or more severe potassium deficiency, affected leaves will have a scorched appearance where the chlorotic or reddened zones progress to brown, necrotic tissue. Leaf margins will curl either upwards or downwards (figure 8.11). Severe potassium deficiency also reduces shoot growth, vine vigor, berry set, and crop yield. Fruit quality suffers from reduced soluble-solids accumulation and poor coloration.

The symptoms described can also appear under conditions of extreme drought or extreme moisture. Furthermore, some pesticides can cause leaf scorching under certain conditions. Such phytotoxicity is generally most acute on the younger leaves, and shoots soon develop newer, unaffected leaves.

Causes of Potassium Deficiency

As described earlier (page 146), vines grown in soils that are very high in exchangeable calcium and magnesium, and low in exchangeable potassium may require periodic

Figure 8.11 ▪ Potassium deficiency

Potassium deficiency on Concord *(V. Labruscana)*. Potassium deficiency is first evident on the basal leaves of a shoot *(top)* with marginal and interveinal leaf chlorosis, or yellowing *(bottom)*.

potassium application. Potassium absorption may also be limited under very basic soil pH (greater than 7.0) or acidic soil pH (less than 4.0). Potassium availability will vary greatly throughout the East. In the mid-Atlantic region, excess potassium absorption is much more common than potassium deficiency. By contrast, Finger Lakes and Great Lakes region growers may routinely apply potassium. High foliar concentrations of potassium can be associated with elevated potassium levels in maturing fruit. Such fruit may, in turn, have undesirably high fruit pH, which can negatively affect wine quality. Thus, aside from the cost, there is good reason not to apply potassium unless it is needed.

Assessing the Need for Potassium Fertilizer

Visual observation of vine performance and foliar symptoms should be coupled with routine leaf petiole sampling to determine the potassium status of vines. Research in New York State indicated that late-summer (70 to 100 days after bloom) tissue sampling was superior to bloom-time for accurately gauging the vines' potassium status. Thus, if visual observations or the bloom-time tissue analysis used for other nutrients indicates a marginal potassium level, additional tissue sampling in late summer should be done to confirm the need for added potassium. Petioles of recently matured leaves (about the fifth to seventh back from the shoot tip) are collected for late-summer samples. As with other nutrients, samples should be collected separately from regions of topographic or soil variation.

Correcting Potassium Deficiency

Potassium deficiency is corrected by applying potash fertilizer. Short-term correction can be achieved with foliar-applied potassium fertilizer; however, the less expensive and longer lasting remedy is soil application. Two commonly used potash fertilizers are potassium sulfate (K_2SO_4) and potassium chloride (KCl) (also called muriate of potash). Potassium chloride is generally much less expensive. Potassium may also be applied as potassium nitrate, but this fertilizer is usually very expensive with low commercial availability because of its explosive properties. As with other essential nutrients, K application rates can be calculated from soil test results or from plant tissue analysis results (appendix 1, page 296). For the latter, rates will vary with the severity of potassium deficiency (table 8.6).

Potassium fertilizers should be banded under the trellis rather than broadcast over the vineyard floor. Banding assures that a major portion of the fertilizer will be available for root uptake and will minimize the amount fixed by soil colloids. Potassium can be applied anytime, but maximal uptake will probably occur between bud break and veraison, and immediately after fruit harvest.

Magnesium

Role of Magnesium in the Plant

Magnesium is the central component of the chlorophyll molecule—the green pigment responsible for photosynthesis in green plants. Magnesium also serves as an enzyme activator of a number of carbohydrate metabolism reactions. In addition, the element has both structural and regulatory roles in protein synthesis.

Symptoms and Effects of Magnesium Deficiency

Magnesium deficiency is usually expressed in midsummer to late-summer when basal (older) leaves develop an interveinal ("between the veins") chlorosis (or yellowing). The nature of the chlorosis is variety-dependent, but generally the central portion of the leaf blade loses green color to a greater extent than the leaf margins. Tissue near the primary leaf veins remains a darker green. As symptoms progress, the yellow chlorosis can become necrotic and brown (figure 8.12). Magnesium deficiency of red-fruited cultivars can cause leaves to turn reddish rather than chlorotic. Because magnesium is mobile within the vine, younger leaves are supplied with magnesium at the expense of older leaves. Magnesium symptoms are, therefore, usually more pronounced in the older leaves, except in cases of severe deficiency.

Insufficient magnesium will cause an impairment of protein synthesis and chlorophyll production, both

Table 8.6 ▪ Applying fertilizers to correct potassium deficiency

Vine potassium deficiency classification [a]	Application rate, per vine (pounds)		Application rate, per-acre equivalent (pounds, rounded to nearest hundred) [b]	
	Potassium chloride (KCl)	Potassium sulfate (K_2SO_4)	Potassium chloride (KCl)	Potassium sulfate (K_2SO_4)
Severe	1.5	2.0	900 [c]	1,200
Moderate	1.0	1.3	600	800
Mild	0.5	0.7	300	400

[a] Potassium deficiency is determined by visual symptom expression or on the basis of plant tissue analysis results.

[b] Based on approximately 600 vines per acre.

[c] Single applications of more than 600 pounds of KCl per acre are not recommended because of the potential to induce salt burn. Split greater rates into two or more applications over three or more months to allow leaching of chloride ions by rainfall.

Figure 8.12 ▪ **Magnesium deficiency.**
Magnesium deficiency on Concord *(top)* and Traminette *(bottom)*.

of which reduce photosynthesis and sugar production. There is also evidence that under some conditions, deficiencies of magnesium may be associated with an increased tendency of some varieties to exhibit bunch stem necrosis (BSN), a physiological disorder that affects fruit set and fruit ripening. However, magnesium applications to BSN-sensitive vineyards have provided inconsistent results, suggesting that the problem is much more complex than that of a single nutrient deficiency.

Causes of Magnesium Deficiency

Grapevines express magnesium deficiency symptoms because they are not obtaining sufficient magnesium from the soil. Magnesium comprises approximately 0.25% to 0.75% of the dry weight of non-deficient, bloom-sampled grape petioles. Research has shown that vines that have petiole magnesium concentrations of less than 0.25% at bloom will typically develop magnesium deficiency symptoms by midsummer to late-summer. Magnesium deficiency is often observed where vines are grown in soils of low pH (less than 5.5) and where potassium is abundantly available. The likelihood of magnesium deficiency appears to increase when petiole potassium-to-magnesium ratios exceed 5 to 1. Ratios of up to 20 to 1 are not unusual in petiole samples from Virginia vineyards. Soils formed from sandstones or granite and from coastal sands are relatively low in magnesium. Soils developed from limestone generally have higher magnesium levels. Plants grown on soil high in available potassium often express magnesium deficiency even though soil magnesium levels test relatively high.

Assessing the Need for Magnesium Fertilizer

As with most other nutrients, leaf petiole sampling at bloom time can be used to determine the vines' magnesium status. Tissue analysis results, coupled with visual observations, should indicate whether magnesium should be applied.

Correcting Magnesium Deficiency

Magnesium deficiencies can be corrected with either foliar or soil applications of magnesium fertilizers (appendix 1, page 296). Foliar application is appropriate for corrections of a mild deficiency or for short-term correction, but soil application offers a more long-term remedy.

If a foliar application is chosen, spray the foliage with 5 to 10 pounds of magnesium sulfate ($MgSO_4$) in 100 gallons of water per acre. This rate will assure uniform coverage of leaves. Apply the $MgSO_4$ three times, at two-week intervals, in the post-bloom period. This timing is significantly more effective than waiting until deficiency symptoms are evident in midsummer to late-summer. Magnesium sulfate can be purchased as sprayable $MgSO_4$ from fertilizer dealers in 50-pound bags, or it can be purchased at drug stores as Epsom salts in smaller quantities. The $MgSO_4$ can be mixed with most fungicide or insecticide sprays, unless the pesticide label cautions against this combination.

Long-term correction of magnesium deficiency is achieved by periodic soil applications of magnesium-containing nutrients. If the soil pH is also low (less than 5.5), high magnesium-content limestone (dolomitic lime containing 20% magnesium) is the preferred magnesium source and should be applied at 1 or 2 tons per acre. Dolomitic lime may not be readily available in areas where magnesium deficiency occurs. Fertilizer-grade magnesium sulfate or other fertilizers containing some percentage of magnesium oxide (MgO) are, however, generally available and are sold either in bulk or in bags. Magnesium sulfate is applied at 300 to 600 pounds per acre; magnesium oxide, 50 to 100 pounds per acre. To be most effective, $MgSO_4$ or MgO should be banded under the trellis rather than broadcast over the vineyard floor. With small plantings, the fertilizer can be hand-applied 12 to 18 inches from the trunks of individual vines.

Boron

Role of Boron

Boron is an essential micronutrient; very small quantities are required for normal growth and development. Boron has regulatory roles in carbohydrate synthesis and cell division. Deficiency of boron can disrupt or kill cells in meristematic regions of plants (regions of active cell division such as shoot tips). Boron deficiency also reduces pollen development and pollen fertility. Reduced fruit set is thus a common occurrence with boron-deficient vines.

Symptoms and Effects of Boron Deficiency

Boron deficiency symptoms can be easily confused with other vine disorders and must be confirmed by tissue analysis before attempting corrective measures. California literature distinguishes early-season boron deficiency symptoms from symptoms that develop later in the spring or summer. The early-season symptoms appear soon after bud break and appear as retarded shoot growth and may include death of shoot tips. Shoots can also exhibit a zigzagged growth pattern; have shortened internodes; and produce numerous, dwarfed lateral shoots. Those early-season symptoms are thought to be more severe after dry weather occurs in the fall or when vines are grown on shallow, droughty soils; either situation reduces boron uptake.

A second category of boron deficiency develops later in the spring and is marked by reduced fruit set. The nature of the reduced set can range from the presence of a few normal-sized berries per cluster, to a condition where numerous BB-sized berries are also present. The "shot" berries lack seeds and often have a somewhat flattened shape, as opposed to the normal spherical to oval shape. A note of caution: poor fruit set is not necessarily due to boron deficiency. Other factors, such as tomato ringspot virus (page 239) and poor weather during bloom, can reduce fruit set. Furthermore, the application of boron can lead to phytotoxicity if the boron concentration is already sufficient (figure 8.13).

Foliar boron deficiency symptoms may accompany the reduced fruit set if boron deficiency is severe. Foliar symptoms begin as a yellowing between leaf veins and can progress to browning and death of these areas of the leaf. Boron is not readily translocated throughout the vine. Thus, the foliar symptoms develop first on the younger, more terminal leaves of the shoot. As with early-season deficiency symptoms, primary shoot tips may stop growing, resulting in a proliferation of small lateral shoots.

Causes of Boron Deficiency

Grapevines are considered to have higher boron requirements (on a dry-weight basis) than many other crops. For bloom-sampled vines, petioles that contain less than

Figure 8.13 ▪ Boron toxicity.
Boron toxicity caused by excess rate of foliar boron application. Note the misshapened, younger leaves.

30 parts per million (ppm) are considered marginally deficient, although clear boron deficiency symptoms may not appear until the boron level drops to 20 ppm or lower. Soil pH, leachability of the soil, frequency of rainfall, and the amount of organic matter in soil affect the availability of boron. A soil pH of less than 5.0 or greater than 7.0 reduces the availability of boron. Boron is actually very soluble at low soil pH; but in sandy soils the increased solubility, if coupled with frequent rainfall, can lead to a leaching of boron from the root zone. Vines grown on sandy, low-pH soils subjected to frequent rainfall are, therefore, prime candidates to express boron-deficiency symptoms. Boron forms complexes with soil organic matter. Topsoils, which generally contain more organic matter than do subsoils, provide vines with the bulk of their boron needs. If the topsoil of the vineyard is eroded, the availability of boron may be reduced. Furthermore, droughts intensify boron deficiency, probably because the topsoil dries sooner than the subsoil. This drying pattern reduces the vines' ability to extract nutrients from the topsoil even though moisture and some nutrients can be obtained from the relatively moist subsoil.

Assessing the Need for Boron Fertilizer

The foremost consideration in correcting boron deficiency is to determine whether the vines are actually deficient. Excess boron uptake leads to pronounced leaf burning and leaf cupping. Therefore, it is imperative not to apply boron unless it is needed. Routine, bloom-time petiole sampling should be used to determine the vines' boron status.

Correcting Boron Deficiency

If plant boron levels are low, corrective measures can be made in the following season (appendix 1, page 296). Confirmed deficiencies are corrected by making two consecutive foliar applications of soluble boron fertilizer. The first application is made about 2 weeks before bloom. The second is made at the start of bloom but no earlier than 10 days after the first application was made. Apply no more than 0.5 pound (one-half pound) of actual boron per acre in each spray using enough water to cover the flower clusters thoroughly. It is important not to exceed this rate of application or to reduce the 10-day interval between consecutive applications, as a phytotoxic response to the applied boron might be observed (figure 8.13). Solubor-20 is a borate fertilizer that contains about 20% actual boron. Thus, 2.5 pounds of Solubor-20 should be applied per acre to provide the 0.5 pound of actual boron.

The water-soluble packaging of certain fungicide and insecticide formulations reacts with boron to produce an insoluble product. Therefore, boron should not be tank-mixed with pesticides packaged in this manner or with any pesticide that cautions against boron incompatibility. Foliar application of boron is a temporary solution but has the advantage of avoiding a potentially excessive soil application; however, with proper calibration, boron can be applied in soluble form to the soil with irrigation equipment or with an herbicide sprayer. In this case, the material should be applied at a rate of 2 to 3 pounds of actual boron per vineyard acre. Soil applications can be made at any time of the season, but their effect will be delayed until the boron reaches the root zone. Dry formulations of boron, such as borax, are difficult to uniformly apply to the soil because of the small quantities used.

Other Nutrients

Other essential elements are generally found at or above sufficiency levels in established vineyards and are usually of minor concern. If low, soil levels of the trace elements iron (Fe), manganese (Mn), copper (Cu), zinc (Zn), and molybdenum (Mo) can be raised to optimal levels in the preplant phase. Many of the trace elements can also be applied to foliage following guidelines provided in appendix 1 (page 296). Iron deficiency (figure 8.14, page 168) occurs in some isolated situations in the East, typically where soil pH exceeds 7.0, as in certain marine deposits around the Great Lakes, and where soils remain cold and wet well after bud burst in the spring. Correction of iron deficiency can be approached in several ways (appendix 1), including acidification and improved drainage of the soil. Foliar, chelated Fe fertilizers have a limited, short-term effect.

Calcium is rarely deficient in plant tissue analyses; but if so, soil Ca and soil pH can both be increased through the addition of calcitic lime. If a soil pH increase is not

Figure 8.14 ▪ Iron deficiency.

Iron deficiency in Concord. Leaves turn yellow starting with the shoot apex.

desired, Ca can be increased by soil application of gypsum (appendix 1, page 299).

Occasionally, tissue analysis results will show excessive levels of certain micronutrients such as iron, manganese, zinc, or copper. Those elevated levels are usually caused by residues of fungicides that contain those elements, not by excessive root absorption. Some tissue analysis labs also include other elements, such as sodium (Na) and aluminum (Al), on test results. Those values have little meaning, however, for most vineyards.

CONCLUSIONS

Achieving and maintaining adequate vine nutrition is one component of sound vineyard management. If one or more nutrients are deficient, vines will not achieve optimal yields and fruit quality, and the grower will not maximize returns from his or her investment. Good vine nutrition starts in the preplant phase and extends through the productive years of the vineyard. It requires recognition of visual deficiency symptoms and the use of specialized soil and plant tissue analysis techniques. Ideally, a grower should apply fertilizers when needed on a maintenance schedule and should not wait until a nutrient deficiency is expressed. The producer must also be willing to apply lime and other fertilizers efficiently where they are needed. Considering the low cost-to-benefit ratio of most fertilizers, that should not be a difficult management decision.

References

Christensen, L. P., A. N. Kasimatis, et al. (1978). *Grapevine Nutrition and Fertilization in the San Joaquin Valley*. Berkeley, CA, University of California.

Goldspink, B. H., N. Lantzke, et al. (2001). *Fertilisers for Wine Grapes*. South Perth, Western Australia, Western Australia Department of Agriculture.

Marschner, H. (1986). *Mineral Nutrition of Higher Plants*. San Diego, Academic Press Limited.

Taiz, L. and E. Zeiger (1991). *Plant Physiology*. Redwood City, CA, Benjamin/Cummings Pub. Co., Inc.

Tisdale, S. L., W. L. Nelson, et al. (1985). *Soil Fertility and Fertilizers*. New York, Macmillan Publishing Company.

Vos, R. J., T.J. Zabadal, and E.J. Hanson. 2004. Effect of nitrogen application timing on N uptake by Vitis labrusca in a short-season region. *American Journal of Enology and Viticulture*. 55:246–252.

—9—

Grapevine Water Relations and Irrigation

Authors
David S. Ross, University of Maryland
Tony K. Wolf, Virginia Tech

Water comprises over 90% of grapevine green tissue fresh weight and is essential to the biochemical (photosynthesis and respiration, for example) and biophysical (such as stomatal conductance and turgor pressure) processes that drive grapevine growth and development. Few would dispute the idea that water is essential to grape developmental processes. The necessary quantity of water, and the point at which that moisture is available to the grapevine, however, represent active areas of viticulture research and industry debate. Mature grapevines that are grown in deep soils have extensive root systems and are fairly tolerant of mild droughts that might affect shallow-rooted, annual plants. Nevertheless, a certain amount of moisture is necessary to support growth and development. Lacking sufficient moisture, vines will suffer drought stress, which can reduce productivity and fruit quality, and predispose the vine to additional stresses, such as winter injury. Irrigation can be used to avoid such stresses. The quantification of optimal irrigation and the timing of irrigation have been extensively studied in arid grape regions, but irrigation has received scant attention in more humid regions, such as eastern North America.

The decision to irrigate vineyards in the East is complicated by two issues. One issue is the unpredictable nature of natural rainfall weighed against the cost of irrigation system installation. The second issue is the perception by some that any irrigation is detrimental to wine quality. In reality, droughts of some duration occur in most years, and severe droughts are not uncommon in the 25+ year life of a vineyard. Drought of any magnitude may increase the risk of vine stress and may therefore justify the use of irrigation to mediate that stress. And, while irrigation represents a substantial input in an already expensive enterprise, the benefits can outweigh the costs, especially for vineyards that are chronically at risk of drought stress.

The goals of this chapter are to examine the factors that govern water supply and demand in vineyards; to assess risk components associated with drought stress; to discuss irrigation systems and preparation for system design; to describe the principal components of vineyard irrigation systems; to explain methodologies for calculating water needs in the established vineyard; and to discuss methods for soil moisture determination and irrigation scheduling.

WATER SUPPLY AND WATER DEMAND

Vineyard water inputs and outputs comprise the vineyard water balance which is influenced by competing processes of supply and demand. Water enters the vineyard as rainfall or through irrigation (see figure 9.1). Some of this moisture drains out of the root zone into deeper soil layers and some runs off the soil surface. Water that remains in the root zone is available for absorption by the vine roots. This water supply is affected by several factors, including the water holding capacity of the soil, effective rooting depth of soil, competition with weeds, and evaporation of water from the soil surface. Water demand by grapevines, in turn, is affected by meteorological conditions, leaf area of the vine, crop level, and perhaps, by some cultural practices such as training system and shoot orientation.

The amount of water held by soils is determined by their textural properties. (See chapter 2, page 14 and Scheduling Irrigation, page 187.) Briefly, clays and clay loams have greater water-holding capacity than do sands and sandy loams. The amount of water available to plants, typically expressed in the United States as inches per foot of soil depth, is a function of both the water holding capacity of the soil and the rooting depth of the soil. For example, a deep sandy loam soil may contain the equivalent of 1.3 inches of plant available water per foot of depth. Recall from chapter 2 that plant available water (PAW) comprises water available to plants in the range from field capacity to the wilting point. If the effective rooting depth of the soil series was 4 feet, the amount of PAW in that soil would be 5.2 inches (1.3 inches/foot x 4 feet). Later we will discuss irrigation water requirements on a volume basis (gallons) by converting inches of water to acre-inches. One acre-inch is the volume of water (27,154 gallons) contained in one inch of water spread over an acre (43,560 square feet).

The rooting depth of a soil may be very deep, as in the case of some alluvial and colluvial deposits. The volume of PAW in such cases can be substantial, and the need for supplementary moisture through irrigation may be relatively low. At the other extreme, shallow soils overlaying impenetrable rock or other textural discontinuities may have extremely limited PAW. Vines planted to such restrictive soils tend to suffer drought stress sooner than those planted to deeper soils. Even in deep soils, however, young vines have limited rooting and, therefore, are subject to greater drought stress than are older, more extensively rooted vines. The volume of soil exploited by grapevine roots may also vary with choice of rootstock; some rootstocks tend to be deeper rooting than others.

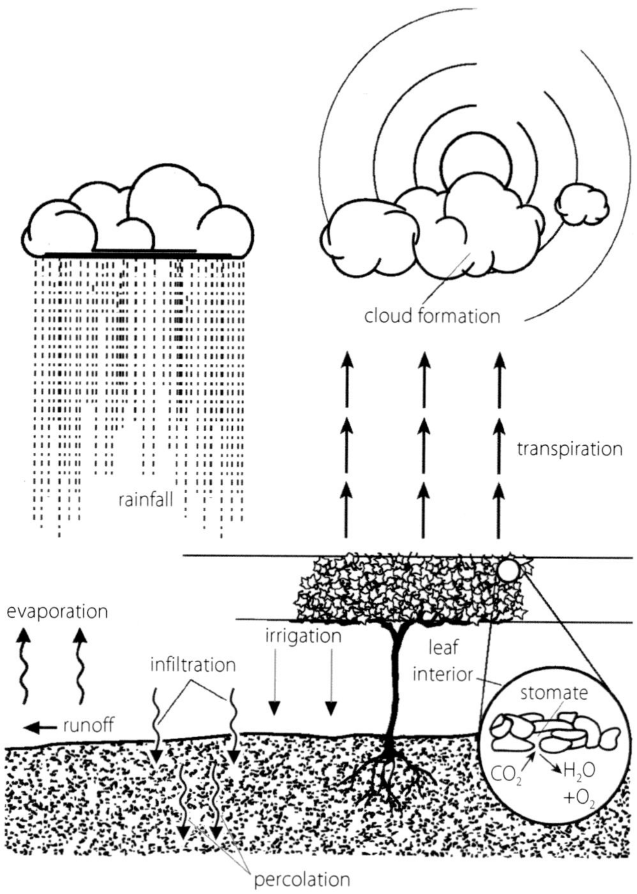

Figure 9.1 ▪ The vineyard hydrologic cycle.
Water enters the vineyard as rainfall or irrigation and is removed through gravity, runoff, evaporation, and transpiration through plant leaves.

Weeds and intentionally planted ground covers compete with vines for available moisture and consequently affect the water supply. While permanent cover crops may be desirable in situations of high vine vigor, the same cover crops may exert an undesirable level of competition during periods of drought. Under western New York State growing conditions, row middle cover crops were estimated to use 2.0 to 2.5 inches of water per month (0.50 to 0.63 inches per week) in midsummer (Lakso and Pool, 2001). To put this into perspective, if our hypothetical sandy loam soil described above had 5.2 inches of PAW at field capacity, the cover crop could be expected to use from 2.0 to 2.5 inches of that soil moisture per month during midsummer.

We've discussed the water supply issues; now let's look at water demand. A vineyard soil at field capacity (the amount of water that the soil can hold after gravitational drainage occurs) will lose moisture in two principal ways:

through direct evaporation into the atmosphere, and through transpiration from the leaves of the vines and any ground cover (see figure 9.1). Water moves out of the leaves through stomata, the small pores that admit carbon dioxide and release water vapor and oxygen. The driving force for this loss of plant moisture is the sum of those meteorological conditions that promote plant water loss, especially high temperature, wind, incident radiation, and low humidity. Collectively, transpiration and evaporation are referred to as *evapotranspiration* (ET).

In addition to being affected by meteorological conditions, the amount of water transpired by grapevines will be affected by vine leaf surface area and training system. Training systems that increase the surface area of canopy exposed to direct sunlight, such as Geneva Double Curtain, will have higher water demands than will single canopy systems. Actual water use by vines is estimated at 0.7 to 0.9 inches per week for vinifera and Concord vines, under New York State conditions. Vine water use is probably somewhat higher in warmer areas of the East—perhaps on the order of 1.0 to 1.1 inches per week in southeastern Pennsylvania and Virginia and as much as 1.3 or more inches per week further south. The amount of crop will also affect water demand. Under mild drought stress, Concord grapevines cropped at 8.0 tons per acre showed a more pronounced depression in sugar accumulation than did vines cropped at 4.0 tons per acre.

If we add evaporative soil moisture loss and the transpirational loss of moisture from weeds, cover crops and vines, the midsummer ET rate from Eastern vineyards is on the order of 1.3 to 2.0 inches per week. Evapotranspiration rates will vary from a minimum early in the growing season to a maximum in midsummer, at or before veraison. The increase in ET is due to increasing leaf area (and transpiration) of the vines, higher evaporative losses from soil, and increased transpiration from ground cover during the hottest period of the season.

SUMMER CLIMATE AND THE POTENTIAL FOR DROUGHT

Agricultural meteorologists and climatologists use the expression *potential evapotranspiration*, or *PET*, to compare the water loss potential of different regions. PET is a combined measure of the potential evaporation of water from soil and from leaf transpiration. Evapotranspiration rates for vineyards vary according to the development of the vine canopy, presence or absence of ground cover, cultivation, and atmospheric conditions. Monthly precipitation is less than PET losses during summer months for many eastern locations, as illustrated by the data for Dulles Airport in northern Virginia (see figure 9.2). Averaged across all of Virginia's National Weather Bureau stations, PET values exceed rainfall by an average of 1.5 inches per month during July.

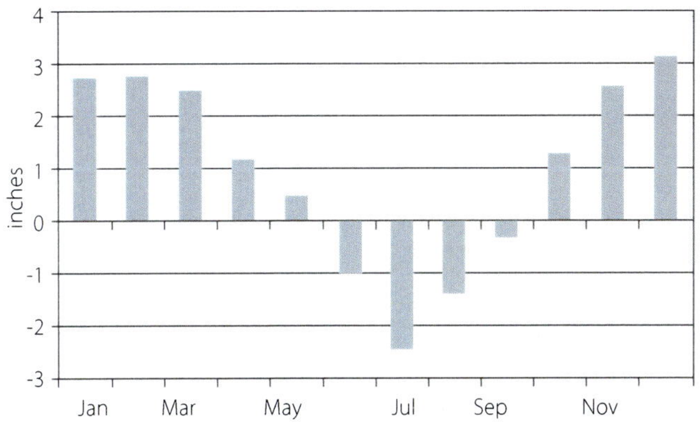

Figure 9.2

The imbalance between precipitation and potential evapotranspiration (PET) is illustrated for Washington Dulles International Airport in northern Virginia. *Source:* HTTP://CLIMATE.VIRGINIA.EDU

Precipitation records indicate that many eastern weather stations record 40 to 60 inches of precipitation per year; however, these averages do not reflect the frequency of rainfall. Even monthly precipitation averages can give a misleading impression of moisture availability. Summer precipitation in this region often results from thunderstorms. These storms are usually restricted to small areas, and significant precipitation might only affect a 10- to 50-square mile area. Furthermore, because rainfall during thunderstorms is intense, less water is absorbed by the soil than if an equal amount of precipitation fell over a longer period. Thus, infrequent summer downpours may not satisfy the vines' critical need for moisture that will develop during extended hot, dry periods. Given high PET rates and the spotty nature of summer precipitation, summer droughts are not uncommon in this region. Depending on soil moisture supply, the droughts can lead to vine stress.

The interested reader can monitor the severity of drought in the eastern United States through WWW sites operated by the National Oceanic and Atmospheric Administration (NOAA) and by following links for National Weather Service and Crop Moisture Index (CMI). The CMI is a short-term drought index that factors available soil moisture holding capacity, rainfall, evapotranspiration, temperature and other factors into a weekly, graphic assessment.

THE ROLE OF WATER IN THE VINE

To an extent, all physiological processes in the plant are dependent upon water. Most grape growers recognize that grapevines absorb carbon dioxide and liberate oxygen, but they may not recognize that most of the liberated oxygen is derived from water. In the larger scheme of plant processes, water plays a pivotal role in driving growth. The cells of adequately watered vines exert an outward hydrostatic pressure, which is termed "turgor pressure." This pressure causes cell enlargement, which, in turn, leads to an increase in tissue and organ size, as exemplified by the lengthening of shoots. The loss of cell turgor pressure results in a flaccid or wilted appearance. Wilting occurs when the transpiration rate of leaves exceeds the ability of the vine to absorb water from the soil and conduct it to the leaves. Water stress or, more accurately, *drought stress* occurs when growth processes are impaired. As we'll see shortly, some degree of stress, imposed at the appropriate time, may be desirable from the standpoint of optimizing fruit and wine quality. The practical challenge is to know how much stress is appropriate and when it should occur.

SYMPTOMS OF DROUGHT STRESS

One of the first signs of drought stress is a change in the appearance of the vines. Rapidly growing shoot tips of well watered vines appear soft and yellowish or reddish green (see table 9.1). As soil moisture becomes limiting, shoot growth slows and shoot tips turn grayish-green, newly formed nodes appear compressed, and shoot tips abort. Tendril drying and abscission serve as a useful early indicator of vine drought stress. As stress intensifies, leaves appear wilted, particularly during midday heat. Under prolonged and severe stress, leaves curl, brown, and eventually abscise. Vines that suffer severe drought stress begin to defoliate, exposing more of the fruit that had

Table 9.1 ▪ Visual Indications of Increasing Drought Stress in Grapevine

Observation	Surplus moisture	Slight to moderate drought stress	Severe drought stress
Tendrils	Turgid, extending well beyond shoot tip horizontally or upright	Drooping or wilted	Yellowed, dried, or abscised
Shoot tips	Actively elongating	Compressed	Aborted
Leaf orientation to mid-day sun	Blade is perpendicular to incident sunlight, receiving full sun	Leaves appear to droop, blades not oriented to receive full, direct sunlight	Leaves may be curling or actually dried
Leaf temperature (check with infrared thermometer or simply press between palms of hands)	Cooler than our body temperature, even at mid-day (at or below ambient temperature)	Warm to touch at mid-day (> 100ºF)	Much greater than 100ºF
Leaf color (basal to mid-shoot leaves)	Vibrant green	Grayish-green to light green	Light green or yellowing
Fruit cluster	Normal berry set and turgid berries	Set may be reduced	Cluster rachis tips may dry if stress occurs during bloom; fruit set may be reduced; berries may become flaccid if severe stress occurs post-veraison

been shaded by foliage. Time and severity of the water shortage can affect berry size, and drought-stressed fruit exposed to the sun can shrivel and experience sunburn. Water shortages also reduce the vine's ability to absorb nutrients from the soil (see chapter 8, page 141). Symptoms of nutrient deficiencies are therefore more apparent during prolonged dry periods.

In addition to visual indicators, special instruments can be used to measure vine water status. Some instruments measure the water status of vines, whereas others measure the moisture status of the soil. Hand-held infrared thermometers can measure the temperature of vine canopies. The leaves of water-stressed vines are often warmer than the surrounding air because of reduced transpirational cooling. Leaves of well watered vines are generally cooler than the air, even during the hottest period of the day. The moisture status of the soil can be determined with instruments that range from simple tensiometers to sophisticated neutron probes (see Soil Moisture-based Measurements, page 190).

The water status of vines also can be measured by determining how much pressure is required to force water from a detached leaf. A wilted leaf will hold its remaining moisture with more tension (negative pressure) than will a fully hydrated leaf. The tension with which a leaf holds water is expressed in units of negative pressure called bars (1 bar = 14.5 pounds per square inch (psi)), or kiloPascals (kPa) (1 bar = 100 kPa), or megaPascals (MPa) (1 bar = 0.1 MPa). Leaf water tension may approach 1500 or more kPa (1.5 MPa) before the plant suffers irreversible injury.

Leaf water potentials become more negative throughout the course of a day as the leaves lose moisture. The leaf water potential is generally most negative during the hottest part of the day and then decreases (becomes less negative) as vines regain their hydrated status in the cool of the night (see figure 9.3). When leaf water potentials reach about −1200 kPa, most grapevine leaf stomata will be closed. This closure conserves the remaining water in the leaf, but the "cost" of this water conservation is decreased sugar production. With stomata closed, carbon dioxide cannot enter the leaf and the photosynthetic conversion of carbon dioxide into sugars will not occur. Extended periods of drought prevent the vine from regaining its hydrated status. Drought-stressed leaves remain at or below −1200 kPa (−1.2 MPa) for much of the day, and, consequently, photosynthesis is greatly reduced. The impairment of the photosynthetic processes will generally occur before leaves are visibly wilted. Reduced photosynthesis can explain why fruit fails to increase in soluble solids during periods of water shortage; little or no sugar is being manufactured. At some point during a drought the daily stress of insufficient water will have an irreversible impact on the vine's performance. By the time leaf wilting occurs, vines are severely stressed.

Many processes are disturbed or impaired by drought stress. The impairment of those processes depends on the severity of stress and can be characterized as either reversible or irreversible. Reversible effects include:

- decreased cell turgor pressure
- reduced stomatal conductance (that is, less carbon dioxide enters the leaf)
- reduced photosynthesis (sugar production)
- decreased shoot growth rate
- reduced berry size

These events are "normal" occurrences in the day-to-day cycle of growth and development even with adequately watered vines. As drought stress intensifies, however, irreversible effects occur. Irreversible effects, in order of increasing stress and severity, include:

- decreased fruit set
- irreversible reduction in berry size
- *delayed sugar accumulation in fruit*
- *reduced bud fruitfulness in the subsequent year*
- reduced fruit coloration

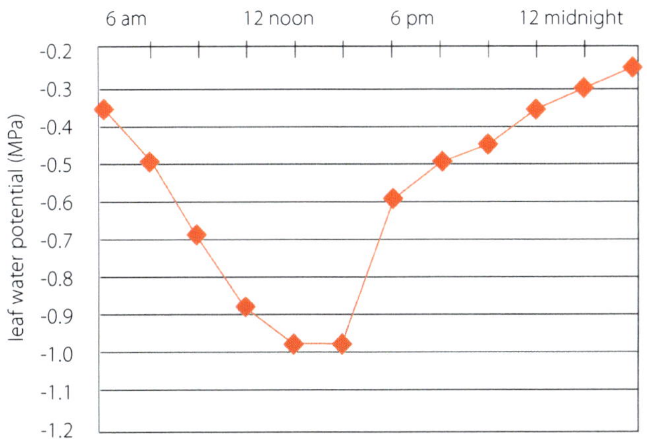

Figure 9.3 ▪ Changes in leaf water potential throughout the course of a day.

Degree of leaf stress is indicated by increasingly negative values.

- leaf chlorosis (yellowing) and eventual burning
- berry shriveling
- reduced wood maturation and possibly reduced vine cold hardiness
- defoliation
- vine death

The two observations indicated by italics are of special interest because their occurrence is variable. Slight drought stress may indirectly hasten sugar accumulation and increase bud fruitfulness by creating a somewhat more open, light-porous canopy. Exposed fruit tends to accumulate sugar at a faster rate than shaded fruit, especially in cool grape regions (see chapter 6, page 124). Furthermore, slowed vegetative growth reduces the "sink" strength of shoots and roots. Thus, more of the vine's carbohydrates are directed to fruit "sinks." Slight drought stress, therefore, might result in hastened fruit maturation. However, with greater stress, sugar, flavor and aroma accumulation is impaired and fruit does not attain optimal quality. Buds exposed to sunlight during their development are more fruitful than those that are shaded; however, severe drought stress reduces the fruitfulness of developing buds and thus reduces crop yields in the subsequent season. As we'll explore in the following section, the optimal vineyard water status seeks a balance between providing sufficient water to meet basic growth requirements but not so much water that vegetative growth is excessive.

WATER BALANCE AND FRUIT QUALITY

It should be apparent that we want to optimize the balance of water available to vines so as to achieve a desired balance between leaf area and crop, as defined in the Canopy Management chapter (page 124). At one extreme, grape growers in the East often deal with excessive rainfall; vegetative growth is unrestrained, requiring the use of remedial canopy measures of shoot hedging, selective leaf removal, and elaborate training systems. On the other extreme, vines that are subject to drought stress suffer a range of problems including reduced yields, inferior vine size, and poor fruit composition. Wines produced from grapes of drought stressed vines may exhibit an increased likelihood of atypical aging aroma. All of these features depend somewhat on the timing of drought stress. An ideal strategy of water supply is shown in table 9.2.

Again, we are relatively powerless to avoid problems with excessive rains; however, a functional irrigation system does at least empower us to avoid the problems caused by severe droughts.

AVOIDING VINE DROUGHT STRESS WITH IRRIGATION SYSTEMS

A properly functioning irrigation system ensures that vines have adequate moisture when they need it (see table 9.2). As stated earlier, the objective of irrigation is to supplement natural precipitation so that vines achieve adequate vegetative growth and berry development. However, the irrigation system must be designed and built to be able to supply all the plant water requirements in an extended drought. Moisture monitoring methods and tools and irrigation scheduling are required to monitor the soil moisture and to provide irrigation sufficient to avoid drought stress. It is important to avoid applying too much irrigation water, as this is both costly and may be counterproductive (see table 9.2).

Drip irrigation systems are the most common systems for providing irrigation to eastern North American vineyards. Overhead sprinklers are much less efficient and, for various reasons, are seldom used (see table 9.3).

As shown in table 9.3, drip irrigation is the preferred irrigation system for eastern vineyards; however, there are some disadvantages to drip irrigation. Moisture distribution may be restricted in some soils, but grapevines and other plants can gather the water they need from a partially wetted root zone. Water distribution depends on the type of emitter and the soil type. Water travels downward more quickly in sandy soils, but the water spreads out more horizontally in heavier soils such as clay or clay loam. Frost protection is not provided by most drip systems. Rodents, insects, and equipment may damage the drip emitters or laterals; damage is dependent on the placement and type of the unit. Some drip systems may require a higher investment than that represented by an overhead sprinkler. This is particularly true with an automated system. The advantages of drip, including cost, however, will normally out-weigh those of overhead sprinklers if the primary purpose of the system is to provide irrigation water.

Table 9.2 ▪ Water Management Strategy for Maintaining a Balance between Vegetative Growth and Fruit Development in Current and Following Years

Stage of growth	Generally desired water status	Consequences of excess water	Consequences of drought stress
Bud break	50% or more of plant available water (PAW) in root zone	Temporary iron chlorosis and reduced nitrogen utilization	Delayed budbreak, slow shoot development
Pre-bloom through bloom	Minimal stress; 50% or more of PAW in root zone	Excessive shoot growth rate and canopy volume with shaded flower clusters; poor fruit set; increased likelihood of powdery mildew infections of clusters	Insufficient leaf area development and/or impaired photosynthetic rates; poor fruit set and poor flower bud development for following year
Post-bloom to veraison	Sufficient water to allow full canopy development, then mild to moderate stress to slow shoot development; leaves function at full capacity	Shoots require frequent hedging; lateral shoots abundant and vigorous; fungal diseases aggravated by dense canopy	Impaired photosynthetic function, reduced berry size and sugar production; reduced crop in following year
Post-veraison	Minimal stress; sufficient moisture to retain leaf function but not stimulate shoot elongation; ripening advanced and fruit/wine quality optimized	Shoot growth continues, fruit ripening delayed, disease incidence may be increased; berry splitting may occur	Fruit desiccates, leaves non-functional and fruit ripening impaired; canopy yellows or leaves are shed; winter cold hardiness impaired
Post-harvest	Mild stress, no new shoot development; leaves normally senesce with decreased temperatures; plant cold hardiness optimized	Shoot growth may continue until cold weather or frost; periderm maturation is delayed; vines may be predisposed to winter cold injury	Severe stress may impair acquisition of vine cold hardiness.

Drip systems can include different materials and placement for the laterals along the row of grapevines. A lateral is the drip line that runs along each row of grapes and is attached to a header or manifold at one end of the field. The options for laterals are described below.

Temporary Laterals

Drip lines can be placed on the ground next to the row of vine trunks to get plants started before a trellis system is installed. These temporary lines are usually a 1- or 2- year service thin wall vegetable row crop tubing of lighter weight and lower cost than the products installed for the life of the grapevines. A row crop tape is a thin-wall tape with the emitter built in at regular spacing as the tape is manufactured. The tape walls may be 4 to 18 mil thick (1 mil = 1/1000th of an inch); the heavier the wall, the longer the tape is expected to last. The emitter usually is formed of the plastic sheet, but individual emitters are also inserted into thin wall tubes as they are extruded.

Emitters are the small orifices through which the water is discharged. An emitter contains a designed flow path that regulates the flow of water. Some are pressure-compensating, but for many the discharge rate is influenced by the water pressure. Pressure compensating emitters use an internal diaphragm that moves with pressure to vary the water pathway to regulate the flow. Higher pressure causes more restriction of the pathway, but the same amount of water passes through.

With the emitters at regular intervals, the tape or tube can be installed along the row of grapevines quickly to give a wetted strip down the row. The tape can be removed when trellis posts are set. Temporary lines are, however, subject to more mechanical damage from rodents, insects, and people, and may expand and contract significantly during changes in air temperature. These are used when drip irrigation is intended only to establish the planting. Most of these drip tapes are not pressure compensating; however, there are relatively thin-walled tubes with pressure compensating emitters. Pressure compensation is important on slopes greater than 3% to 4%.

Table 9.3 ▪ Advantages and Disadvantages of Drip Irrigation Compared to Overhead Irrigation for Vineyards

Water storage capacity	Drip is more efficient and requires less water than an overhead system. A smaller water supply will be required and a smaller flow rate will be adequate.
Water quality	Drip requires extremely clean, particle-free water. Surface water needs good filtration. Newer emitters do have more ability to pass particles and self-clean than older drip products.
Pump and pipe network	Drip requires a lower operating pressure and lower flow rates; equipment will be sized smaller and less energy will be used. Polyethylene pipes distribute the water along each row.
Water use efficiency	Drip applies the water along the crop row in the root zone, rather than broadcast over the entire field. Much less water is lost to evaporation in the air or from plant foliage than from overhead sprinklers. Overhead sprinklers may lose up to 25% of applied water to evaporation.
Disease/weed management	Diseases that develop when free water is on leaf and fruit surfaces do not develop when drip irrigation is used because the water is applied to the soil. Weeds are less problematic between rows and in non-irrigated areas. However, weeds can occur near emitters.
Labor/automation	Drip irrigation generally requires less labor and can be extensively automated. Field operations can continue during irrigation.
Wind	Water can be applied by drip under windy conditions.
Fertilizer	Drip can distribute soluble fertilizer. Broadcast fertilizer may not be efficiently moved into root zone by drip.
Moisture distribution	Drip has more limited distribution, but for grapevines a portion of the roots can retrieve enough water to maintain whole plant moisture status.
Frost protection	Drip is not effective for frost protection. Overhead sprinklers need to be designed to provide proper frost protection. Microsprinklers may be available to help give some coverage.
Rodents/insects/people damage	Both systems can have trouble with clogging. Drip systems can be damaged more easily, but they are relatively easy to repair.
Cost	Drip systems may require more investment, particularly when automated.

Above-ground Laterals

After the trellis system is in place, many growers like to place a permanent drip lateral on a trellis wire about 18 inches above the ground. A low-density polyethylene pipe or heavy-wall tubing is used for this purpose, and it is fastened to the trellis wire. The pipe looks like the polyethylene pipe one can buy in hardware stores, but it has a different chemical makeup that resists the effects of heat and freezing. Also, it has more carbon black in it to resist the effects of the sun. The pipe may have emitters built into it at regular intervals, or individual emitters can be installed at the necessary spacing. The water drips down onto the ground so the wetting pattern can be observed. This method allows the trellis wire to hold the pipe up out of the way of cultural practices, and allows for observation of the emitter operation. In addition, with this method there is less danger of damage by rodents, insects and machinery operation.

Underground Laterals

Subsurface drip irrigation is popular for many crops and is used in vineyards in more arid regions. The drip line with emitters is buried to one side of the vineyard row. Water is delivered underground, thereby eliminating the loss of water to evaporation. There is always some danger of hitting the buried lateral with tillage equipment, and it is more difficult to visually monitor the operation of the lateral, although pressure gauges and flow meters serve as monitoring tools.

Microsprinklers

Small sprinklers that throw the water within a circle of only a few feet can be placed on posts above the crop for the purpose of applying water to the foliage for crop cooling or for frost protection. This has been done in some parts of the country on a limited basis. There is still the problem of icing of the sprinkler and a flow rate inadequate for the coverage needed for frost protection. The water protects the tissue from freezing by a continuous process of ice formation, which is an exothermic, or heat-releasing, process. If the formation of ice ceases at subfreezing temperature—if water is no longer available—the plant tissue may quickly freeze. Sprinkler protection against frost injury, once started, must be continued until air temperatures rise well above freezing. The volume of water needed can be significant.

IRRIGATION WATER QUANTITY AND QUALITY CONSIDERATIONS

The first step in planning an irrigation system is to determine a source and volume of high quality water. Water can come from a pond, spring, stream, river, or a well; each source has unique characteristics to consider. Surface water sources (ponds, streams, or other impoundments) are acceptable but can contain suspended inorganic and organic matter that might cause problems. During summer months, when water is most needed, a stream or spring may be at low flow and inadequate for irrigation. Wells are generally the cleanest water source and are fairly consistent in flow but may also dry up in droughty years. It is important to check on the flow level during the dry summer months before planning to use this water source. Another source is municipal water close to the field. It is a clean source but does have a more direct cost that is noticeable when the water bill arrives.

The legal right to withdraw large quantities of water must be examined. In the eastern United States water rights are called riparian, or landowner's rights, because anyone is entitled to use any water associated with land ownership. However, landowners downstream also claim rights to water flowing past their property, so rights are subject to reasonable use interpretations. In the western states water is only available through appropriative rights. Permits may be required before surface or ground water can be withdrawn for irrigation. Check on local permit requirements.

Water Quantity

An estimate is needed of the amount of water that will be available for a growing season before an irrigation system can be designed. During a drought a stream or well flow may be reduced considerably so a mid- or late-summer flow record is important. An estimate of the amount of water available is compared to what is needed for irrigation.

A professional irrigation dealer or designer should design the system, but since one might consider planning for their own system, the procedures for estimating water quantity will be given here. One should observe well yield, pond recharge, or stream flow in the summer to know its seasonal characteristics. If a designer gets involved, he or she will need to know the summer flow or yield characteristics.

A well driller or well records can provide the yield and the maximum rate at which water can be pumped from the well. The flow rate of water available may be the limiting factor in the irrigation system design—water cannot be used faster than it's available. However, water can be stored or accumulated for 24 hours a day for one day's use. "Intermediate storage" of water in a large tank, lined reservoir, or pond can be used to take advantage of small wells or streams in this way. Be careful to check state law before taking all the water in a stream because neighbors downstream may be legally entitled to their share of the water.

Pond water volume can be estimated by taking some measurements and calculating the volume of stored water. The flow of water into the pond might also be estimated unless subterranean springs are the main source. For a pond, determine the surface area by using geometric shapes, such as squares, rectangles, and triangles, to describe its surface area. Take measurements, in feet, of sides and use the area formulas. Total the surface areas to get an estimate. Then determine an average depth. Pond depth can be determined with a boat and by taking depth measurements at regular intervals along a line across the pond. Average the depth readings. Multiply the average depth in feet times the surface area in square feet to get cubic feet. Multiply by 7.5 gallons per cubic foot to get gallons. On a pond that has not been silted-in, the average depth can be estimated by multiplying 0.4 times the maximum depth.

A stream or river flow can be estimated by determining its cross-sectional area and the average velocity of the water in the stream or river. Start by taking water depth measurements, in feet, across the waterway at regular intervals. Select an area where the stream is fairly uniform for a short distance. The average of these depth measurements gives the average depth. Multiply the average depth by the width to get a cross-sectional area, in square feet. Then, determine the stream velocity by throwing a wood chip or stick into the middle of the stream and measuring the distance, in feet, the object floats in a minute. Estimate the average stream velocity by multiplying this number by 0.7. Now, multiply the average stream velocity times the cross-sectional area to get the stream flow in cubic feet per minute. Convert stream flow to gallons per minute by multiplying by 7.5.

Water Quality

Water quality is important, particularly for drip irrigation. Drip emitters can be clogged by particulate or organic matter, such as algae and bacterial slimes. Dissolved minerals can form precipitates when exposed to air or temperature changes. Iron and manganese are a problem in some areas. Hard water that contains high levels of calcium and/or magnesium can cause white film deposits such as calcium carbonate in pipes. A standard irrigation water quality test should be performed early on to determine potential problems with the water source. An annual water analysis during periods of low flow (supply) is suggested to verify the quality and to catch any changes. Most private water testing companies offer this service. Surface waters can be sampled and allowed to settle out any silt or debris before testing. Surface water quality is subject to storms that can cause silt and debris to be added to the flow, so note the weather when sampling.

Water quality is more important when using drip irrigation because of the small orifice size of the emitters. A good filter, 100 to 200 mesh or as required by the drip product manufacturer, is essential for keeping the irrigation water free of debris and silt. Filters are discussed in more detail below under Drip Irrigation Components, page 181.

CROP IRRIGATION REQUIREMENTS

For system design purposes, it is important to determine the required maximum capacity of the irrigation system for delivering water. This is different from the task of irrigation scheduling that will be discussed later. Some of the terminology is the same but the goal is different.

While the humid East frequently averages adequate annual rainfall, there are often periods of drought during the growing season and an irrigation system must be sized to supply the water requirements of the crop. As stated earlier, the midsummer ET rate from eastern vineyards is on the order of 1.3 to 2.0 inches per week.

The water requirements of specific crops, including grapevines, can be calculated as a fraction of a reference Potential Evapotranspiration, ET_o, which is the potential evapotranspiration of a reference crop. By convention, the reference crop used is a cool season grass with a 3- to 6-inch sward grown on deep soil and under well watered conditions. For various reasons, the actual ET rate of our vineyard will normally be somewhat less than the ET_o rate. The vineyard ET rate is therefore adjusted for grapevines by multiplying by a crop coefficient (K_c) that reflects the specific vineyard conditions. Grapevine K_c values, which do not have units of measure, will vary according to the extent of seasonal canopy development, which increases during late-spring and early-summer, and also to the presence or absence of a row middle ground cover and weeds or other vegetation under the trellis. The K_c values increase with increased ground cover and grapevine canopy because, in either case, there's more leaf area present to transpire soil moisture. Accordingly, the crop coefficient will generally range from $K_c = 0.51$ (no ground cover) to 0.80 (40 to 50% ground cover) under eastern conditions. Stated another way, our vineyard ET rates will generally range from 50% to 80% of potential evapotranspiration (ET_o) rates. A method to determine K_c is described in appendix 2 (page 304).

Calculation of Required Irrigation Water Volume

To illustrate the calculation of vineyard water use, we will use a $K_c = 0.60$. One can estimate the vineyard water usage, ET_c, from equation 9.1.

Where ET_c is the estimated daily water use (inches of water) by the grapevines (also simply stated as "ET" in

$$ET_c = K_c \times ET_o$$

Equation 9.1

following text); K_c is the crop coefficient; and ET_o is the reference daily potential evapotranspiration rate (inches of water). Equation 9.1 is expressed in units of inches of water per day. A site-specific, electronic weather service provider or a local weather station might be able to provide reference potential evapotranspiration (ET_o) data; however, this information may be difficult to obtain in some areas of the East where irrigation is less commonly used than in the West. An ET_o in the humid East of 0.20 inch per day may be used as a midsummer approximation in the absence of any local data. Use the crop coefficient (appendix 2) to estimate the maximum water requirement in July or August and use that value for developing a water supply and for designing the drip irrigation system.

Assuming a K_c of 0.60 and ET_o to be 0.20 inches per day, we can use Equation 9.1 to derive a water consumption rate of 0.12 inches per day as follows:

$$ET_c = 0.60 \times 0.20$$
$$ET_c \text{ (or ET)} = 0.12 \text{ inches per day}$$

To estimate water requirements for a week, one can multiply the daily ET above by 7 to get 0.84 inch per week. While rainfall is normally expected to provide some of this water demand, it is best to plan for an extended drought and find the total quantity of water to be used by the crop for a production season or at least for several weeks of dry weather.

Pumped Water or Source Supply Volume

One must also account for losses in delivering the water to the crop in the water use estimates. If delivered to the crop by drip irrigation at 90% efficiency, the water source must pump 0.84 inches per week divided by 0.90 to give 0.93 inches of pumped water per week for planning purposes (equation 9.2). Overhead sprinkler irrigation systems have an efficiency of 70% to 80% and require enough water to cover all of the ground. Pumped water reflects the efficiency of the irrigation system in delivering the water to the root zone.

Pumped water = ET / Irrigation system efficiency
Pumped water = ET / 0.90
Pumped water = 0.84 inches per week / 0.90
Pumped water = 0.93 inches per week

Equation 9.2

Converting to volume of water required per crop acre, use the conversion factor of 1 acre-inch of water is 27,154 gallons of water. Weekly crop water requirements = 0.93 inches per week x 27,154 gallons per acre-inch = 25,253 gallons per week per crop acre. Multiply by the number of acres to get total irrigation water requirements for the entire vineyard.

Flow rate of water to satisfy gallons per minute (gpm) demand of irrigation system is another number to calculate, particularly when the water supply comes from a well which has a fixed maximum yield in gallons per minute. This information becomes more of a concern for an irrigation system designer but will be shown here to give a perspective on the volume of water required. To apply this amount (25,253 gallons per acre) in one 8-hour work day, divide the volume by time to apply. This would be 25,253 gallons per week/(8 hours x 60 minutes per hour) = 52.6 gpm for 8 hours of application time per acre, once each week. The time to apply must also consider the number of acres and total time in a week to run the irrigation system. Obviously, the volume of water from Equation 9.2 indicates the amount of water that a pump must be able to obtain water from a well or other source.

By way of comparison, a clean-cultivated vineyard might use a K_c = 0.51 (appendix 2, page 304). Equation 9.1 would then yield an ET of 0.71 inches of water per week. This translates to a water volume of 21,421 gallons per acre per week. The pumping rate to supply this volume in 8 hours would be 44.6 gpm.

For water consumption planning, consider soil moisture conditions at the start of the growing season and the normal rainfall distribution over the season. The number of weeks that you plan to irrigate will depend on how much insurance you want to provide and the period of time when droughts are most common. Twelve weeks of irrigation water reserve (June to August) might be a minimum starting point. The amount of water available from a well, pond, stream or spring is the ultimate limitation on developing an irrigation plan.

Deep loamy soils will store several days of moisture but sandy, shaley, or shallow soils store less water and will require more frequent irrigation to maintain the moisture level in the soil. Soils vary in their water holding capacities (see Water Budget, page 187) but 1 to 2 inches of water may be held in each foot of soil in the root zone. So, for example, in 5 feet of soil there may be 5 to 10 inches of water stored for the crop under good rainfall conditions. It is important to learn soil types, water holding capacities, and rooting depth from soil maps at the local county soil conservation district office or from a Natural Resource Conservation Service (NRCS) office.

Soil moisture sensors can be used to monitor soil moisture as an aide to irrigation scheduling (see Irrigation Scheduling) and new technologies make this less expensive and give real time information. Soil moisture storage is an advantage with a deep-rooted crop but it is important to know if there is water down 1 to 5 feet in the ground for the crop. Normally, the goal of irrigation is to bring the soil moisture up to within 50% to 80% of field capacity so the crop has adequate water for growth. When mild stress is desired (see table 9.2), the entire rooting depth should be monitored and the vines should be visually monitored for drought stress (see table 9.1), because grapevines can extract moisture from depth even if the upper soil layers are dry.

IRRIGATION SYSTEM DESIGN PREPARATION

The irrigation system is best designed by a competent dealer or other irrigation specialist. A grower may want to use his or her own time and labor for some of the installation, but it is emphatically recommended that an experienced person design the system. It is not the intention of this chapter to teach system design, but rather to give the grower an understanding of what information a system designer needs from the grower and to give the grower a basic understanding of an irrigation system.

The grower must make decisions on the vineyard layout and should compile certain information for the system designer. A grower with a 5- to 10-year plan for his or her operation is in the best position to help the irrigation designer put together a good system with expansion possibilities. The vineyard will last longer than 10 years but we assume that the irrigation system will be installed within the first 10 years.

A suggested planning process is described below.

Aerial Photograph or Field Sketch

If available, an aerial photograph can be useful for identifying the fields, water source, power source, future expansion areas, and other details of the area. The Natural Resources Conservation Service (NRCS) often can provide a copy of an aerial photograph with a normal scale of 1 inch equal to 660 feet. An alternative is a farm sketch to show the locations and dimensions of fields and distances to power and water sources. A topographic map is a handy addition for establishing elevations that will affect water pressures. A local library, your NRCS office, or web-based services may help you get a topographic map of the property. If the vineyard is on relatively flat ground, then a topographic map is unnecessary. Row orientation, row length, and field widths should be overlain on the photograph.

A detailed sketch of the planned vineyard blocks and other site improvements will assist the design engineer (see figure 9.4). Specific information should include:

- a diagram showing the vineyard blocks with their dimensions and location relative to water and power supplies
- topographic information to show contours and elevations
- general soil type, texture, and characteristics
- general information on vine water requirements. We suggest a system capacity of about 1.0 acre-inch per week for a minimum of 12 weeks as discussed earlier
- length of rows
- row direction and spacing between rows
- number of rows per block. For drip, the total number of lateral rows and total length of lateral line is needed
- vine spacing in the row. For drip systems, the emitter spacing is often a function of vine spacing in the row

As noted earlier, a grower should have a 5- to 10-year business plan developed so that the growth of the operation is defined. This allows the irrigation designer to plan expansion for the system's principal components.

Water Supply Information

Information on the water supply is important. Using the earlier discussion under Irrigation Considerations, one should determine the quantity of water available and

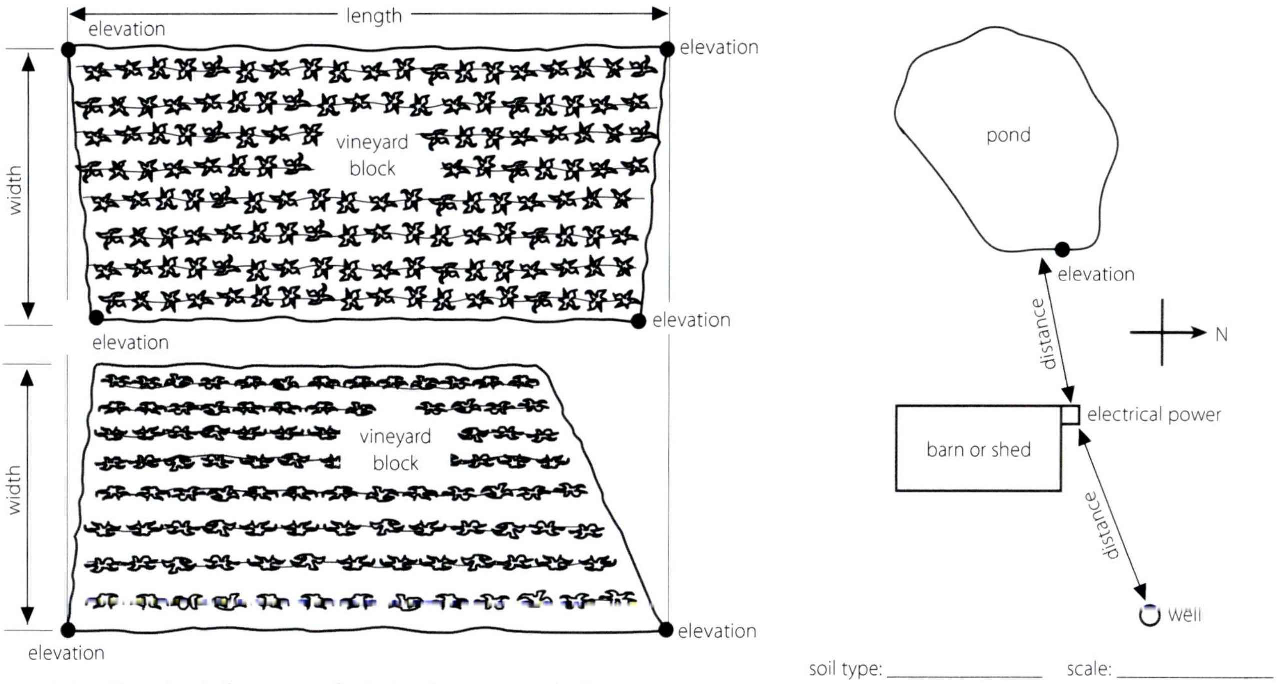

Figure 9.4 ▪ Planning information for irrigation system designer.

Adapted from: Production of Vegetables, Strawberries, and Cut Flowers Using Plasticulture, *NRAES 133.*

have an irrigation water analysis to give to the designer. If a well is to be used, find out from the well driller the pumping yield in gallons per minute. The irrigation system should be designed to provide 100% of the crop water requirements because droughts of six weeks or more can occur.

Existing Pump Details

Sometimes there is an existing pump that one wants to consider for use in the irrigation system. If so, learn as much about the pump as possible. The owner's manual may include a pump chart showing the flow rate and pressure capabilities of the pump. Note that pumps are designed for specific flow rates and pressure heads. A pump that was designed to provide high pressure for overhead sprinklers might not match the requirements for a drip irrigation system. Appendix 3 (page 305) describes some of the factors involved in sizing pumps for irrigation systems.

For small wells and other sources, a water pump controlled by a pressure tank and pressure switch may be used. This system provides continuous water available for many uses and the pump is turned on and off to maintain pressurized water. For domestic and irrigation uses, one must know the total daily yield of the well and try to schedule irrigation when domestic requirements are low. Provide backflow protection when potable water is used for irrigation so that chemicals are not drawn into a domestic water supply.

Electrical Power Availability

If possible, note the voltage and phase available, the distance of the electricity from the site, and the name and phone number of the regional power company.

DRIP IRRIGATION COMPONENTS

A drip irrigation system has many components that may be foreign to the beginner. This section will give some background on components and will serve to call attention to the need for specific items. Consultation with an irrigation designer is recommended. Seek referrals from local growers or NRCS offices. A designer should be able to explain the purpose and need for each of these items. Learn to maintain and use each component correctly. Appendix 3 reviews some of the irrigation math and physics appropriate to the discussion that follows. See figure 9.5 (page 182), which shows an illustration of the components of a typical drip irrigation system. A further description of each component follows.

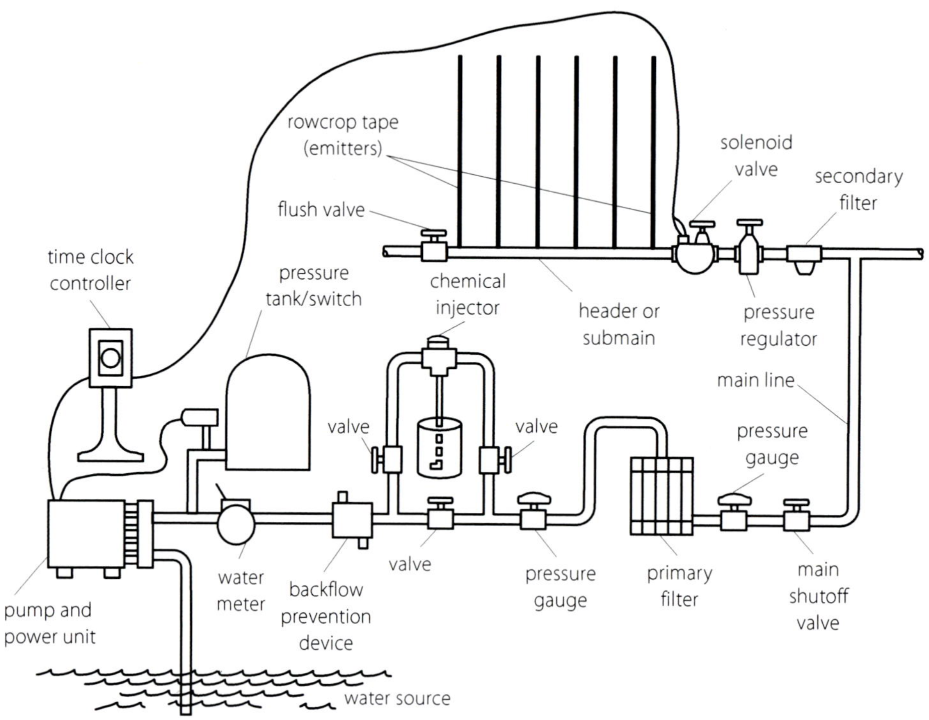

Figure 9.5 ▪ Components of a typical drip irrigation system for grapevines.

Source: Production of Vegetables, Strawberries, and Cut Flowers Using Plasticulture, *NRAES 133.*

Emitters

Emitters deliver the water to the vine from the lateral lines. Emitters come in several forms: imprinted in the forming of thin-walled rowcrop tape from sheet plastic; as individual emitters inserted into thin- or regular-walled tubing as it is extruded; and as stand alone emitters that are punched into polyethylene pipe (PE). The very thin rowcrop tapes are commonly used for only one year, as in the case of vegetable production, but heavier walled tapes may be used for several years in small fruits and vineyards. A heavy-walled rowcrop tape of approximately 8 to 25 mil may be used if the owner wished only to have irrigation during the vineyard establishment period. One disadvantage of rowcrop tapes, as reported by some growers, is that heat expansion of the tape can cause it to buckle into the row, displacing the emitters into the row middle where they may be ineffective at meeting grapevine water needs.

For permanent irrigation systems, polyethylene tubing of about 45 to 50 mil thickness is recommended and is normally installed on an irrigation wire, 18 to 24 inches above the ground. The tubing has integral emitters pre-inserted at regular intervals during manufacture. These emitters can be pressure compensating. Another option is to use regular low-density polyethylene pipe (PE) and insert individual emitters into the pipe after positioning on the irrigation wire. Use low density PE that is made for outdoor exposure in drip systems. Do not use common polyethylene found in hardware stores.

Most of these emitters operate at 8 to 10 psi and are affected by water pressure changes. Thin-walled, rowcrop tape and many individual emitters are very sensitive to pressure changes and can be used only on flat or gently sloped (1% to 3%) fields. Otherwise, water application rates will not be uniform. Pressure-compensating emitters are designed with a diaphragm that responds to pressure and changes the size and/or shape of the orifice so that water is metered out at roughly the same flow rate for water pressures ranging between 10 to 60 psi. Only pressure-compensating emitters should be used on rolling terrain where slopes are greater than 3%. Emitter manufacturers can provide data showing the coefficient of uniformity (CU) for different lengths of lateral tubing runs and for changes in elevation. A chart may suggest the maximum length of run for a given slope in order to maintain a CU of 90% or more. Discharge rates for rowcrop tape as a line source are given in gpm per 100 feet of tape. Individual emitters show their discharge rates in gallons per hour (gph) or in liters per hour (lph). The number of emitters multiplied by the discharge rate gives the total water required for a system zone. Either way, one can calculate the total discharge for a lateral or for a zone.

In permanent systems, emitters are normally spaced at regular intervals that place the water 12 to 24 inches from the base of vines. The spacing is not critical, as vine roots will extend some distance from the trunk; however, avoid dripping directly onto trunks if possible.

Main

A main line carries the water from the pump out to the field and distributes it to submains, headers or manifolds. For a long term crop such as grapevines, the main pipe line is best installed underground with provisions for it to be drained in the fall. If the pipe line is buried a safe depth below the frost line, the drainage is not as critical. Installing drain valves at low points allows for draining and flushing as needed. Any future expansion of the irrigation system should be considered when sizing the main line. Buried polyvinyl chloride (PVC) pipe (schedule 40) is suggested for this service. Risers should be installed at regular intervals where water will be taken off for irrigation zones via submains, headers or manifolds.

Submain

A main line may divide to carry water to irrigation zones away from the main line. These branched extensions of the main distribution line are called submains (see table A3.1, page 306 for more detail on main and submain pipe sizing).

Header or Manifold

A header, or manifold, is the pipeline across the end of the field to which the drip rowcrop or polyethylene tubing laterals that run down each row are attached. An irrigation zone may be one or more manifolds, each controlled by a valve near the main line. It is important to design the layout to provide the correct operating pressure to each lateral.

A flexible pipe called "layflat" can be used for carrying water to lateral lines. Layflat is similar to a fire hose; its flat design allows equipment to drive over it easily. For a permanent irrigation system on grapes, a polyethylene or PVC pipe can be placed underground with a riser at the end of each row. This reduces maintenance around surface pipes. To allow drainage of the system, slope these underground pipes and install drains at low points. (see Irrigation System Maintenance, page 186).

Pump

Centrifugal pumps are commonly used for shallow well or surface water applications. They may be powered at ground level by electric motors, dedicated gasoline or diesel engines, or tractor power take-off (PTO) attachment; however, if water must be lifted more than 20 feet from the source to the pump, then a deep well pump is required. A well driller who has the specifications on the well (such as the yield in gpm and the total dynamic head pressure, in feet of head or psi), can select an efficient pump for the project. Total dynamic head pressure includes the energy needed to lift water out of a well, to push it through a filter and a fertilizer injector, to push it through a mainline and up a hill, to replace friction loss in the pipeline, and to operate all emitters at proper pressure in the field. A pump is specifically selected to match the total head pressure and the total flow requirements of the irrigation system.

The total flow or pump output cannot exceed the yield of the water source unless water is also pumped from a holding tank. If the total vineyard to be irrigated requires more water flow than available from the source, the vineyard must be divided into irrigation zones. Zones are vineyard blocks with irrigation water requirements that match the gallons per minute that are available to be pumped. Zones can be watered one at a time.

Pressure Tank and Pressure Switch

In small systems the domestic water supply may be used for more than just irrigation, so a continuous supply of water must be available. In this case, the dealer or well driller can advise on the pressure tank water-holding capacity required to supply water for at least a minute of "off" time for the pump. The pressure tank provides a continuous availability of water but gives the pump a rest so it can cool down between restarts, or cycles. A minute or so rest period before recycling is recommended for the pump.

A pressure switch operates the pump between two pressure set points; the pump turns on at a low pressure point and turns off at a high pressure point as required to maintain a pressurized water supply, as water is used for irrigation or for other purposes.

Vacuum Breakers/Air Vents

Air vents are placed at high points in the system to release air (air vent) as the pipes fill or to admit air (vacuum breaker) as the pipes drain. A drip emitter on the ground sucks in air and dirty water if draining water creates suction in the drip line. The vacuum breaker reduces this problem. An air vent located at high spots on an irrigation main or lateral will prevent air locks from forming in the pipe where it goes over a hill or high point. Air can accumulate at the high point and could block water flow if not released. As a vacuum breaker, a vent is used next to a backflow preventer near the pump to reduce the chance of a vacuum forming and pulling water and chemical past the check valve.

Backflow Prevention

When the irrigation system is connected to a potable (drinking) water supply or when chemicals are to be injected into the irrigation water, it is essential that a backflow prevention device be installed. A "reduced pressure backflow preventer" represents a minimum level of protection for a potable water source. For nonpotable water supplies, a backflow prevention device includes: (a) a check valve that closes when water tries to flow back toward the source, (b) an air vent or vacuum breaker on the source (pump) side of the check valve. and (c) a drain on the source side to remove any water that gets past the check valve.

For household use, a simple hose bibb backflow preventer can be purchased and installed to keep water from being drawn into the house from a hose on the hose bibb outdoors. This represents a minimum level of protection when only water (no chemical injection) is involved.

Shutoff Valves

The type of shutoff valves used in the system is an important consideration. Typical household valves significantly restrict the flow of water. To reduce this restriction, use ball or gate valves that open to about the same diameter as the pipe to which they are attached.

Pressure Gauges

A pressure gauge is like the speedometer or odometer on a car or truck. Naturally, no one should operate a car or truck without the proper safety equipment. In an irrigation system, the pressure gauge tells the operator if the system is functioning properly; the proper pressure of a new system should be recorded and the system monitored for any deviation from this. A low pressure may mean a break in a pipe or damaged emitters. A high pressure may mean a blockage or valve not open to allow the full irrigation zone to operate. If you see that the pressure is wrong, check the system. A gauge reads most accurately at about its midpoint, so select gauges that have adequate range for anticipated pressure extremes. Note that liquid-filled gauges give longer service than those that are gas-filled.

Place pressure gauges on the pump discharge, on both sides of filters, and on both sides of solenoid valve/pressure regulator combinations used on manifolds. These gauges will permit the irrigator to monitor the functioning of the system.

Pressure Regulators

A drip irrigation system works properly at a low pressure of usually 8 to 10 psi. Rolling terrain and hillsides can cause the pressure to vary. Each single foot of elevation means 0.433 psi change in the pipeline pressure; thus, 2.31 feet of elevation change means 1 psi pressure change. When the operating pressure is only 8 to 10 psi, small changes in elevation translate into large pressure changes. Uniform applications of water depend on the correct water pressure in the drip lines. Main lines can run up slopes under high pressure and pressure regulators can be used to maintain proper pressure in laterals laid out on the contour from the main. Use pressure regulators after filters and just before the emitters.

Filters

Filters are the most important component in a drip irrigation system. A filter is critically important because the emitters have small passageways that can be easily fouled. Some emitters are designed to open to flush, but this is a bonus feature. Selection of the correct filter is very important. An irrigation dealer or designer can explain the filtration requirements for a given irrigation emitter product.

A filter will clog over time and this clogging will restrict water flow, resulting in an eventual pressure drop. For this reason it is wise to install a pressure gauge on both sides of a filter. When the pressure difference is about 5 psi, it is time to clean the filter. A back-flushing or self-cleaning feature can reduce maintenance time when dirty water conditions cause fairly frequent clogging.

Pressure regulators should be installed after the filter so that pressure is adjusted just before the lateral line and water pressure is not affected by a clogged filter; full line pressure should be available at the filter for allowing a pressure drop over the filter as it clogs. Also note that full pressure is good for filter cleaning purposes.

The appropriate type of filter depends on the water source condition. Well water is usually fairly clean and free of particles except sand or grit, so a screen filter of 100 to 200 mesh (0.15 to 0.08 mm openings) is used to catch these particles. Spring, stream, and clean pond water carries some debris and, possibly, a bit of organic matter. A disk filter has more storage space for debris and is a little stronger, to withstand pressure and more debris. A disk filter is a round flat disc with grooves across it to allow water to pass through while debris is caught. A disk filter, a backflushing or oversized screen filter, or a back-flushing disc filter is better for moderately dirty water. For very dirty water conditions, including heavy organic loads, a "media" or "sand" filter is recommended to remove algae, fine silt, or other fine organic or inorganic particles. A pair of sand filters is best so that one unit can filter the water used to backflush or clean the other unit. Swimming pool filters have been used in some operations, but, in such cases, the filter must be sized adequately. It is best to seek advice from an irrigation dealer or designer.

Electric Solenoid Valves and Controller

Automatic control of the system is a good management tool. Electric solenoid valves can be used to open the valve on the manifold or laterals automatically rather than by turning a manual valve. A timer is used to operate the solenoid valves. To irrigate each zone, the operator can determine the soil moisture and set the clock for the appropriate opening of each valve. The operator can spend his or her time doing other tasks while the irrigation system runs the proper length of time and then shuts down. Automation means not worrying about turning valves on or off or failing to water a zone, although some system monitoring is of course necessary.

Fertilizer Injectors

An important component of a drip irrigation system for nutrient management is a fertilizer injector. An injector can deliver nutrients through the irrigation system into the wetted zone where plants are actively removing both water and nutrients. The major types of injector includes positive displacement (hydraulic or electric powered) pumps and pressure differential systems such as the venturi injector.

Water-driven positive displacement injectors use the irrigation water flow to drive a piston pump that is coupled with an injector pump, which pulls in the chemical. The hydraulic pump must be sized to the water flow but its speed of operation is proportional to the flow rate so that the chemical is pumped in at a proportional rate.

The pressure differential system using a venturi device relies on a large pressure drop created by a pressure regulator or a centrifugal booster pump. The venturi device is placed in a bypass line parallel to the irrigation line. A pressure drop over the device creates a suction that pulls the chemical into the line. More energy is used in this system and the pressure drop determines the chemical flow rate. The choice of a particular injector system depends on required accuracy of injection, desired longevity of the piece of equipment, cost, and how corrosive the chemical is to the injector.

Backflow prevention is very important for most injection situations. Potable water supplies must be carefully protected. Flow controls to turn on and shut off the chemical flow are very important as a means of ensuring that the chemical does not flow after the irrigation system is shut down. Calibration of the injector is critical to proper application of the chemical.

Moisture Sensors

Moisture sensors serve as an important tool for judging the soil moisture situation and for helping to determine the irrigation schedule for maintaining a good soil moisture condition. Sensors and scheduling methods are discussed below under Scheduling Irrigation.

IRRIGATION SYSTEM MAINTENANCE

Irrigation system maintenance is important for maintaining the system at its initial operating condition. Maintenance tasks can be grouped into three seasonal time periods: startup in the spring; growing season maintenance; and winterizing. Each period is important to successful operation and system longevity. Drip irrigation systems are assumed to be the system used in this discussion.

Startup in Spring

Inspect the system in the spring for damage caused by weather or pests before pressurizing the system. If there are water quality issues, a good flush of the system with water or with an appropriate chemical treatment (acid, chlorine, or other product) to loosen and remove any residue in the pipeline is recommended as a way to reduce potential clogging problems. Small manual drain valves or automatic flush valves on the ends of laterals facilitate this flushing. Residues should not be pushed out through emitters; instead, residue in emitters can be flushed out after first cleaning the lateral lines to which they are attached.

Reinstall any components that were removed during winterization. Replace any broken components. Pressurize and check the system's integrity.

Retension lateral lines that have been displaced from their position in a row.

Seasonal Maintenance

Visually inspect the system on a regular basis to determine if any emitters are not functioning or to spot any breaks in the lines.

Monitor the pressure each operating day by viewing a pressure gauge at the pump, on either side of the filter, or in the irrigation zone to be sure the water is properly applied. Expect the pressure to drop several psi across the filter as it clogs. A pressure gauge is the tool that warns the operator of problems in the delivery of water to the plants.

A water meter measures the water used. This can be useful in monitoring water use to show variations due to a broken line, incorrect pressure, or changes in length of irrigation.

Flushing is perhaps the best maintenance for a drip irrigation system if there is a need to treat the water for chemical or organic problems. The filter should be selected based on water quality and then properly maintained. Pressure gauges on either side of the filter allow the operator to monitor the pressure drop across the filter and to (a) see that the filter self-flushes at the proper pressure point, or (b) inform the operator when to manually flush non-automatic filters.

Air vents and vacuum breakers are important components for a drip system and their proper operation is critical to proper line filling and to avoid having dirt pulled into emitters as lateral lines drain, as described earlier.

Calibration of fertilizer injection systems is important to ensure that the proper amount of fertilizer is being delivered. Check the calibration on a regular basis.

Pressure regulators are important for regulating the water pressure at an irrigation zone. The water pressure in lateral lines should be checked several times each season to ensure that the drip emitters are functioning correctly.

Winterizing

If the system is subject to algae, iron deposits, carbonate deposits, or other chemical problems, the pipelines might be treated with chlorine or other appropriate chemicals, allowed to set, and then flushed to remove residues from the lines.

Draining the system and using compressed air to blow the water out of pipe and the components that do not drain are the most important practices in preparing the system for winter. Water in any pipe or component that is above local frost depth is subject to freezing and breaking pipes and other system components such as valves, filters, and meters. Drains can be installed on pipelines that are sloped to be drained at low points; an access box to a drain valve on the pipeline is an option. An air compressor that can generate 50 to 75 cubic feet per minute (cfm) is adequate for smaller systems. A 125 cfm compressor is adequate for up to a 3-inch diameter pipe. Air pressure should be limited to less than 50% of system operating pressure. Some systems may be adequately drained by gravity and limited system disassembly.

Drain filters, fertilizer injectors, and other frost-sensitive equipment and protect from winter weather.

SCHEDULING IRRIGATION

There are several approaches to determining the irrigation needs of established vineyards. While none is perfect, using at least two in combination is infinitely superior to not using any. We will discuss each in the order of lowest to highest level/cost of technology. The approaches are based on: (a) visual basis, (b) water use "budgeting," and (c) soil moisture-based measurements. Other methods are briefly mentioned but will be beyond the practical needs of most readers.

Visual Bases

With visual bases, the vineyardist simply monitors the growth of the vines throughout the season, paying attention to the plant indices and responses listed in table 9.1 (page 172). While there is no formula to state what response is needed or tolerable at a given time, irrigation could be applied if mild stress symptoms appeared in the prebloom through immediate postbloom period (see table 9.1). The amount of water to apply (for example, 10 gallons per vine per week) would be rather empirical unless some measures of soil moisture were included. Indeed, to fully meet the vine's water requirements, the volume of water might be closer to 30 to 40 gallons per week based on calculations using ET and foliage surface area.

At minimum, the soil "squeeze" test can be used to gauge the effects of added irrigation water (see page 190). As the season progresses, some degree of stress (reduced shoot elongation) may be tolerable or even desired, but irrigation water should be metered out in a fashion that balances that reduced growth with sustained leaf function; leaves should remain relatively cool (transpiring) for most of the day. A relatively inexpensive infrared thermometer can be purchased to read the canopy temperature on sunny afternoons to facilitate the temperature measures.

Pros and cons

Visual clues represent a low-tech, and therefore inexpensive approach. By the time visual clues are apparent, some negative consequences of stress (such as reduced fruitfulness in the following year) may have already occurred. Using visual clues requires conscientious monitoring throughout the vineyard.

Water Budget

Weather records are useful for monitoring the plant available water supply; in particular recording and tracking the daily ET_o if it is available to you. The ET water budget approach to irrigation scheduling is based on the premise that we can accurately estimate the volume of water removed by a vineyard and, by irrigation, replace that amount of water either at 100% of loss, or at a desired deficit level (<100%). A bookkeeping ("checkbook") method uses more information, accounting for rainfall and ET along with using knowledge about the water storage capacity of the soil in the root zone to estimate water availability for the crop. Both methods are described here.

ET Water Budget Approach

The ET water budget approach uses equations 9.1 and 9.2 to track vine water use as described above in "Crop Irrigation Requirements," (page 178). Equation 9.1 requires the availability of daily reference evapotranspiration rates (ET_o) specific for the local conditions of the subject vineyard and a crop coefficient (K_c). Equation 9.2 requires knowledge of the irrigation system efficiency. Crop coefficients adjust the ET_o for degree of grapevine canopy development (appendix 2, see page 304). The results of equations 9.1 (page 178) and 9.2 (page 179) are an estimate of the amount of water it would take to replace soil moisture lost by evapotranspiration by grapevines and any ground cover on a daily basis. Irrigation water (in the absence of rainfall) can then be applied to the vineyard to replace moisture lost by evapotranspiration. Of course, the water can be applied at some other frequency (such as weekly) but one would sum the daily losses for whatever period is used, as in the following example:

Cumulative one-week reference ET_o (obtained locally):
1.40 inches (0.20 inches per day x 7 days)

Crop coefficient (K_c):
0.51

Estimated water use for one-week period (ET) =
cumulative one-week ET_o X K_c

ET = 1.40 x 0.51 = 0.71 inches of water
(This value goes into the budget ledger)

The amount of water to apply must account for the irrigation efficiency so when it is time to irrigate one must do a calculation to get the amount of "pumped water."

System efficiency for drip system = 90%
Pumped water = (1.40 x 0.51) / 0.90
Pumped water = 0.79 inches for the week
(This value is used to set irrigation application time)

This method makes one aware of the estimated water use of the crop but does not account for rainfall or soil moisture in guiding one to schedule irrigation. This method is more suited to an arid climatic region with a high ET.

Bookkeeping or Checkbook Budget Method

The bookkeeping budget method of scheduling irrigation might be called the checkbook method. One can look at the balance of one's checkbook to find one's financial status. Over time there are checks written that take money out and, conversely, deposits are made to add money to the account. Likewise, the soil in the root zone of the crop has a moisture status. Rainfall and irrigation add water to the soil while evapotranspiration takes water out. To use this method one needs to have some understanding of soil water holding content, the role of evapotranspiration, and amounts of daily rainfall or irrigation. This method will be illustrated here for a warm summer week. ET_o values may vary from 0.10 inches per day to 0.25 inches per day over a season, crop coefficient K_c will vary as the grapevine canopy develops, and soil moisture conditions will vary with ET and rainfall. Thus, over a growing season one should monitor and adjust these values to one's own crop situation for increased accuracy of the budget.

A bookkeeping (checkbook) budget method is easiest to do if started out with a soil at field capacity. And, when the soil capacity is refilled to field capacity in the future, one can reset the budget to start over. Field capacity is the soil moisture condition after rainfall has filled the soil profile and gravitational drainage of excess water has occurred. One can maintain a record of the daily soil moisture status on paper or on a computer spreadsheet by making daily entries of ET, rainfall and irrigation, as appropriate. In the absence of rainfall or irrigation, soil moisture will decrease daily due to evapotranspiration. In the absence of local ET_o data, use ET = 0.20 inches per day in summer. Soil moisture will, conversely, increase with rainfall and irrigation. When the soil moisture drops to about 50% available moisture, then irrigation is used to bring it back up to a desired level, such as field capacity.

Recording the daily weather along with the soil moisture status gives data on which to make future decisions and provides data for reviewing a season. The budget is only as good as the data put into it, but it provides some basis for decision making. Perhaps the most important factor is the existence of a weather record, which can improve the soil moisture accounting. A sample of soil can be checked visually, by the squeeze method (page 190), or by other means to help one make judgments.

Knowledge of the soil water-holding capacity and depth of rooting is essential. For each foot of soil depth, there would be an estimated number of inches of water in storage. Table 9.4 provides the range of Plant Available Water (PAW) holding capacities for several soil textures. Local soil maps at a county soil conservation district or Natural Resource Conservation Service (NRCS) office can be consulted to determine the soils in the vineyard. Knowing the soil type and estimated rooting depth, one can estimate PAW. Allowing depletion of 50% of PAW provides the number of inches of water to use as a maximum available water amount for the budget method. A sandy loam soil with a 3-foot rooting depth may pro-

Table 9.4 ▪ Approximate Plant Available Water (PAW) of Soils as a Function of Soil Texture

Soil texture	Inches available water per foot of soil depth
Coarse sand	0.50
Fine sand	0.75
Loamy sand	1.00
Sandy loam	1.25
Loam	1.50–2.00
Clay loam, silt loam	1.75–2.50
Clay	2.00–2.40

vide 1.25 inches of water per foot of soil depth, or 3.75 inches of PAW. With a 50% to 55% depletion allowed, 2.0 inches goes into the budget when the soil is at field capacity after a good rainfall or irrigation. This is resetting the budget to full or 2.0 inches of water.

The 2.0 inches of water would be the maximum water available for the plants in the budget; this gives a margin of safety over using all available water for this scheduling purpose. For each day an estimate of ET (in inches) would be made and entered into the spreadsheet or chart for water consumption (as negative quantity). Each day for which there is rainfall, the amount of rainfall, or at least the amount that would be held in the soil, would be entered in the chart as rainfall (as a positive quantity). An irrigation event would also add water to the budget (as a positive quantity). Table 9.5 represents a budget sheet for irrigation scheduling, showing a few days of the record for August. A long, slow rain that fills the soil profile would reset the budget to the maximum soil moisture value (2 inches of PAW).

Note that it is not desirable for the PAW to drop below 0.0 inches. One should allow for rainfall storage by not re-filling the PAW capacity to full with the irrigation system. Applying the irrigation water over two or more days allows the moisture sensors, if used, to reach equilibrium so that soil moisture can be brought up to the desired point without over-application of irrigation water. In this situation, 2.0 inches in the right-hand column would indicate soil at field capacity. The goal is to maintain the soil between 0.0 and 2.0 inches of PAW.

Pros and Cons

Water budget calculations can be computerized with the use of Excel™ or other spreadsheet programs to speed up calculations and avoid computational errors. The bookkeeping method gives a method for tracking soil moisture, weather, and irrigation events. On the negative side, it is hard to know the best estimate of ET in many areas that do not have local reports or to know an accurate soil moisture holding capacity to use. Without precise measurements it is an estimate only. This method should ideally be used in concert with visual clues and perhaps with other soil moisture determinations, as described in the following section. This method, at a minimum, provides a visual record of weather data, rainfall, and irrigation, enabling the grower to see the trends that are occurring. At some point a good rainfall or heavy irrigation will bring the available moisture up to 100% and the record starts afresh so errors are not carried forward.

Table 9.5 ▪ Worksheet Example of Irrigation Scheduling by the Soil Moisture Budget Method

Water budget — Field 4: Root depth = 3 ft; sandy loam soil; 2.0 inch PAW						
Date	Maximum temperature (°F)	Weather	ET (inches)	Rainfall (inches)	Irrigation (inches)	Soil moisture balance (inches)
August 1	85	Cloudy	0.2			0.4
2	87	Cloudy	0.2	0.3		0.5
3	84	Partly Sun	0.2			0.3
4	82	Partly Sun	0.2			0.1
5	80	Cloudy	0.2		0.5	0.4
6	84	Sunny	0.2		0.5	0.7
7	90	Sunny	0.2		0.5	1.0
8	82	Partly Cloudy	0.2			0.8
9	78	Cloudy	0.2	0.8		1.4
10	80	Partly Sun	0.2			1.2

PAW = Plant Available Water

Soil Moisture-based Measurements

The basis for using soil-based measurements is that we want to provide some level (for example, 50% to 75%) of PAW to vines throughout the growing season. Rarely would we need to have 100% PAW. The use of this approach therefore requires some basic knowledge of the PAW for your site/vineyard block. This can be achieved with knowledge of the soil type and texture so an idea of the water holding capacity for the depth of soil is known. Knowing the soil texture should give an idea of the depth of available water as well as the amount of water between field capacity and the permanent wilting point (see table 9.4).

Simple techniques and special instruments may also be used to measure soil moisture. These include a hand squeeze method, water-filled tensiometers, gypsum blocks, and time domain reflectometry and capacitance probes.

Squeeze or Ribbon Test (Feel and Appearance)

A simple manual squeeze test gives a reasonable estimate of soil moisture and requires no equipment other than a shovel or auger to dig a hole. There are several variations of this test, but in essence it is based on the ability to form either a ball or a ribbon of soil—the more moisture in the soil, the easier the task. Appearance is also a factor. Table 9.6 gives the feel or appearance for each of four soil textures at five different moisture levels. With experience a person can become reasonably accurate at this method. Be sure to check a couple depths in the root zone.

Procedure:

- Choose an area to sample; perhaps near a vine in a representative area of the vineyard block.
- Dig a hole with a tractor-mounted auger or by shovel to a depth of at least 30 inches.

Table 9.6 ▪ Estimating Soil Moisture Levels Using the Soil Squeeze Test[a]

Plant available water	Coarse-textured soils (sand); gritty when moist, almost like beach sand	Moderately coarse-textured soils (sandy loam); gritty when moist, dirties hand; some silt and clay	Medium-textured soils (silt loam); sticky when wet	Fine- and very fine-textured soils (clay loams and clay); sticky when moist, behaves like clay
0–50%	Dry, loose, and single-grained; flows through fingers. Appears to be dry; does not form a ball under pressure.	Dry and loose; flows through fingers. Appears to be dry; does not form a ball under pressure.	Powdery dry; in some places slightly crusted but breaks down easily into powder. Somewhat crusty but holds together under pressure.	Hard, baked, and cracked; has loose crumbs on the surface in some places. Somewhat pliable; forms a ball under pressure.
50–75%	Appears to be dry; does not form a ball under pressure.	Forms a ball under pressure but seldom holds together.	Forms a ball under pressure; somewhat plastic; sticks slightly under pressure.	Forms a ball; ribbons out between thumb and forefinger.
75% to field capacity	Sticks together slightly; may form a very weak ball under pressure. Upon squeezing, no free water appears.	Forms weak ball that breaks easily; does not stick. Upon squeezing, no free water appears.	Forms ball, very pliable; sticks readily if relatively high in clay. Upon squeezing, no free water appears.	Ribbons out between fingers easily; has a slick feeling. Upon squeezing, no free water appears.
Field capacity (100%)	Same as above and upon squeezing, no free water appears on soil, but wet outline of ball is left on hand.	Same as above and upon squeezing, no free water appears on soil, but wet outline of ball is left on hand.	Same as above and upon squeezing, no free water appears on soil, but wet outline of ball is left on hand.	Same as above and upon squeezing, no free water appears on soil, but wet outline of ball is left on hand.
Above field capacity	Free water appears when soil is bounced on hand.	Free water is released with kneading.	Free water can be squeezed out.	Puddles, free water forms on surface.

Source: National Engineering Handbook, US Department of Agriculture, Natural Resources Conservation Service.

Note: Depending on the stage of growth, irrigation method, and soil type, irrigation should be started when the soil is at 50%–75% field capacity in the root zone. Soil moisture should not drop to less than 50% field capacity, or wilting can occur.

[a] A ball is formed by squeezing a handful of soil very firmly.

- Remove a handful of soil from the wall of the hole near the bottom.
- Squeeze the soil in your hand(s) to try and form a cohesive ball.
- Toss the ball from hand-to-hand to test how cohesive the ball is.
- Compare your results with those in table 9.6.
- Use additional soil bores and the squeeze test to monitor the penetration depth and persistence of applied irrigation water.

Tensiometers

A tensiometer (see figure 9.6) is a water-filled tube with a hollow porous ceramic tip on the bottom and a vacuum gauge and airtight seal on the other end. The device is installed into the soil so that the porous ceramic tip is in tight contact with soil at some depth. Water moves through the porous ceramic tip to achieve equal tension with the soil particles. The reading on the vacuum gauge on the tensiometer equals the tension (attractive force) of soil particles. If the soil is dry, the attractive force (tension) is high. If the water column is intact, a zero reading indicates saturated soil. At about 70 to 80 cb (cb = centibar, a measure of tension; 100 cb = 1 bar = 14.5 psi = 100 kiloPascals [kPa]) the tensiometer breaks tension with the soil particles and air enters the water column. It is time to service the tensiometer. A low reading means a good moisture level and readily available water. Tensiometers should be placed in strategic locations within the vineyard row, preferably at a minimum of two rooting depths at each location. For example, one might be placed at a depth of 9 to 12 inches and another at a depth of 30 to 36 inches, or deeper, depending on the soil and root depth. When the shallow instrument says the soil is dry, irrigate. When the deep instrument says it is wet after an irrigation application, do not apply more water as it is moving below the root zone. Water lightly if the top is dry and the bottom is wet. Tensiometers should be located under the trellis and in an area that would normally be wetted by the irrigation system but not immediately beneath an emitter. Drip emitters wet a bulb-shaped volume of soil with a shape dependent on several factors, including soil texture, depth of soil and hardpan, and length of irrigation application. A moisture sensor should be placed 1 to 2 feet away from the point where water enters the soil from the emitter. In a fine textured loamy-to-clay soil the water will spread more laterally than in sandy soil. Tensiometers should be well flagged or otherwise marked to avoid damage by cultivation or other equipment.

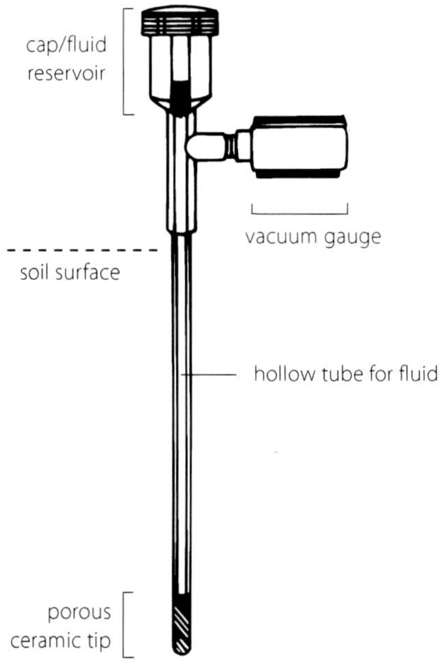

Figure 9.6 ▪ Typical tensiometer. *Source:* Production of Vegetables, Strawberries, and Cut Flowers Using Plasticulture, *NRAES 133.*

Tensiometers require frequent monitoring and must be correctly installed and serviced in order to obtain reliable results. They work best in coarse soils because they measure soil tensions up to about 70 cb which is near the point of available water depletion in sandy soils. Clay soils hold water tightly and require higher tensions than sandy soils to remove available water.

Grapevine roots are generally distributed as shown in figure 9.7 with about 40% of the roots found in the top one quarter of the root zone, 30% in the next one quarter, 20% in the next one quarter and 10% in the bottom one quarter of the root zone. Placing one tensiometer about one foot deep puts it into or under the top quarter of the root zone but the tensiometer may function better at this depth because the soil may not dry as frequently down below 70 to 80 cb.

Tensiometers are best read at the same time each day, preferably in the morning, and the readings plotted on a chart for reference. Plot the centibar readings on a vertical axis and place days of month on a horizontal axis. The

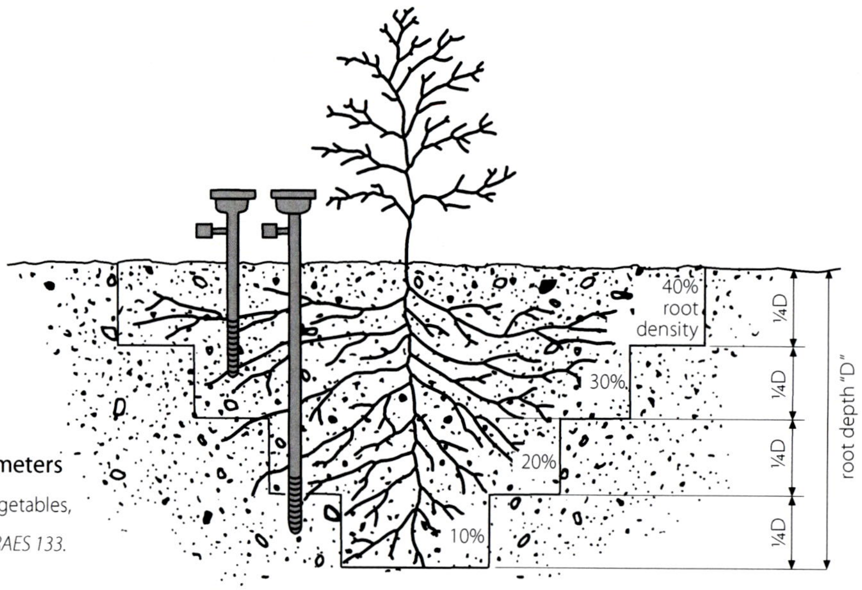

Figure 9.7 ▪ Placement of a pair of tensiometers in the crop root zone. *Source:* Production of Vegetables, Strawberries, and Cut Flowers Using Plasticulture, *NRAES 133.*

Table 9.7 ▪ Soil Tension Guidelines for Using Tensiometers

Soil moisture and irrigation status	Soil texture	Soil tension (centibar)
Soil at field capacity—no irrigation required	Sand, loamy sand	5–10
	Sandy loam, loam, silt loam	10–20
	Clay loam, clay	20–40
50% of plant available water depleted—irrigation required	Sand, loamy sand	20–40
	Sandy loam, loam, silt loam	40–60
	Clay loam, clay	50–100

trend in the daily readings will indicate the trend in the soil moisture level and will indicate when to irrigate. It is not a hard chore, compared to digging down to the same depth to get a soil sample to feel or dry. Two tensiometers are used to monitor both shallow and deep, to show the change in moisture status at top and bottom depths of the root zone. The movement of rainfall or irrigation water through the soil profile should be detected by the readings. Irrigation should move water to the lower depth when that soil is dry but water should not be moved deeper and out of the grape root zone. Table 9.7 gives an indication of the readings in three soil textures for field capacity and for 50% of available water depleted. These values show that tensiometers have limited value in fine-textured soils when the soils are allowed to dry out.

Careful installation and maintenance are required for reliable readings. The ceramic tips must be in firm and complete contact with the soil particles. To achieve this, use an auger to make a hole to the proper depth, making a soil water slurry with the soil that was removed, and place the slurry back into the hole. Push the tensiometer down into the hole and slurry. Place soil around the tensiometer at the surface so the water drains away from the hole. Allow the tensiometer a few hours or several days to equilibrate with the soil. Use the maintenance vacuum pump to remove any air bubbles in the tube. When high readings have occurred, use the vacuum pump to remove any air bubbles and refill the reservoir with distilled water. Costs of tensiometers are in the range of $60 to $80 each, plus a maintenance kit for $35 to $40.

Gypsum Block Moisture Sensors

Gypsum block moisture sensors are electrical resistance blocks. Electrodes inside the block are used to measure the resistance of electrical current flow between electrodes. The electrical current flow is affected by the moisture content of the block, which is a function of moisture tension. Higher resistance readings mean lower water content and, thus, higher soil water tension. The advantage to these blocks is the working tension range from 100 to 200 centibars so they are suggested for fine textured soils that hold available water at higher tension. The sensors require that a meter be connected to them in order to read them. The gypsum block is more sensitive in the heavier soils that hold more water. Read the meter in the morning and note values in a chart or graph so trends can be seen.

Watermark™ Blocks

Watermark™ blocks, or granular matrix sensors, are a relatively new style of electrical resistance block. The electrodes are embedded in a granular matrix material similar to a compressed fine sand. A gypsum wafer is embedded in the granular matrix near the electrodes. The granular material enhances the moisture movement to and from the soil, so the block is more responsive to soil tensions in the range of 0 to 100 cb. Watermark™ blocks have good sensitivity from 0 to 200 cb, making them more adaptable to a wider range of soil textures.

Readings are taken using a special electrical resistance meter that is connected to wire leads, and the estimated soil temperature is set. The blocks have a longer service life than gypsum blocks and can be left in the soil under freezing conditions. Soil salinity affects the electrical resistivity of the soil water solution but the gypsum offers some buffering of this effect.

Time Domain Reflectometry (TDR) and Capacitance Probes

Time Domain Reflectometry (TDR) units measure the dielectric constant of a medium by finding the time taken for an electromagnetic pulse to traverse the distance between two parallel conducting probes in the soil profile. The dielectric constant of a material is a measure of the capacity of a nonconducting material to transmit high frequency electromagnetic waves or pulses. Research has shown that measurement of the dielectric constant of the soil is a sensitive measurement of soil water content.

The TDR propagates a high frequency electromagnetic wave along a cable attached to parallel conducting probes (waveguide) inserted into the soil. The signal is reflected from the end of the waveguide back to a meter where the time between sending the pulse and receiving the reflected wave is accurately measured. By knowing the length of the transmission line and waveguide, the propagation velocity of the signal in the soil can be computed. The dielectric constant is inversely related to this propagation velocity; that is, a faster propagation velocity indicates a lower dielectric constant and thus a lower soil water content. Or, as soil water content increases, propagation velocity decreases, and the dielectric constant increases.

TDR has been shown to be very precise, if accurately calibrated for the soil in question. Waveguides inserted into the soil usually consist of a pair of parallel stainless steel rods spaced about 2 inches apart. A shielded, coaxial cable connects the waveguide to the cable tester. The TDR soil water measurement system measures the average volumetric soil water percentage along the length of the waveguide. The volume is measured in a cylinder surrounding the waveguide with a diameter about 1.5 times the spacing of the parallel rods.

Due to the high cost of the cable tester and associated electronics, TDR soil water measurement system costs usually begin in the range of $6,000 to $7,000. Currently, the length of cable is limited so the sensors cannot be placed very far from the meter.

Capacitance Probes

Capacitance probes, or the Frequency Domain Reflectometry method, uses radio frequency (RF) capacitance techniques to measure soil capacitance, which is then related to soil water content. A pair of electrodes is inserted into the soil. Two commercially available instruments using this technique are the Troxler Sentry 200-AP ($4,000 to $4,500) and the Aquaterr ($400 to $500).

A new sensor that is now available is the ECH_2O probe (Decagon Devices, Inc.) which is a capacitance type probe. The probe measures the dielectric constant or permittivity of the material in which it is embedded. The capacitance technique determines the dielectric permittivity of a medium by measuring the charge time of a capacitor that uses that medium (soil) as a dielectric. This probe is relatively small and can be read using a small handheld meter or a datalogger. The advantage of this probe is that it can be buried to any depth. Each probe can measure up to a 20-cm rooting depth. Two probes could monitor soil moisture at the surface and at any depth. Up to five probes can be connected to a datalogger from up to 250 feet without signal attenuation, using extension cables. Dataloggers can connect wirelessly to a central computer using a wireless relay, to provide a network of sensors that gives continuous soil moisture data from the vineyard. (A wireless network with computer, irrigation management software, 20 soil probes, 4 field dataloggers and a wireless relay station costs about $7,000, as of 2006.) Technology is rapidly changing in this field and wireless transmitters at remote locations are sending data to a central computer to show soil moisture data in real time.

Pros and Cons

Vine growth is generally well correlated to PAW, and knowledge of the percent PAW at any given time provides a predictive measure of vine drought stress. The limitation of this approach is that instrumentation or "squeeze test" methods must be routinely used to monitor soil moisture, and this requires a conscientious approach by the vineyard manager. As with all approaches, the greater the variability in soil type within the vineyard monitoring unit, the more variable will be the vine response to soil moisture differences. Soil moisture probes (tensiometers, gypsum blocks, and TDR or capacitance sensors) cost from hundreds to thousands of dollars and generally only reveal the soil moisture in an 8 to 10-inch cylinder of soil. This may be entirely satisfactory if that cylinder of soil is representative of the vineyard block; however, a number of moisture sensors (or sensor access tubes) will be required to accurately sense the water status of a vineyard comprising different soil types.

Other methods

There are other techniques for measuring the moisture status of the plant or the soil. Some are used for research and some have application in larger vineyards. Their use, however, may involve expensive equipment, user certification for handling of radioactive materials, or specialized skills in interpreting the results. The methods include using pressure bombs to measure the degree of plant tissue hydration, neutron probes for measuring soil moisture, and carbon isotope discrimination for evaluating the history of plant drought stress. These techniques might be used by eastern North American vineyardists in the future, but the more commonly used techniques described here will have more practical use for the foreseeable future.

CONCLUSION

Water is essential for grapevine growth and productivity. Rainfall can be unpredictable in the East so an irrigation system provides some insurance against drought stress. Depending on location the grapevines transpire about an inch of water per week. In areas where the grapevines, weeds and row middle cover crops coexist, the combination of transpiration and evaporation may amount to 1.3 to 2.0 inches of water per week. During a drought the soil moisture reservoir can be depleted to a point that grapevines are drought stressed. A drip irrigation system can maintain adequate soil moisture levels.

Plant water requirement can be estimated by determining daily evapotranspiration, ET, under local conditions in mid-summer. Knowledge of the local Potential Evapotranspiration, ET_o, from a weather station and the crop coefficient, K_c, for one's grapevines allows the plant water requirement to be calculated. In the humid East a mid-summer ET_o of 0.20 is a default value to use. The crop coefficient, K_c, varies over a season but default mid-summer values range from 0.51 for bare ground to 0.80 when there is 40 to 50% ground cover in the row middle. These values allow one to determine the ET or water lost daily from the vineyard and, subsequently, the amount of water needed to be supplied by irrigation to avoid water deficits.

An irrigation system must be designed by a qualified person to achieve adequate system capacity and high application uniformity.

Scheduling irrigation can be done using many methods. The simplest may be using visual indices coupled with the soil squeeze test to determine soil moisture content but that works well for only shallow depths. A bookkeeping method tracks available soil moisture using estimates of soil moisture storage capacity, rainfall, irrigation, and evapotranspiration. Tensiometers are relatively inexpensive, but have maintenance issues in drier soils and will be time-consuming to maintain. TDR probes are too expensive and do not compare to the ease of capacitance probes. Capacitance sensors give accurate readings of soil moisture condition for scheduling irrigation. Sensor technology has been improving in recent years with more capability at lower cost. Some sensor systems can be used for irrigation system control or to give real time data on soil moisture for management purposes. A combination of several techniques will provide a cost-effective and reasonably accurate means of assessing vine water status.

References

Allen, R.G., L.S. Pereira, D. Raes, and M/ Smith. 1998. Crop evaporation—guidelines for computing crop water requirements. *FAO Irrigation and Drainage Paper 56*. Food and Agriculture Organization of the United Nations. Rome.

Goulart, B. Water Management. Chapter 6. pp. 38–51. IN: Pritts, M. and D. Handley, Ed. *Strawberry Production Guide For the Northeast, Midwest and Eastern Canada*. NRAES-88. Natural Resources, Agriculture and Engineering Service. Cooperative Extension. Ithaca, NY.

Hellman, E. (undated). Irrigation scheduling of grapevines with evapotranspiration data (HTTP://WINEGRAPES.TAMU.EDU/GROW/IRRIGATIONSCHEDULING.PDF)

Lakso, A.N, and R.M. Pool. 2001. The effects of water stress on vineyards and wine quality in Eastern vineyards. *Wine East* Nov.–Dec., 12–20, 51.

Martin, D.L., E.C. Stegman, and E. Fereres. *Irrigation Scheduling Principles*. Chapter 7, pp. 155–203. IN: Hoffman, G.J., T.A. Howell and K.H. Solomon, Ed. *Management of Farm Irrigation Systems*. ASAE monograph Number 9, American Society of Agricultural Engineers. 1990.

Plant and Irrigation Water Requirements, Chapter V, pp. 106–180. IN: Pair, C.H., W.H. Hinz, K.R. Frost, R.E. Sneed, and T.J. Schiltz, Ed., *Irrigation, Fifth Edition*, 1983. The Irrigation Association. Alexandria, VA Library of Congress Card No. 83–81489.

Smart, R. E. and B. G. Coombe. 1983. Water relations of grapevines pp. 137–196 In: T.T. Kozlowski (ed) *Water Deficits and Plant Growth* Vol VII. Academic Press, New York.

Ross, D.S. 2004. Drip Irrigation and Water Management, Chapter 3, pp. 15–35. In: Lamont, W.J., Ed., *Production of Vegetables, Strawberries, and Cut Flowers Using Plasticulture*. NRAES-133. NRAES, Cooperative Extension, Ithaca, NY.

— 10 —

Spray Drift Mitigation

Author

Andrew J. Landers, Cornell University

Methods of reducing drift and improving deposition during the application of pesticides have been of concern for many years. Pesticide drift, as defined by the National Coalition on Drift Minimization, is "the physical movement of pesticide through the air at the time of pesticide application or soon thereafter from the target site to any non-target site." Ganzelmeier (1993) describes direct drift as the amount of active ingredient which, during the application process, is borne beyond the treated area due to atmospheric currents (horizontal and vertical) and settles outside that area (ground sediment) or is driven as solid or liquid particles in suspension, over longer distances (atmospheric drift).

Spray drift is an important and costly problem facing growers. Drift results in damage to susceptible off-target crops, environmental contamination to watercourses, and an unintentionally reduced rate of application to the target crop, thus reducing the effectiveness of the pesticide. Pesticide drift also affects neighboring properties, often leading to public outcry and conflict. As more people choose to live in or near the picturesque setting of a vineyard, the public's concern with spray drift will continue.

WHAT CAUSES SPRAY DRIFT?

Traditional airblast sprayers direct the air from a single axial flow fan, mounted directly behind the sprayer, in an upward and outward direction (see figure 10.1). Axial fans are designed to move large volumes of air at low pressures, and an increase in fan diameter, rather than fan speed, is an efficient way of increasing airflow rate. Traditional advice to growers has been to use small adjustable deflector plates, fitted at the top and base of the air outlet, to direct the air toward the target canopy in an attempt to confine it. Unfortunately many growers appear to be using sprayers with no deflectors, and many manufacturers offer optional deflectors that serve little or no purpose.

Most growers use older, traditionally designed airblast sprayers fitted with hollow cone or air-shear nozzles that provide a large amount of air to penetrate the canopy and beyond, often resulting in a vast plume of spray drifting above the target row(s). Air-shear nozzles rely on high air velocities [160 to 180 miles per hour (mph)] to create fine droplets, often exacerbating the drift problem, particularly when a sparse canopy exists in early season.

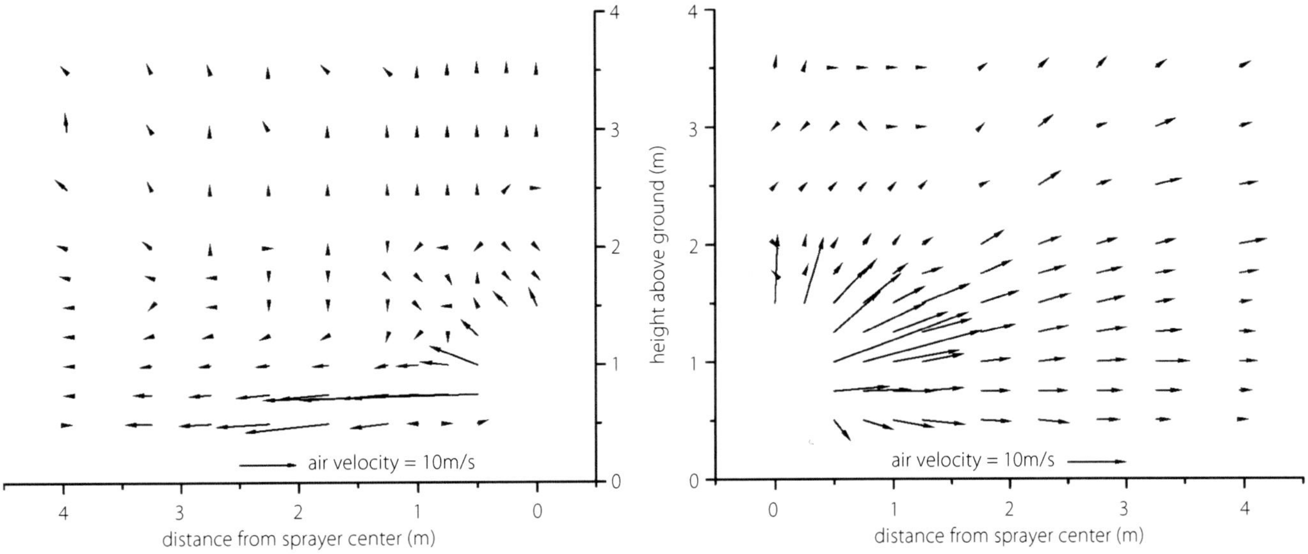

Figure 10.1 ▪ Direction and velocity of air from airblast sprayer fan.
Air vector diagram showing air velocity and direction on left and right side of counter-clockwise-rotating axial sprayer fan, when viewed from the rear. Size of arrows indicates air velocity. 10 meters per second (m/s) = 22 mph; 1m = 39 inches.

Many growers choose not to replace old, mechanically reliable sprayers; as a result, many vineyard sprayers are in excess of ten years old and lack the finer adjustments found on modern sprayers such as directed deposition machines.

A current spray practice is to use the same settings on an airblast sprayer in the vineyard from the first application through to the last, regardless of changes in canopy volume or density. Many growers change application spray volume per acre but changes to air flow (speed, volume, or direction) or ground speed are rarely made. As the season progresses, and as the canopy fills, growers frequently drive too fast and pay little attention to drift, deposition on the leaves, and clusters of grapes where disease or insects may occur. The objective in spray drift mitigation is to find the optimum combination of application parameters for different stages of canopy development in order to reduce drift while improving deposition.

Two types of drift occur during or after pesticide application: Vapor drift, which is the airborne movement of evaporated chemical (highly volatile materials), can occur even after the droplet is deposited on a leaf surface. The second, and more prominent, droplet drift is the movement of spray droplets in liquid form. Smaller droplets, those less than 100 microns (25,000 microns per inch) may be projected 33 feet or more vertically. For hydraulic nozzles found on airblast sprayers, 45% of droplets may be in the 30 to 100 micron range.

The Spray Drift Task Force (1998) measured the droplet size spectrum from airblast sprayers and found a Volume Median Diameter (VMD) in the range of 138 to 210 microns and a VMD of 73 to 110 microns from mistblower sprayers. Volume Median Diameter is the droplet diameter at which half of the spray volume is composed of larger droplets and half is composed of smaller droplets. The percentage of droplets less than 141 microns ranged from 26% to 52% for airblast and 65% to 90% for mist blower sprayers. Both the VMDs and the percentage of droplets less than 141 microns confirmed that mist blowers produce finer droplets and a higher volume of drift-prone droplets. The majority of growers use traditional airblast sprayers with hollow cone nozzles.

What happens to these droplets that are sent skyward in an airstream in excess of 50 mph? The size of a droplet strongly influences its trajectory after being emitted from a hydraulic nozzle. The droplet rapidly decelerates until it attains a velocity that is solely a function of its diameter. Air movement in which the droplet descends also influences its trajectory. The major drift factors are droplet size, wind velocity, relative humidity, ambient temperature, droplet discharge height, and initial droplet velocity. On a typical spraying day in the East, high temperatures and low humidity can result in fine droplets becoming smaller and drifting further off-target, particularly when the droplets are launched over 30 feet above the target in an airstream of 180 mph.

Droplets smaller than 150 microns generally pose the greatest drift hazard; droplets less than 50 microns have insufficient momentum for impact as they remain suspended in the air indefinitely or until they evaporate. Research has shown that a 100-micron droplet takes approximately 11 seconds to fall ten feet in still air. A similar sized droplet will drift about 75 feet before hitting the ground when it is released into a wind of 5 mph. Imagine the distance it could travel in a 180-mph airstream! The higher the operating pressure, the smaller the droplet. Conversely, low pressure produces large droplets that may bounce off the target. Deposition efficiencies are typically only 55% of the applied spray from an airblast sprayer; this suggests that 45% of pesticide either hits the ground or goes up into the air. Much of the ground losses are spray droplets returning to the ground from airborne drift. Depending on the stage of canopy development, losses that occur during spraying are: evaporation (4% to 6%), drift (10% to 15%), and ground deposit (30% to 50%).

The Spray Drift Task Force, a consortium of 38 agricultural chemical companies, conducted tests with airblast sprayers to provide data to support registration of products belonging to their member companies. Results from spraying vines trained to 6-foot tall trellises, with full canopy development, showed that most of the spray that passed the first 2 rows moved above the top of the vines. Results show that the highest amount measured was less than 0.75% of the total applied active ingredient. The amount of spray moving through the vineyard decreased rapidly and never exceeded 0.06% of the active ingredient at any height beyond the tenth row. This was due to a combination of droplet settling and the filtering effect of the foliage. Care must be taken when interpreting these results. It is very important to note that the percentage of drift is based on the amount drifting from the total area to which pesticides were applied, and that in practice the percent drift can be much higher, particularly from the edge rows of a vineyard and particularly under moderate wind speed, low humidity, and high temperature conditions.

Disease and insect management is a critical factor in vineyards and, while such management may represent a small proportion of crop value, there is a demand for increased spraying efficiency. Attention to detail is necessary to reduce drift, improve efficiency of deposition, and increase sprayer output. There are many interrelated factors affecting drift and deposition, and all factors should be addressed if droplets are to be placed into the target canopy and not drift past the grapevines.

Canopy

The vine canopy should intercept the spray plume as it passes outward from the sprayer. The amount of droplet capture will depend on the volume, speed, and direction of the spray cloud. The canopy also plays a major role during the first prebloom spray in early season, when there is very little target area to capture the droplets and most of the spray passes through and is subject to drift. Research was conducted in New York State using an LIPCO™ tunnel sprayer to measure spray retention in the canopy over the growing season. The LIPCO™ uses a recycling feature and any spray not captured in the canopy is recycled. The quantity being recycled was measured and in early season (May 22) Cabernet Franc and Riesling canopies required only 17 gallons per acre. In mid-July the full canopy retained 50 gallons per acre. Research has shown that early season losses to drift and ground deposition can be as high as 45%, depending on the volume applied, the variety, and the extent of canopy development.

If average losses from a traditional airblast sprayer amount to 50% of sprayed material, the tunnel sprayer results clearly show the need for reduced rates in early season spray. Growers should consider reduced volumes along with lower fan speeds, low-drift nozzles, and correctly oriented nozzles.

One of the major areas of concern is alternate-row spraying in early season. Many growers have observed a large spray plume going past the target row, onto the second and third rows. As they are under pressure to improve their timeliness, they choose to spray alternate rows. When a disease outbreak occurs, questions are raised! Our research shows that sometimes the sprayer is calibrated for approximately 50 gallons per acre, but growers frequently forget that it was calibrated for single-row spacing at 9 feet. Driving alternate rows results in only 25 gallons per acre being applied, with very erratic spray deposition on the second row. A disease outbreak is likely to occur with this situation.

Growers often say they haven't time to spray every row. Using a computer model to study sprayer output, we note that a grower spraying 50 gallons per acre, using alternate rows on a 25-acre vineyard, has an output of 3 acres per hour and will take 8.3 hours to spray the vineyard. If the grower set the sprayer up correctly, in terms of nozzle orientation and reduced airflow, and targeted the canopy with 25 gallons per acre, all other factors being equal, the output would be 2.5 acres per hour, or ten hours to spray the entire vineyard. To spray every row, the difference per acre is only 4 minutes.

Many pesticides are applied at one of two rates, early season application being slightly less than full-canopy spraying. Grapevine canopy volume dramatically increases as the season progresses. In Australian vineyards, there is a move to adopt application rates based on a unit canopy row (UCR) basis (volume of pesticide per cubic volume of canopy), omitting the inter-row spacing, but making adjustments for canopy density and foliage type. This method is similar to tree row volume (TRV) used in spraying apple trees. The concept of UCR is currently being tested in Australia, Europe, and the United States. Unit canopy row measurement offers many benefits, including a reduced amount of spray material applied per acre, which in turn, reduces drift.

Meteorological Conditions

The weather plays a critical role in drift management. Wind velocity, relative humidity, and temperature all affect the characteristics of the spray plume. As wind speed increases, spray droplets will drift farther. Humidity and temperature determine how quickly the spray will evaporate. A higher temperature means the droplets are more apt to dissipate and, in some cases, completely vaporize. Evaporation occurs much more rapidly at low humidity. Larger droplets (>200 microns) are affected the least, due to their shorter travel times, but these may still evaporate after depositing on the canopy. Atmospheric inversion is another problematic condition that occurs when the temperature, instead of decreasing, rises with altitude. With cooler air at lower altitudes, there is no tendency for upward air currents to form; this causes pesticide droplets to become trapped at a lower altitude. Spraying in ideal, neutral conditions greatly reduces drift.

Wind speed, temperature and humidity

Wind speed and direction are often the most important variables governing the potential for spray drift. Spraying should only be done when the wind is moving away from sensitive crops, neighbors, and watercourses at a wind speed of 2 to 10 mph. At speeds less than 2 mph, spray can dissipate unpredictably through evaporation; the operator has no idea where the airblast from the fan will drive the plume. By contrast, a slight breeze will serve to indicate the direction of any potential drift and the operator can work accordingly. At speeds greater than 10 mph, medium and large droplets begin to drift off target. Practicality must be a major consideration, particularly in windy locations.

Droplet size is important in determining the amount of drift in conjunction with the wind, temperature, and relative humidity. The size of a droplet strongly influences its trajectory. The droplet rapidly decelerates due to friction until it attains a velocity that is solely a function of its diameter. Air movement, in which the droplet descends, also influences trajectory of the droplet. Droplets with a diameter of less than 200 microns are transported considerable distances, especially under low-humidity conditions (see figure 10.2). Temperature affects the extent

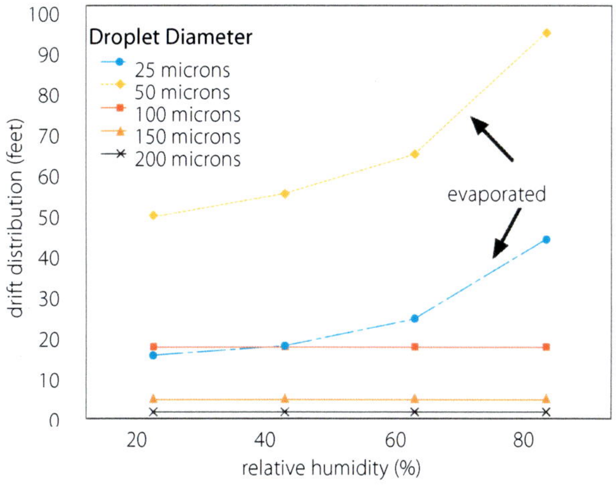

Figure 10.2 ▪ **The effect of relative humidity on mean drift distances.**

Ten mph wind velocity. Ambient temperature = 65°F.

Source: Ozkan (1998).

of drift with small droplets (< 50 microns), but it is difficult to isolate the effects of temperature and humidity, as relative humidity decreases with decreasing air temperature. Training the operator to recognize conditions that lead to excessive drift, such as high winds, fine spray, and inversion layers, is important.

Unstable conditions—do not spray

Warm air rises because it is less dense than cold air. An example of this is can be seen on a sunny afternoon when the vineyards are warm. The air near the ground warms up and rises into the cooler air above. This is referred to as an "unstable surface layer." Air rises and mixes with the air above, carrying pesticide particles with it.

Stable conditions—do not spray

At night the vineyard land gives up its heat rapidly (a process known as "radiant cooling"); the air nearest the ground cools and, being heavier than the warm air above it, remains close to the ground. This situation is referred to as a "stable surface layer," or "temperature inversion." The cold air is unable to mix with the warm air above, causing small suspended droplets to form a concentrated cloud that can lead to long-distance migration of pesticide drift extending for several miles. Clear skies favor radiant cooling and, therefore, favor inversion layers; this can occur in early morning and late afternoon.

Neutral conditions—spray

Cloudy skies and 3 to 10 mph winds are neutral conditions, ideal for spraying.

Changes in the weather may occur as the day progresses. It is often possible to identify an inversion layer by watching smoke rise. If an inversion layer is present, the smoke rises vertically for a short distance and then turns, moving horizontally.

Growers who are located near lakes should be aware that the effect of open water on the dispersion of droplets can be pronounced. The temperature of the lake water rises more slowly than the land around it during the daytime and lowers more slowly than the surrounding land during the night. On clear, sunny days, cold air moves off the water and replaces warm air rising off the land, causing an on-shore breeze. The reverse occurs during clear nights. There should be no doubt that the dispersion of pesticide droplets will be influenced by atmospheric stability. It is clear that the effects of changes in atmospheric stability will be greater on finer droplets; therefore, changes in applicator practice intended to reduce the number of fine droplets will reduce the effects of changing atmospheric stability.

REDUCING DRIFT

Barriers to Drift

Pesticide drift can be dissipated with buffer zones before it reaches sensitive areas, or the potential for off-target drift can be reduced by use of wind screens or shelter belts. Both options are considered here.

Buffer zones

In an ideal world, growers would avoid planting grapevines too close to sensitive areas. A buffer zone should generally be 50 to 150 feet wide, but the distance required will depend on the nature of the buffer zone, the type of sprayer being used, toxicity of the pesticides, trellis design, and prevailing wind direction. Buffer zones are usually located on the downwind side of the vineyard. Vegetative buffer zones may include plants such as trees or shrubs deliberately planted to filter pesticides from the air.

Shelter belts

Augmenting the buffer zone with a natural or artificial shelter belt is an effective means of further reducing drift, and providing some visual screening of vineyard operations. Fast-growing evergreens such as Leyland Cypress may be considered, but severe damage may occur if these are subjected to a heavy snow fall. The primary purpose of a shelter belt or windbreak is to reduce wind speed across the vineyard, reducing drift from the vineyard and, in some cases, allowing the spraying operation to continue when it would be impossible on a more exposed site. Research has shown that most deciduous windbreaks are insufficient to reduce drift by more than a factor of 2 during early season spraying when drift potential is at its greatest. Windbreak efficiency is a function of the height as well as the porosity of the barrier. Droplets and particles that are not trapped by the windbreak can be blown over the barrier and deposited on the leeward side. A porous buffer that slows air movement but does not redirect it up and over will remove more spray droplets than will a solid wall of vegetation or a tall fence. The minimum height of the shelter belt should be at least double the release height of the pesticide. For example, a poorly directed airblast sprayer discharging a spray plume 20 feet into the air will require a tree barrier at least 40 feet high. Mixed plantings of trees with thin, rough foliage and small pubescent leaves are best for providing a porous barrier and entrapping the spray drift.

Improving Existing Application Techniques to Reduce Drift

Weather monitoring equipment

Growers should purchase and use high quality instruments for measuring wind speed, temperature, and humidity. Small hand-held anemometers cost around $100 and provide fairly accurate information on wind speed. Spraying with no wind present is dangerous due to potential problems with vaporization and inversion layer entrapment. Spraying when conditions are too windy is equally hazardous.

Sprayer check

One of the simplest ways to help reduce drift is to maintain and calibrate the sprayer; equipment should be maintained and operated according to the manufacturer's instructions. Studies have shown that many growers apply pesticides with worn-out nozzles, inaccurate pressure gauges, and poorly trained operators. Pressure is one of the main factors affecting application rate; pressure that is too high results in too many fine drift-prone droplets. A well trained operator will notice a spray plume drifting away as wind speed increases or changes direction. Nozzle calibration is a very beneficial practice for growers, as it takes very little time to perform and it ensures that all nozzles are discharging at the correct amount for the target. Choice of nozzle can affect drift by governing particle size. Select low-drift nozzles that minimize the number of drift-prone fine droplets; those less than 150 microns in diameter. Replacing worn nozzles when they exceed the manufacturer's flow rate by 10% and matching the correct nozzle to the type of application are fairly inexpensive practices that help to reduce drift. Application rates will be more constant and easier to adjust. Once the sprayer is properly calibrated, the operator can start positioning the nozzles to the target canopy.

Nozzle orientation

Orientation of the nozzles affects the spray pattern from an airblast sprayer. Traditionally, growers have positioned nozzles radially around the air outlet of the sprayer. The direction of the airstream on a sprayer with a counter-clockwise rotating fan (as viewed from rear of sprayer) carries the droplets upward, over the canopy on the right-hand side of the sprayer, and predominately downward on the left-hand side. Where the air flows, the droplets will surely follow. At Cornell University, an Italian vertical tray patternator (MIBO, Milano, Italy) is used to simulate the canopy at the center of the vine row (see figure 10.3). The patternator comprises a series of collecting trays that are situated at 10-inch intervals above the ground; the sprayer is turned on and the water plume collects in the trays. Water then runs down pipes

Figure 10.3 ▪ The MIBO vertical patternator.

into a row of measuring cylinders, indicating the spray profile or vertical distribution pattern. Different nozzle configurations (number and orientation) are selected and vertical spray patterns determined.

Results from the patternator (see figures 10.4 and 10.5) show not only the great variability in spray pattern produced according to nozzle orientation but also the lack of symmetry between each side of the sprayer. Note that 33% of the spray was overshooting the canopy before a change in nozzle orientation was made. On the Berthoud™ S600EX sprayer used as an example, nozzles set in the "typical growers" pattern (see figure 10.5a) pointing radially outward resulted in a large amount of spray being blown above the target row. The best spray pattern for the grape zone (see figure 10.5b) occurred when the right-hand side nozzles were pointing horizontally and the top two nozzles were 20° below horizontal on the right side, to counteract the upward movement of the air from the fan. Best results occurred with the left side nozzles pointing 45° upward to counteract the downward direction of the air from the fan. The results show the importance of correct nozzle orientation if pesticides are to be effectively applied on target.

To see where the spray is actually going, a grower can create a simple system to check the spray plume. One method is to use a 16-foot high pole with a paper tape

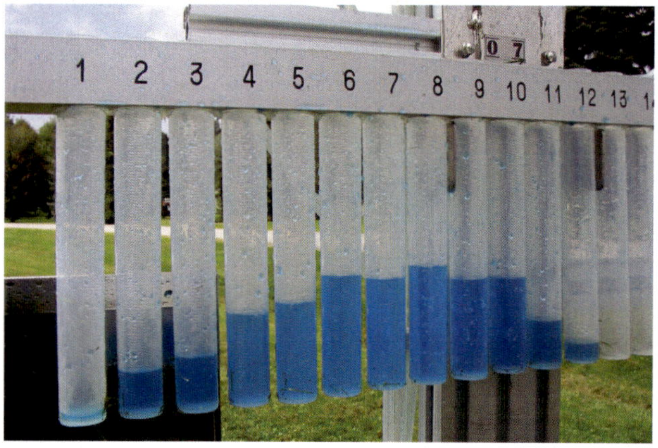

a. Before nozzle adjustment—all right side nozzles pointing radially outward.

b. After nozzle adjustment—all right side nozzles pointing horizontally except the top two are pointing 20° below horizontal.

Figures 10.4 ▪ **The MIBO collecting tubes before and after nozzle adjustment.**

Results shown are for a Berthoud S600EX sprayer, individual sprayer models will vary.

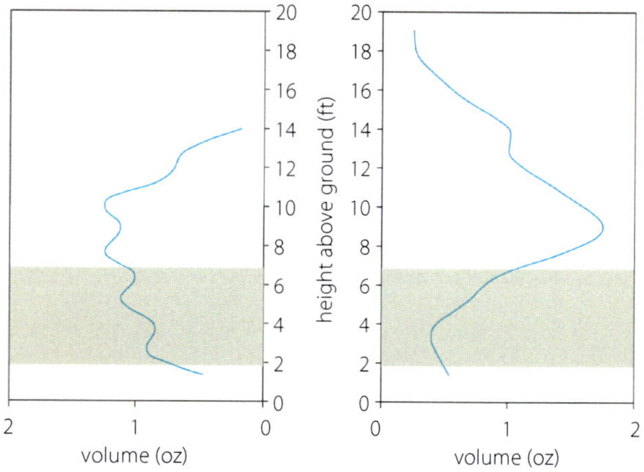

a. Typical nozzle setting — nozzles pointing radially outward.

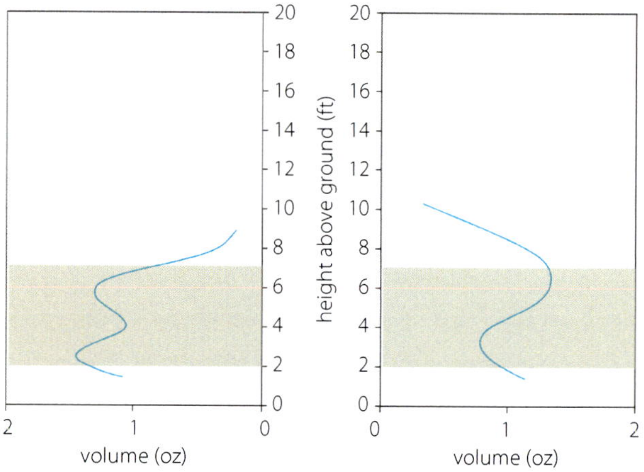

b. Improved nozzle setting—right side, nozzles pointing horizontally with top two 20° pointing below horizontal; left side—nozzles pointing 45° above horizontal.

Figure 10.5 ▪ **Spray pattern for two nozzle settings.**

Shaded area indicates canopy area 2 feet through 7 feet above ground level.

stapled along the leading edge. Place the pole between two vines within the row and spray a mixture of clean water and food coloring. Travel between the rows, spraying out the mixture. The spray will stain the paper where it hits. By looking at the colored spray droplets on the paper, the grower can alter the orientation of the nozzles or deflectors until the spray is only hitting the portion of the vertical pole that corresponds to canopy height.

Air induction nozzles

Air induction (AI) nozzles, when properly used, can reduce drift by at least 50%. The principle behind AI involves creating a larger droplet that won't drift as far but that will still maintain good leaf and fruit coverage. Air induction nozzles are available either as flat fan or hollow cone nozzles where an internal venturi creates negative pressure inside the nozzle body. Air is drawn into the nozzle through one or two holes in the nozzle side, mixing with the spray liquid. The emitted spray contains large droplets (approximately 375 microns) filled with air bubbles and virtually no fine, drift-prone droplets. The droplets explode on impact, dissipating the energy, dispersing the liquid, and producing coverage similar to that achieved by conventional, finer sprays. Figure 10.6 shows the results of a drift comparison trial at Cornell University, where a Berthoud™ airblast sprayer was fitted with flat fan air induction nozzles and compared to the same sprayer fitted with hollow cone nozzles. The results clearly show that the conventional hollow cone nozzles created plenty of drift, which was detected up to 6 rows (54 feet) from the target row. When AI nozzles were fitted, the larger droplets came back down and droplets were recorded up to 3 rows from the target, demonstrating that it is crucial to orientate all nozzles to reduce drift, even from a drift-reducing AI nozzle. Balsari et al. (2005), in trials in Italian vineyards, concluded that AI nozzles, in comparison to conventional hollow cone nozzles, not only reduced ground deposits by 20% (see figure 10.7) but also reduced potential drift by up to 70% (see figure 10.8). They found no significant difference in the vine incidence of downy mildew (*Plasmopara viticola*) between nozzle types.

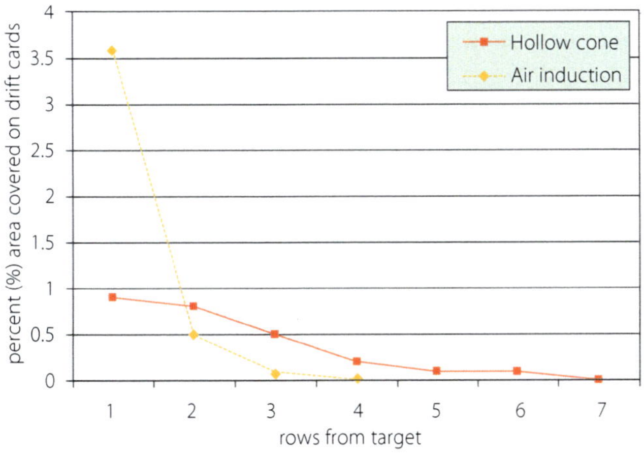

Figure 10.6 ▪ Drift comparison: flat fan air induction nozzle versus hollow cone nozzle (Berthoud™ sprayer).

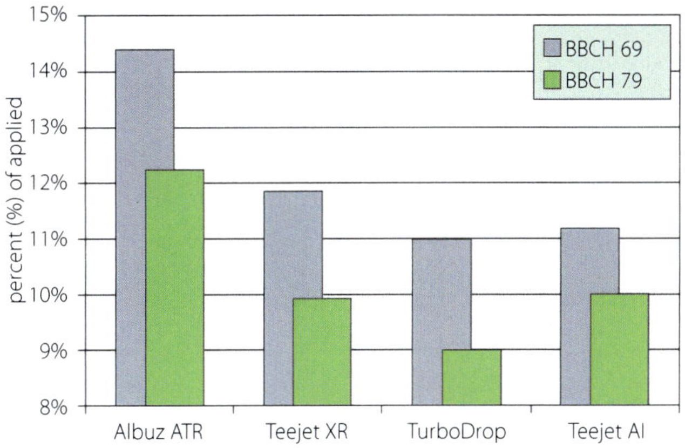

Figure 10.7 ▪ Spray ground losses for different nozzles.
See key below. *Source: Balsari et al. (2005)*

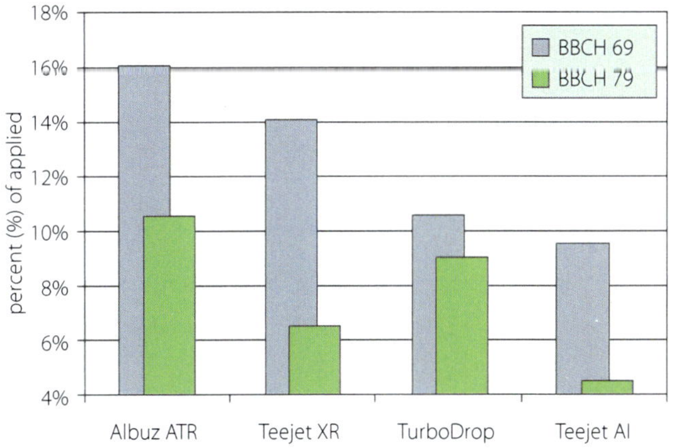

Figure 10.8 ▪ Potential spray drift for different nozzles.
See key below. *Source: Balsari et al. (2005).*

Key to Figures 10.6 and 10.7:
BBCH 69: growth stage: end of flowering
BBCH 79: growth stage: majority of berries touching
Albuz ATR: hollow cone nozzle
Teejet XR: a flat fan nozzle
Turbodrop: an air induction hollow cone nozzle
Teejet AI: an air induction flat fan nozzle

Regulating Air Speed

PTO speed

Regulating the power-takeoff (PTO) speed of the tractor is an inexpensive way to reduce drift. Early- to mid-season sprays are frequently applied at full fan speed, resulting in a mighty plume of pesticide going toward a small leaf area or up into the air. Is it necessary to have the fan rotating at all? Air helps transport droplets to the target, protects against the effects of wind blowing across the vineyard, and can shake the canopy to aid penetration. Unfortunately, too much air can transport droplets through the target canopy, leading to excessive shaking of the leaves, which may result in blowing off droplets already deposited. Excessive air can "shingle" leaves, creating an impenetrable wall for the spray plume, and resulting in the spray cloud going up and over the canopy. Air-assistance aimed at directing the spray to the canopy and improving penetration into the canopy has shown to be effective, but over and under supply of air energy can result in inadequate deposition and poor control of the target pest or disease as well as potential for off-target drift. Drift and deposition efficiencies are often erratic, and research shows that when the volume of air is increased there is no corresponding improvement in the quality of pesticide application. Increasing air volume reduces deposition on the spray side of a fully developed canopy while it doesn't affect deposition on the opposite side of the canopy. Lowering the PTO speed reduces fan speed, preventing excessive amounts of air from blowing pesticides through the target canopy and allowing improved deposition to occur.

In trials with an air-shear type sprayer (AgTec™), drift was detected on water-sensitive cards fitted to 14-foot tall poles that were spaced at 20-foot intervals up to 80 feet from the target row. When the tractor PTO speed was reduced by 25%, drift was reduced by 75%. The reduced air speed also increased droplet size (due to the operating characteristics of air-shear nozzles), further reducing the effects of drift. Reducing PTO speed with a conventional airblast sprayer will have a similar effect.

Hydraulic drive

Using an hydraulic motor to drive the sprayer fan will allow the operator to regulate air velocity. An hydraulic control valve can be fitted in the tractor cab to allow the operator to infinitely vary the fan input speed from 0 to 540 revolutions per minute (rpm). The operator can reduce air speed during early- to mid-season applications and increase it for full canopy applications. When approaching the end of a block of vines near neighboring properties, drift-sensitive crops, roads, or watercourses, the operator can easily adjust the airflow and, therefore, reduce the spray plume. Care should be taken to ensure the tractor has a high enough oil flow (gallons per minute) to drive the hydraulic motor at speed.

Cornell doughnuts

Round, doughnut-shaped attachments (see figure 10.9) restrict air intake and therefore reduce air flow through the airblast sprayer. A simple plywood doughnut can be constructed using a jigsaw. This can then be attached to the sprayer air intake using bolts. The diameter of the intake is measured and a hole is created in the center of the plywood circle. A safety grill is fitted to prevent fingers and large debris from entering the fan. For early season, a ½ diameter air intake doughnut can be used, allowing enough air to penetrate just the target row. A ⅔ air intake hole can be used for early- to mid-season to allow more air to flow. Finally in a full canopy, no doughnut is required. Results of trials indicate excellent drift reduction from a simple yet effective device.

Directing Airflow

End plates

In many traditional situations, such as spraying the last row of a block of vines, only one side of the airblast sprayer is required. In such cases, the operator switches off one side of nozzles, resulting in the spray plume moving inward toward the block. Unfortunately, many fine droplets (< 150 microns) can be sucked back in by the fan, resulting in fine droplets blowing out in the airstream of the non-spraying side and away into the sky, perhaps over neighboring properties or drift-sensitive crops. A quick inspection of the fan blades on an airblast sprayer (see figure 10.10) shows the recirculation effect of the air intake and small droplets. A metal shroud can be used to block any air travel on the non-spraying side of the sprayer (see figures 10.11, page 206). Using the end

a. Doughnut hole of ½ of the diameter of the sprayer intake — for early season.

b. Doughnut hole ⅔ of the diameter of the sprayer intake — for early to mid-season.

Figure 10.9 ▪ Cornell doughnut fitted onto the air intake of an airblast sprayer to reduce airflow.

plate on the last couple of rows in the vineyard allows the operator to spray inward, reducing drift by 50%.

Tower sprayers and deflectors

Tower sprayers (see figure 10.12, page 206) and deflectors (see figure 10.13, page 206) are better than a conventional airblast sprayer at reducing drift and targeting the spray in the canopy; these are further described below in the section "Sprayers Designed to Reduce Drift and Improve Deposition." The conventional airblast sprayer sends droplets in an airblast from a central fan upward and outward into the canopy; whereas the addition of a tower, or deflectors, creates a horizontal air curtain. In

Figure 10.10 ▪ Air intake and fan blades covered in recirculated spray.

some cases, the horizontal airflow can be adjusted by internal deflectors to direct the air toward the fruiting zone. A number of manufacturers sell "grape towers" as a retrofit to existing airblast sprayers.

Axial Fan Size, Speed and Air Induction Nozzles

The axial fans of airblast sprayers typically move about 12,000 cubic feet of air per minute (approximately 20,000 m^3 per hour). The use of air induction nozzles with airblast sprayers will result in a 75% reduction of drift. In order to accommodate various crops and seasonal canopy development, a number of modern sprayers are fitted with adjustable pitch propellers to provide a variable airflow. Operators can manually adjust blade pitch either by turning a handle or altering individual blades. Care must be taken when adjusting individual blades to ensure that all blades have the same pitch. It is important to note that, as pitch is altered, power requirement is also altered. Many modern airblast sprayers are fitted with a gearbox but, unfortunately, they offer only slow, fast, and neutral settings. The slowest setting speed may be too fast for most canopies, particularly in early- to mid-season applications. Growers should assess air volume requirements by observing spray penetration into the

a. End plate open to allow spraying of both sides.

b. End plate closed to block spraying from that side of the sprayer.

Figures 10.11 ▪ End plate to reduce drift at the end of a block.

canopy and the amount of spray going up and over, or through the canopy.

Foliage Sensors

Infrared or ultrasonic sensors detect the absence or presence of a vine or canopy. A sensor emits a ray of light or sound toward the canopy and any resulting reflection, due to the presence of vines or canopy, is picked up by the sensor and automatically opens or closes a valve near the nozzles allowing spray to be switched on or off. The use of sensors reduces overspray of the canopy and can reduce drift by 50%.

Tunnel Sprayers

Tunnel sprayers represent the best method for reducing drift. Tunnel sprayers are fully described in the section below titled, Sprayers Designed to Reduce Drift and Improve Deposition. Very little spray gets out of the tunnel spraying system; this allows for a 90% reduction in drift. If AI nozzles are used with the tunnel sprayer, 99% drift reduction can be attained.

Drift-Reducing Additives

A number of manufacturers supply drift-reducing agents. Most of these agents work by increasing droplet size, as larger droplets are less prone to drift. Be aware

Figure 10.12 ▪ An example of a tower sprayer.

Figure 10.13 ▪ An example of deflectors.

that not all drift-reducing additives can withstand the higher pressures associated with fruit sprayers and, interestingly, often create more drift than would be present without the use of agents. Growers should seek independent verification of the effectiveness of drift-reducing additives.

Drift is impossible to eliminate but can be minimized. Implementing some of the above methods will greatly reduce drift and improve the efficiency of the spray application, thus saving time, money, and future problems.

SPRAYERS DESIGNED TO REDUCE DRIFT AND IMPROVE DEPOSITION

The following sprayers are available to growers in the eastern United States. Some of these reduce drift when compared to the traditional airblast and mistblower sprayers. In many cases, their novel design concomitantly improves deposition. It should be noted that drift will still occur in early season due to the sparse canopy to intercept the spray; results with a tunnel sprayer are a possible exception to this tendency. The sprayers described here direct the spray cloud into the canopy, while a traditional sprayer directs air upward and outward. Generally, sprayers that direct air into the canopy are considered superior to those of traditional design.

Developed in recent years, the tower, or "grape tower," is an option available from most airblast sprayer manufacturers, including Rears Manufacturing of Oregon and Durand-Wayland Inc., of Georgia (see figures 10.12 and 10.14). Towers comprise a steel volute, a chamber that prevents the air escaping from the top 25% of the fan outlet and redirects that air horizontally toward the canopy. Hollow cone nozzles are fitted in a vertical plane along the air outlet. The operator is able to manually switch individual nozzles on and off to match canopy height. An example of variation in the design of the tower is the air deflector manufactured by Slimline Engineering and sold as the Major™ sprayer. The Paris™ sprayer from France is a traditional airblast sprayer imported by Lakeview Harvesters, Ontario, Canada. This sprayer has an hydraulically driven fan encased in a chamber; the chamber outlet uses a grid to vector the air into a horizontal pattern.

Trials at Cornell University with the Rears, Slimline Major, and Paris sprayers show that the internal deflector system directs the spray-laden air in a horizontal pattern, as opposed to the conventional airblast, which sends the spray plume upward and outward. Owners of airblast sprayers may be able to purchase "grape towers" from some manufacturers as a retro-fit attachment; otherwise, they can consider the use of deflectors.

GreenTech

The GreenTech™ sprayer (see figure 10.15), imported from Australia by OXBO International, comprises a series of fan units mounted on a gantry that straddles the vine canopy. Hydraulic fan or cone nozzles are mounted in front of the fans and the airflow directs the droplets toward the canopy. The diagonally opposed fans create good turbulence and deposition within the canopy.

Figure 10.14 ▪ Grape tower mounted on a Durand-Wayland sprayer.

Figure 10.15 ▪ The GreenTech sprayer.

Research at Cornell University indicates that maximizing the convergence angles gives the best penetration and deposition, with the units pointing backward by 10°. The fans are driven by electric motors, with power supplied from a PTO-driven generator.

Italian CIMA Sprayer

The Italian CIMA™ sprayer (see figure 10.16) uses multiple outlets ("fingers") to direct air and spray toward the canopy. The basic model comprises outlets mounted at the base of the sprayer, moving air upward and into the canopy. The individual units comprise 4 or 5 fingers and can be moved into various positions, up or down and forward or backward. The next model comprises upper and lower units; the upper units are mounted at the top of an air column and can be positioned to spray downward onto the canopy. The CIMA uses a tractor PTO-driven, high-speed, low-volume fan to produce the airflow and air-shear nozzles to create the droplets. Trials at Cornell University show that using the top set of units doubles the spray deposition in the top third of the canopy.

Danish Hardi Tango Sprayer

The Danish Hardi Tango™ sprayer (see figure 10.17) comprises a small PTO-driven high-speed, low-volume fan to produce airflow. The air passes through a series of corrugated pipes that can be adjusted forward or backward, up or down, to direct the spray toward the target. Droplets are created by traditional hollow cone nozzles situated in the outlet hole of the air-carrying pipes. Trials at Cornell University showed that by changing the direction of the air outlets to point backward, drift was reduced and penetration and deposition within the canopy was greatly improved.

Figure 10.17 ▪ The Hardi Tango sprayer.

Berthoud Sprayer

The Berthoud sprayer (see figure 10.18) comprises a PTO-driven fan that transports the air in long pipes to the outlet near the canopy. The outlet is adjustable, moving up and down, forward and backward, allowing the operator to adjust air direction according to variables such as canopy size and training system. The sprayer is available with either air-shear or hydraulic nozzles.

Figure 10.16 ▪ CIMA sprayer.

Figure 10.18 ▪ Berthoud sprayer.

Micronair Cage

The Micronair™ cage (see figure 10.19), manufactured in England by Micron, and the Proptec™ (see figure 10.20), manufactured by Ledebuhr Industries, are very similar. Both use a rotary cage to create droplets that can be directed into the crop canopy. Liquid pesticide is fed into a high-speed spinning cage from which centrifugal forces spread the liquid. The majority of droplets are the same size, depending upon flow rate and cage speed. The fine droplets are then carried toward the target in a directed airstream developed by small rotating fan blades mounted on the same shaft as the spinning cage. The hydraulically driven cage and fan assembly can be located on the end of a gantry overhanging the canopy, while a second unit can be mounted diagonally opposite, resulting in very good turbulence of air and droplets within the canopy.

Figure 10.19 ▪ The Micron controlled droplet applicator.

A number of growers have fitted a third fan/cage unit in the uppermost corner of the gantry to aid deposition. Trials of both machines showed improved deposition at application rates lower than those of a standard airblast sprayer.

LIPCO Tunnel Sprayer

The LIPCO™ tunnel sprayer (see figure 10.21) comprises a plastic tunnel mounted on a frame that encloses the grapevine canopy during spraying. Mounted inside the tunnel walls are two vertical spray-booms fitted with hollow cone or flat fan nozzles that direct spray horizontally into the grape canopy. Any excess spray runs down the inside of the tunnel wall and is collected in a trough at the base of the tunnel, where it is recycled back to the main tank. In early season trials, up to 75% of spray was recycled; at full canopy very little recycling occurs, but average reduction in pesticide use amounted to 35% over the season. Research conducted in Germany showed that the tunnel sprayer reduced drift by 90%. Drift was reduced to 99% when air induction nozzles were used in conjunction with the tunnel sprayer. Lightweight plastics or fiberglass construction and well designed hitches allow a fair degree of maneuverability. In contrast to the large spray material plume created by an airblast sprayer, the lack of visible spray discharge beyond the tunnel enclosure reduces public concerns about pesticide application. Trials at Cornell University showed that the LIPCO™ tunnel evenly deposited spray throughout dense canopies, even without the fan in operation.

Figure 10.20 ▪ Proptec controlled droplet applicator.

Figure 10.21 ▪ LIPCO™ tunnel sprayer.

REDUCING DRIFT DURING HERBICIDE APPLICATION

Some of the commonly used herbicides can damage vines, so care must be taken with application. Herbicides, although relatively inexpensive, require good application techniques to improve deposition and reduce drift. Off-target application wastes money, reduces deposition on the target plant, can cause damage to vines, and can lead to pesticide residue on grapes.

Boom Applicators

Two main types of boom applicator are available. With one type, a tractor-mounted frame straddles the canopy, spraying under the vine on each side of the target row. Alternatively, a boom may be fitted to the front of the tractor, spraying one side only of adjacent rows. Herbicide spray volumes are typically in the range of 20 to 30 gallons per acre. Nozzle hoods and break-back devices are important considerations. A hood will protect the vines from drift created by the small droplets being emitted from a conventional flat fan nozzle (see figures 10.22 and 10.23). A simple hood can either be purchased or made from a used plastic drum cut in half. A break-back device will protect the sprayer boom and nozzle from damage caused by inadvertently striking a trellis post or similar object. Break-back devices normally incorporate a spring-loaded arm.

Figure 10.22 ▪ A full-width tarpaulin cover to reduce drift.

Figure 10.23 ▪ Half a plastic container provides a shield to reduce drift.

Correct Nozzle Selection

Correct nozzle selection is one of the most important, yet inexpensive, elements of drift reduction. A nozzle's droplet size spectrum determines deposition and drift and is referred to as "spray quality." Modern nozzle catalogs provide information on spray quality for each nozzle. Select a MEDIUM quality spray for herbicide application. Conventional flat fan nozzles produce droplets in the range of 10 to 450 microns. Drift is a major problem with droplets less than 150 microns and can affect both foliage and the grapes (see figures 10.24 and 10.25). Increasing the Volume Median Diameter (VMD) will certainly reduce drift, but a droplet that is too large (>300 microns) will bounce off the leaves to the ground, thus causing pollution, wasting money, and resulting in less product on the target. Growers should consider the following when selecting nozzles.

Conventional flat fan nozzles

Nozzles with an 80° angle produce coarser droplets than do nozzles with a 110° angle operating at the same flow rate; however, 80° nozzles require the boom to be set at 17 to 19 inches above the target whereas 110° nozzles can be set at 15 to 18 inches above the target; the lower the boom, the less chance of drift. Spray quality is fine to medium at 15 to 60 pounds per square inch (psi).

Pre-orifice flat fan nozzles

The internal design of this nozzle results in reduced internal operating pressure and coarser droplets than those produced by a conventional flat fan nozzle. Pre-orifice flat fan nozzles are available as 80° or 110° nozzles. Spray quality is medium-to-coarse at 30 to 60 psi.

Figure 10.24 ▪ Spray drift effect on vine canopy.

Figure 10.25 ▪ Spray drift effect on grapes.

"Drift-guard™" nozzles sold by TeeJet are an example of pre-orifice flat fan nozzles.

Turbo-Teejet

A turbulence chamber in Turbo™ nozzles produces a wide angle flat spray pattern of 150°. Spray quality is medium-to-coarse at 15 to 90 psi. Nozzles can be set at 15 to 18 inches above the target.

Air induction nozzles

Air induction, air inclusion, or venturi nozzles are flat fan nozzles in which an internal venturi creates negative pressure inside the nozzle body. Air is drawn into the nozzle through two holes in the side mixing with the spray liquid. The emitted spray contains large droplets filled with air bubbles with virtually no fine, drift-prone droplets emitted. The droplets explode on impact with leaves and produce coverage similar to conventional, finer sprays. Air induction nozzles work best at higher pressures, in the range of 75 to 90 psi. They are available at 110° fan angles, so boom height may need to be adjusted to 15 to 18 inches. The use of adjuvants will certainly help create bubbles. Air induction nozzles work very well for herbicide application.

Sensor-Controlled Applicators

Sensor-controlled pesticide applicators (see figure 10.26) use optical sensors designed to determine where weeds are located. These sensors, coupled with a computer controller, regulate the spray nozzles and apply herbicides only where needed, thus considerably reducing herbicide use. A computer-controlled sensor detects chlorophyll in plants and then sends a signal to the appropriate spray nozzle, applying the herbicide directly to the weed. The operator calibrates the system to bare soil or pavement, allowing the computer to determine when there is a weed present. Sensor-controlled applicators are often mounted on all-terrain vehicles (ATV), John Deere Gators™, and other vehicles including tractors or trucks. Typically, this type of applicator can be used at speeds up to 10 mph. A complete sensor-controlled system consists of a chemical tank, pump, battery power, computer controller, optical sensors and spray nozzles.

Benefits of sensor-controlled applicators

These are benefits of sensor-controlled applicators;

- Reduced amount of herbicide applied
- Reduced potential for groundwater contamination
- Ability to apply herbicides in dark or light conditions
- Reduced herbicide drift, if wind-deflecting shields are present

Figure 10.26 ▪ Sensor-controlled applicator.

Controlled Droplet Applicators (CDA)

Traditional flat fan nozzles produce a wide range of droplets (10 to 450 microns); some drift, some drip off the leaves, and other droplets adhere to the target leaves as intended. A controlled droplet applicator (CDA) herbicide applicator (see figure 10.27) incorporates an electrically driven spinning disc under a large plastic hood, bristle hood, or dome. The perimeter of the spinning disc has small teeth that break up the liquid herbicide into droplets, of which 95% are the same size. The speed of the spinning disc dictates droplet size—the faster the disc, the finer the droplet size. As there are no large or small droplets in the CDA spectrum, all the droplets stick to the plant and reduced rates can be applied (for example, 1 to 8 gallons per acre). Reduced rates improve logistics, allowing for better use of time when spraying conditions are ideal. The smaller application rate means a smaller tank on the sprayer, which, in turn, means less compaction, allowing the tractor or ATV earlier access after rain. Various widths of hood or dome can be selected and are fitted with break-back devices. A bristle skirt may be desirable on rough or rocky ground. An optional plastic cover can be fitted over the bristle skirt for use with young vines. Covers are lightweight, relatively inexpensive, and very maneuverable.

Flame Applicators

Flame applicators simply use a flame to destroy weeds. These are the ultimate in drift reduction! Most flame applicators burn liquid propane gas to create a flame with a temperature near 2000°F. The flame is applied directly to the weeds by using a hand-held wand or boom-mounted torch attached to a tractor or an ATV. The flame is applied to the weed for only a short period of time, usually about 1/10 of a second. The length of time the flame is applied depends on the age, size, and tenderness of the weed. It is recommended that the flame be applied to weeds at 1 to 3 inches in height, typically in the spring and early summer. The weeds are not burned. Rather, the water inside the plant cells boil, causing the cells to burst when exposed to the flame. With destroyed cells, the plant is unable to transport water and continue photosynthesis and is, therefore, likely to wilt and die. Flame applicators should be used only when there is little or no potential for setting fire to dry plant material and trellis posts. Benefits of flame applicators include: (a) no chemical residues on land or in water, (b) no drift, and (c) no exposure of workers to chemicals. Cost of fuel and lack of persistence are the two chief drawbacks.

Figure 10.27 ▪ Controlled droplet applicator with canvas-covered bristles

THE HUMAN FACTOR

The focus of this chapter has been on the machine and the environment. It is important, however, to include a brief discussion on the crucially important human factor that impinges on spray drift mitigation. Here we look at the responsibilities of the operator, the operator's supervisor, and the vineyard manager.

Sprayer Operator

The operator has a direct influence on drift mitigation. The sprayer operator should simultaneously be student, meteorologist, mechanic, diplomat, grower, and well motivated tractor driver. Sprayer operators must always be students, attending training sessions on the safe and efficient use of pesticides. Operators should learn about new techniques to improve deposition and reduce drift, as these techniques and devices are constantly being researched and developed for commercial use. Operators must also keep abreast of all regulatory developments. Operators must monitor and recognize changes in the weather, such as wind speed and direction, humidity, and temperature, and they must be aware of drift-sensitive areas relative to meteorological conditions.

Maintenance of sprayers is crucial if drift and environmental pollution are to be minimized. There are currently no state or federal requirements for sprayer inspections in the United States; however, a properly calibrated and adjusted sprayer could result in a more productive spray season, saving time and money and reducing drift. It is interesting to note that sprayer testing is compulsory in many European nations.

Operators obviously should know exactly what product, or products, they are applying and what the target pests are. This is as much for their own safety and pest management efficacy as for grower-neighbor relations. Increasingly, the public wants to know about pesticide usage on farms. Good neighbors communicate with one another, and vineyard owners, managers, and operators need to be able to address the "Whats?" and "Whys?" of pest management.

Operators must keep records detailing application conditions. These are important in case there is any subsequent claim by a bystander or neighbor against the grower, or in the event of a failure of the product to control a pest. Records should be kept in accordance with state and federal requirements.

Good driving habits are necessary if drift is to be reduced. Switching the nozzles off as a row is finished and not turning them back on until the sprayer is in the next row will dramatically reduce drift. Operators must maintain a constant speed and pump pressure. Driving too fast is a common mistake; the spray plume must be given time to penetrate the canopy. Operators also need to be able to critically adjust the sprayer for seasonal changes and for changes in canopy geometry and leaf area index.

Vineyard Manager and Timeliness

The vineyard manager must be knowledgeable about what the operator needs to know, along with the "Six Ps" of vineyard machinery management: Proper Prior Planning Prevents Poor Performance!

Management of the spraying system has assumed much greater importance due to increased financial constraints, the need for timeliness, operator safety, and increasing legislation regarding the environment—particularly in reference to drift. Frequently, many growers are unable to spray their entire vineyard within two days because their sprayer capacity is insufficient. If spraying conditions deteriorate, the potential for drift will increase. Managers should plan the spraying operation and consider using vineyard field work cards that detail the amount of pesticide to be mixed or sprayed per block. The work cards should be given to the sprayer operator, who can then get on with the job at hand. Timeliness is often lost if the operator is trying to calculate quantities to mix while the sun is shining and the wind is low.

Providing a back-up service to enable the operator to spray in the event of mechanical problems is of great importance in larger vineyards. A crop sprayer should be regarded as part of a spraying system rather than just a machine. Operators should pinpoint any aspect of the system that is impeding the task. System components that should be considered include the following:

Single versus multiple rows

The effect of changing from a single row to multiple rows is only noticeable after other parts of the spraying system have been considered, as spraying width only affects the output.

Tank

A large tank reduces the proportion of time spent traveling to refill the sprayer within the spraying cycle; however, larger tanks mean greater weight and increased potential for soil compaction.

Availability of water

The output of the spraying system can be greatly increased if spray water is close to the vineyard. Large overhead tanks can be used to speed up sprayer filling, with the tank filling over time and quickly filling the sprayer by gravity during refilling operations.

Nurse tanks

The use of a nurse tank for water will dramatically increase sprayer output by decreasing the time spent traveling to the water source. Nurse tanks can be emptied with the sprayer self-fill hose or a centrifugal pump.

Pumps

Centrifugal pumps transfer liquids at high speed and low pressure and are ideal for filling sprayers or nurse tanks. The pump can be driven by a small gas engine. Filling at 100 gallons per minute helps to make a rapid turn-around of the sprayer, thus taking full advantage of good spraying conditions.

Mixer tanks

When a second person is employed, a very high output can be achieved, as the sprayer driver can remain on the tractor or sprayer and the second person can be in charge of measuring chemicals, mixing, and filling. The modern trend toward engineering controls such as a chemical fill probe or a mixing bowl ensures safety because the operator doesn't have to clamber over the sprayer with a container of concentrated pesticide. On large vineyards the mixer tank can discharge mixed spray straight into the sprayer, as it is already mixed and doesn't require measuring.

Application rate

The volume of water required to spray vine canopies varies as the canopy develops over the season. Research shows that early season applications require far less water than mid- to late-season applications. Adjusting the volume-per-acre measurement drastically reduces the amount of water required. A tank of a given volume will provide longer periods of spraying and therefore will reduce the relative amount of time spent traveling to refill.

Vineyard

Vineyard size, shape, position, and topography will affect system output. Small irregularly shaped and undulating vineyards reduce output. Multi-row sprayers require wide headlands on which to turn if output is to be maintained. The distance of the vineyard from a water source will affect output, particularly if it is not possible to use a nurse tank.

Drift and the Man—Machine Interface

Spray drift, besides going off-target, contaminates the tractor. Modern tractors with cabs protect the driver to some degree. Trials were conducted to determine the level of pesticides remaining in the tractor cab after many days of spraying activity. Surface levels of pesticides were measured at various interior and exterior locations on five tractors—four with cabs using carbon-bed air-filtering systems and one without—and three commercial spray rigs. All equipment was used on farms or by commercial spray contractors for both fruit orchard and vegetable crop applications. Sampling was conducted at the end of the growing season. The highest chemical concentration levels were found on steering wheels and gauges and in dust collected from the fabric seats. Contamination of the steering wheel and seat areas could result in operator exposure to pesticides if the tractor is used for other purposes, such as mowing grass, during which the operator will not be using personal protective equipment (PPE). Pesticide residues on the outlet louvers from the air filtering systems of the enclosed cabs often included more compounds, at higher levels, than did samples from the inlet louvers. A carbon filter once saturated may release compounds back into the tractor cab environment. Information on carbon filter efficiency and specificity of chemical removal in relation to filter size and chemical break-through trends is not generally available from the manufacturer; however, this will be important in future studies, to optimize filter replacement schedules. If a tractor cab is not available, personal protective equipment should be worn to prevent operator contamination (see figure 10.28).

Figure 10.28 ▪ Personal protective equipment must be worn to avoid operator exposure.

CONCLUSIONS

The aim of this chapter was to identify the social and environmental problems associated with pesticide spray drift and to illustrate how vineyard owners and operators can take a systems approach to mitigating drift. Starting with an understanding of the factors that contribute to spray drift, we have shown techniques and equipment that can be used to modify the spray particle size, the distribution of particles that contact and are retained by grapevine (or weed) foliage, and the proportion of material that stays in the vineyard, as opposed to moving outside the vineyard. It is incumbent upon the vineyard manager to wisely use the appropriate technologies and practices that are currently available and, at the same time, to be aware of research that is progressing in this area.

References

Balsari, P, Marucco, P. and Tamagnone, M. 2005. Quality of distribution, spray losses and drift during spray application with different nozzle types used in vineyards. In proc. 8th workshop on spray application techniques in fruit growing. June 2005, Barcelona. eds. Gil, E., Solanelles,F., Escobar,C., Planas,S., Rosell,J.R., Val, l. pp. 129–130. Barcelona: Universitat Politecnica de Catalunya.

Ganzelmeier, H. 1993. Drift of plant protection products in field crops, vineyards, orchards and hops. A.W.P.P -B.C.P.C, Second International Symposium on Pesticides, Strasbourg. Farnham, British Crop Protection Council. 1: 125–132.

Ozkan, H.E. 1998. Effect of major variables on drift distances of spray droplets. Extension fact sheet AEX 525-98 Columbus, OH: The Ohio State University, 12pp.

Spray Drift Task Force, 1998. A summary of airblast sprayer application studies. Stewart Agricultural Research Services, Inc. Macon, MO 63552, 10pp.

Further Reading

Landers, A.J. 2000 Farm machinery selection, investment and management. 152pp Sidcup: Farming Press (Miller Freeman).

Landers, A.J. 2002. The answer is blowing in the wind. In: *Aspects of Applied Biology* 66. International advances in pesticide application. pp. 177–184.

Landers, A.J. and Farooq M. 2004. Factors influencing air and pesticide penetration into grapevine canopies. In: *Aspects of Applied Biology 71*. International advances in pesticide application. pp. 343–348.

Landers A.J. 2004. Prevention is better than cure- reducing drift from vineyard sprayers. In Proc. International meeting on spray drift. Kona, Hawaii, 26–29th October. pp.116–124. Pullman: Washington State University.

Further Information

Further information on vineyard spraying can be found at HTTP://WWW.NYSAES.CORNELL.EDU/ENT/FACULTY/LANDERS/PESTAPP

A simple vertical patternator can be constructed in the farm workshop using readily available materials. A build list and photographs can be found at HTTP://WWW.NYSAES.CORNELL.EDU/ENT/FACULTY/LANDERS/PDF/PATTERNATOR.PDF

Videos showing calibration and nozzle selection may be found on the Internet at www.youtube.com.

Type in: Calibration of airblast sprayers for vineyards part 1 selecting and changing nozzles

Type in: Calibration of airblast sprayers for vineyards part 2 measuring liquid flow

—11—

Disease Management

Authors
Wayne F. Wilcox, Cornell University
Tony K. Wolf, Virginia Tech

Grapevines and grapes are subject to attack by numerous disease-causing organisms. Recognizing and understanding the nature of these diseases is essential to the design and implementation of management practices. This chapter describes the major diseases that routinely threaten bunch grapes in eastern North America and reviews management options. Readers are advised to consult state-specific pesticide (insecticide, fungicide, herbicide) recommendations, as legally prescribed products vary from state to state, as well as from year to year. Pesticide recommendations are typically updated each year and issued by the Cooperative Extension division of land-grant universities.

Infectious diseases of grapes are caused by microorganisms that include fungi, bacteria, phytoplasmas, and viruses. Abiotic, noninfectious disorders can be caused by nutritional deficiencies or imbalances, climatic extremes, chemical toxins, or other environmental stresses. This chapter focuses on identifying the infectious diseases common to grapevines in eastern North America, understanding the factors that promote the development of these diseases, and describing general management principles. Specific management programs will vary according to varietal susceptibility, regional importance of the various diseases, and climatic conditions during each individual season, and should be designed accordingly.

FUNGAL DISEASES

With the exception of canker and internal wood diseases, most fungal diseases of grapevines are not perennial, that is, the causal organisms must reinfect the plant each season. Therefore, disease control is aided by limiting the amount of fungal inoculum (spores) capable of initiating new infections once the season begins. Of course, one of the best ways to accomplish this is to provide good disease control the previous year. Additionally, inoculum of some fungal pathogens can be reduced through sanitation programs focusing on the removal of old diseased tissues in which these organisms overwinter. Furthermore, most fungal diseases intensify with the frequency and duration of moisture on the plant tissues, so control is aided by the selection of planting sites and canopy management practices that promote rapid drying of the fruit and foliage. Finally, commercial control of most common fungal diseases also requires the use of fungicides at certain times of the year. Knowing when fungicides are most likely to be needed is the key to providing the necessary level of disease control without making unnecessary pesticide applications. Because grape varieties vary in their susceptibility to individual diseases, spray programs can often be adjusted on a block-by-block basis to account for the partial resistance of some varieties and the high susceptibility of others (see table 3.2, page 62).

In addition to legal (label) restrictions on the use of fungicides for disease control, grape growers and vintners must recognize that excessive residues of certain fungicides such as sulfur, copper-containing compounds, and captan can affect fermentation, impart undesirable aromas, or mute the varietal character of the resulting wine. Fungicide selection within the final 30 to 45 days of harvest must therefore consider not only efficacy and legality, but also potential enological impacts. Grape growers are encouraged to closely cooperate with grape buyers or the winemaker to minimize potential negative effects of spray materials on wine quality (see chapter 15, page 278).

Powdery Mildew

Powdery mildew, caused by the fungus *Uncinula (Erysiphe) necator*, is the most important fungal disease of grapevines worldwide. Generally, varieties of *Vitis vinifera* and some of its hybrids are much more susceptible to powdery mildew than are varieties of native American grape species. If left uncontrolled, powdery mildew can destroy infected berries outright or reduce their quality and predispose them to bunch rot infections. Foliar infections can limit photosynthesis, thereby limiting berry maturation, vine growth, and winter hardiness.

Symptoms and Signs

The powdery mildew fungus can infect all green tissues of the grapevine. On leaves, it appears as a white or grayish-white powder on the upper and lower surfaces (figure 11.1). Heavily infected leaves may turn dull, dry out, and drop prematurely in the autumn, whereas very young leaves that become infected may become distorted and stunted as they expand. When the stems of green shoots are infected, the fungus initially appears white, but the infection site soon turns gray to dark brown or black, with dark patches remaining on the surface of the dormant canes (figure 11.2). Fruit infections appear white and powdery when young (figure 11.3, page 218), eventually turning dark and dusty as they age (figure 11.4, page 218), and sometimes result in shriveling or cracking of the berries depending on variety, climate, and the time of infection. Beginning in mid- to late summer, the fungus produces small black spherical bodies (cleistothe-

a. Powdery mildew colonies appear as white or grayish-white powder, especially noticeable on upper leaf surfaces.

Anton Baudoin, Virginia Tech

b.Young powdery mildew colonies on leaves are best viewed by angling the leaf blade away from a perpendicular view.

Figure 11.1 ▪ Foliar symptoms of powdery mildew.

Roger Pearson, Cornell University

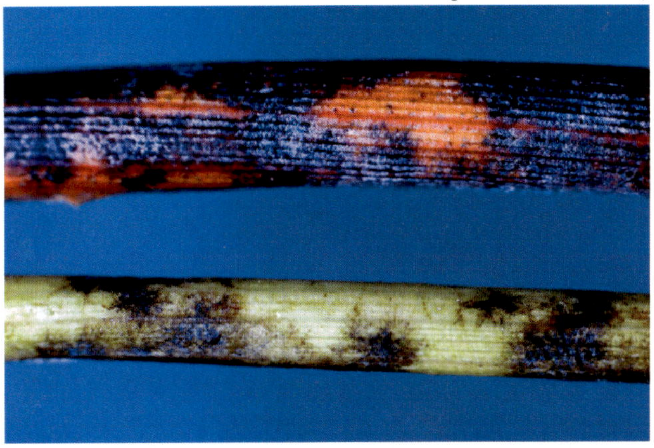

Figure 11.2 ▪ Powdery mildew on dormant cane and stem of a green shoot.

Figure 11.3 ▪ Severe powdery mildew on Chardonnay clusters near the time of bunch closing.

Figure 11.4 ▪ Severe powdery mildew on Chardonnay clusters at the time of harvest.

cia) on the surface of infected leaves, shoots, and berries. When they mature, some cleistothecia are washed into the rough bark of the trunks and cordons of the vines, where they lodge and overwinter.

Disease Cycle and Conditions for Development

In New York, the powdery mildew fungus is known to overwinter only as cleistothecia lodged within cracks in the bark of the vine. When rains of approximately 0.1 inch (2 to 3 mm) or more occur in the spring, with temperatures of at least 50°F (10°C), these structures swell with water and discharge their infective spores (ascospores) into the air. The ascospores are blown to nearby leaves and clusters, where they germinate and begin forming a fungal colony on the surface of the infected tissue. Ascosporic infections are significantly more likely at 59°F (15°C) than at the minimum temperature of 50°F (10°C), and temperatures of 68°F (20°C) or higher are optimal. These first infections of the season (primary infections) are small and inconspicuous, usually occurring on leaves near the trunk of the vine where cleistothecia had overwintered. In general, ascospore discharge starts soon after budbreak and is completed by the end of bloom or shortly thereafter.

In California and other regions with relatively warm winters, the fungus also survives between seasons within infected buds, producing heavily infected "flag shoots" as soon as these buds push in the spring (figure 11.5). Flag shoots have never been observed in the northeast USA. This is presumably because either the infected buds or the vegetative stage of the mildew does not survive cold winters. The precise conditions governing the survival of infected buds are poorly understood; however, they might theoretically survive in some eastern viticultural regions during exceptionally warm winters, particularly in more southerly locations, although this has not been documented. Because flag shoots provide abundant fungal spores for disease spread shortly after they emerge, fungicidal protection during the early stages of vine growth is particularly critical if they are present.

After colonies from primary infections have developed (or flag shoots have emerged), the fungus produces a mass of white powdery "summer" spores (conidia)

Trevor Wicks

Figure 11.5 ▪ "Flag shoot" covered with growth of powdery mildew fungus.

that function to spread the disease. Conidia are blown by wind throughout the vineyard and do not require rain for dispersal or infection. New fungal colonies that develop from these secondary infections produce still more conidia, which can continue to spread the disease. This repeating cycle of infection, spore production, spore dispersal, and re-infection can continue throughout the season so long as conditions remain favorable, causing disease levels to "snowball" at a rate that is determined primarily by temperature (table 11.1). At optimum temperatures in the mid-60s to mid-80s (°F), a new generation of the fungus can develop every 5 to 7 days, resulting in a severe epidemic of the disease on susceptible varieties unless it is effectively managed.

Although spore germination and infection do not require rain and can occur across a wide range of atmospheric moisture levels, all stages in the growth of the fungus are favored by high humidity, with an optimum of approximately 85% relative humidity. Thus, disease severity can be particularly high in vineyards close to bodies of water or in parts of the vineyard where air circulation is poor. Direct exposure to intense sunlight is detrimental to the powdery mildew fungus, hence the disease is strongly favored in shaded sections of the vineyard or within dense canopies where light penetration is poor. Powdery mildew is often a particular problem in seasons with frequent rains, which promote ascosporic infections and result in protracted periods with high humidity and low sunlight intensity.

Table 11.1 ▪ Approximate generation time for powdery mildew at different average temperatures

Temperature (°F)	Temperature (°C)	Days
44	7	32
48	9	25
52	11	16
54	12	18
59	15	11
63	17	7
74	23	6
79	26	5
86	30	6
90	32	—[a]

Source: C. J. Delp, University of California, Davis (1954).

[a] Little or no disease development at or above 90°F (32°C)

Period of Susceptibility

Berries are extremely susceptible from the immediate prebloom stage through fruit set, and severe disease on the clusters is usually a result of uncontrolled infections during or shortly after this period. Berries of the American variety Concord become virtually immune to powdery mildew infections about 2 weeks after fruit set, whereas fruit of *V. vinifera* varieties maintain some susceptibility until 4 weeks postbloom. Infections that occur on such berries during their final stage of susceptibility (near the time of bunch closure) are difficult to see with the naked eye; however, they may have serious consequences, because they promote the growth of fungi responsible for bunch rots and of other microorganisms that cause spoilage of the wines made from affected grapes. Grape leaves are also most susceptible when very young, and become significantly more resistant to infection shortly after they have fully expanded.

Management

Maintaining good control of foliar infections until one month before leaf drop greatly reduces overwintering inoculum and improves the efficacy of management programs the following year. Cultural practices that reduce humidity within the vineyard, promote good air circulation through the canopy, and provide good light exposure to all leaves and clusters greatly aid in the control of powdery mildew. Some specific practices include site and slope selection, orienting row direction to the prevailing winds, vigor control, shoot and cluster thinning, and selective removal of leaves in the cluster zone. However, commercially-acceptable control also requires the application of effective fungicides at the proper times, as affected by variety susceptibility, the availability of fungal inoculum, the stage of crop development, and climate. Sprays are often required as early as the 3- to 5-inch (8- to 13-cm) shoot growth stage on *V. vinifera* varieties. High disease levels the previous year increase the importance of the earliest sprays, whereas excellent control reduces their importance. Similarly, warm temperatures or frequent rains increase disease pressure, affecting both the initiation of spray programs and their necessary intensity (spray intervals, materials, and rates). Spray programs should be at their peak from the immediate prebloom stage through 2 to 4 weeks later (depending on variety susceptibility), after which fruit become highly resistant to infection. Subsequent control of foliar disease is necessary for proper ripening and vine health on *V. vinifera* and susceptible hybrid varieties but may be less important on native American varieties, depending on crop load, climate, and existing disease levels.

Unlike all other pathogens of the grapevine, the powdery mildew fungus grows almost entirely upon the surface of infected tissues rather than within them. Hence, it is vulnerable to topical applications of certain materials such as oils and salts that have no effect on other diseases. The efficacy of these "nontraditional" products is, however, strongly dependent upon complete coverage of all susceptible tissues and relatively frequent spray intervals. In addition to sulfur (phytotoxic on Concord and some other native varieties and hybrids), powdery mildew is controlled by a number of modern fungicides. However, the causal fungus is particularly prone to developing resistance to many of these materials. Therefore, it is important that they be used in accordance with recommended resistance-management practices, such as limiting the number of applications of products in any single fungicide class, and rotating with chemically-unrelated materials during each season.

Black Rot

Black rot, caused by the fungus *Guignardia bidwellii*, is one of the most potentially serious diseases of grapes in eastern North America. Although capable of causing complete crop loss in most regions, it is relatively easy to control when standard spray programs are employed. However, disease pressure is especially severe in more southerly regions, where warmer temperatures are regularly optimal for infection during critical periods of crop development, and management efforts are often most intense in those regions. There is a wide variation in susceptibility to this disease among native and hybrid varieties, whereas all common varieties of *V. vinifera* appear to be highly susceptible.

Symptoms and Signs

All young green tissues of the vine are susceptible to infection. Relatively small, brown circular lesions develop on infected leaves (figure 11.6), and within a few days tiny black spherical fruiting bodies (pycnidia) protrude from them. Elongated black lesions on the petioles may eventually girdle these organs, causing the affected leaves to wilt and fall off. Shoot infection results in large black, elliptical lesions, which often are difficult to distinguish from those caused by the *Phomopsis* fungus (described in a later section). These lesions may contribute to breakage of shoots by wind, or in severe cases, may girdle and kill young shoots altogether.

Infection of the fruit is by far the most serious phase of the disease and may result in substantial economic loss. Infected berries first appear light or chocolate brown, but quickly turn darker brown, with masses of black pycnidia developing on the surface, often in a target-shaped or "bulls-eye" pattern as the rot progresses. Finally, infected berries shrivel and turn into hard black raisin-like bodies called "mummies" (figure 11.7).

Disease Cycle and Conditions for Development

The black rot fungus survives winter primarily in mummies within the vine and on the ground, although it also can persist for at least 2 years within lesions on infected shoots. Spring rains trigger release of airborne spores (ascospores) that form within mummies on the ground and in the trellis, and these can be dispersed for moderate distances by wind. Spores of a second type (conidia) also can form in overwintered cane lesions or in mum-

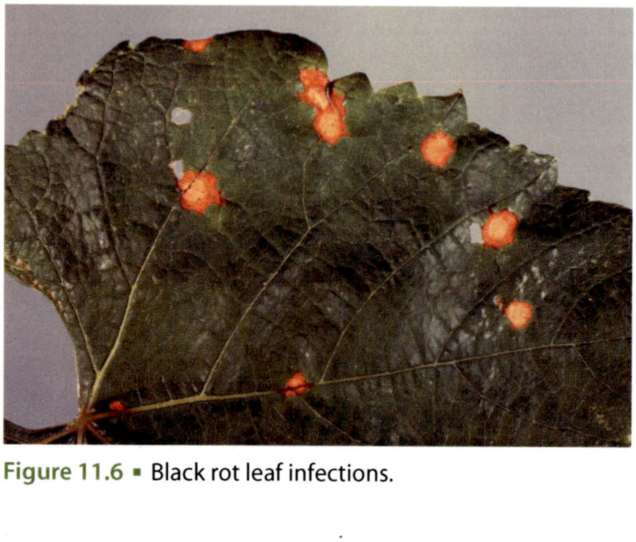

Figure 11.6 ▪ Black rot leaf infections.

Figure 11.7 ▪ Various stages of black rot development on an infected cluster.

mies that have remained within the trellis, and these are dispersed to nearby fruit and foliage by splashing rain drops. Infection occurs when either type of spore lands on susceptible green tissue and it remains wet for a sufficient length of time; the duration of wetness necessary for infection to occur depends on the temperature (table 11.2). Furthermore, the period during which these overwintered spores are available to cause infections depends on their source. Studies in New York and the upper Midwest show that significant discharge of ascospores from mummies on the ground begins about 2 to 3 weeks after bud break and is virtually complete within 1 to 2 weeks after the start of bloom. In contrast, mummies retained within the trellis can continue to release both conidia and ascospores from the early prebloom period through veraison, and conidia can be dispersed from overwintered cane lesions as soon as bud break begins.

The time required for symptoms to appear after infection occurs (the incubation period) depends on both post-infection temperatures and the age of the tissue at the time of infection. In New York, young leaves and fruit generally start showing symptoms about 2 weeks after infection occurs, and the small black pycnidia form within a few days later. The splash-dispersed spores (conidia) produced within these structures can cause substantial spread of the disease under warm and rainy conditions, particularly if berries are still highly susceptible to infection (see below). Most berries that become infected near the end of their susceptible period do not show symptoms until at least 3 weeks later, and the majority do not begin to rot until 4 to 5 weeks after the infection event. These incubation periods should be considered when trying to determine the origin of unexpected disease problems.

Period of Susceptibility

Young leaves are highly susceptible to black rot as they unfold but become resistant about the time that they finish expanding. Berries do not become infected while the caps remain attached but are extremely susceptible for the first 2 to 3 weeks after cap fall. Susceptibility declines progressively after that. In New York, Concord berries become highly resistant about 5 weeks after bloom and immune 1 week later, whereas berries of *V. vinifera* varieties maintain some susceptibility until they become highly resistant 6 or 7 weeks after bloom. The duration of berry susceptibility has not been investigated in other locations. The development of age-related resistance is affected by temperature, and occurs more rapidly in warm seasons than in cool ones.

Table 11.2 ▪ Duration of continuous leaf wetness necessary for infection by the black rot fungus at different air temperatures

Temperature		Hours of leaf wetness required for infection
(°C)	(°F)	
7.0	45	No infection
10.0	50	24
13.0	55	12
15.5	60	9
18.5	65	8
21.0	70	7
24.0	75	7
26.5	80	6
29.0	85	9
32.0	90	12

Source: R. A. Spotts, The Ohio State University.

Management

Black rot should be managed through a combination of cultural and chemical methods. The success of any fungicide program will be greatly enhanced by sanitation practices designed to reduce inoculum of the black rot fungus, and these may be essential for avoiding losses in vineyards where the disease is a perennial problem. Sanitation is a crucial component of black rot management in "organic" production systems: to date, there are no organically-approved fungicides with good efficacy against black rot, and this disease remains the "Achilles heel" of organic grape production in the East. It is critical to remove all mummies from the canopy during the dormant pruning process; because overwintered mummies produce spores immediately adjacent to the current season's crop, even a relative few can produce enough inoculum to cause significant damage. Burying mummies by cultivating beneath the vines also will greatly reduce the number of spores available to initiate the disease cycle for that year; ideally, this should be done by bud break or soon thereafter. Many growers do not need to perform this operation, particularly if black rot was well-controlled the previous year. However, it should be beneficial for those who have difficulty controlling the disease and may be especially important for those who choose to severely limit or avoid the use of conventional fungicides. As with all fungal diseases, control is also improved by canopy management practices that promote air circulation, speed the drying of the leaves and fruit, and improve spray penetration.

Traditional fungicide recommendations have specified regular applications from the early shoot growth stage through veraison. In the Great Lakes region, however, both research and experience have shown that excellent control can be obtained in most vineyards when fungicides are applied from the immediate prebloom stage through 4 weeks postbloom. Sprays should start at least 2 weeks prebloom if disease was severe the previous year. Note also that elimination of sprays in the early and late season has received only limited testing in states further south, where black rot pressure often is greater. Nevertheless, because fruit are most susceptible during the first few weeks after the start of bloom, this is when the fungicidal component of black rot management programs should be focused most strongly, whether additional sprays are applied or not. Fungicides vary considerably in their effectiveness against black rot. Furthermore, some materials, such as certain sterol-inhibitor fungicides, provide significant activity when applied up to several days after the beginning of an infection period, whereas others such as mancozeb, ziram, and the strobilurins are effective only if applied prior to the start of the rain that induces infection. Recognizing these fungicide characteristics will result in more efficient use of the products.

Downy Mildew

Downy mildew, caused by the fungal-like organism *Plasmopara viticola*, is extremely responsive to environmental conditions. As a result, it varies considerably in its intensity among regions and between years. Although downy mildew can be difficult to detect during dry weather, it can suddenly "explode" if conditions become warm and wet. Infections are most common on the foliage. Severely infected leaves typically fall from the vine, reducing the plant's ability to ripen fruit and acclimate for the winter. Premature defoliation from downy mildew is not unusual in wet seasons if spray programs are stopped early. In the absence of effective control programs, serious yield losses can also result from pre- and postbloom infections of the cluster stems and berries, which destroy these organs. All common *V. vinifera* varieties are highly susceptible. Native and hybrid varieties range from highly resistant to highly susceptible (see table 3.2, page 62).

Symptoms and Signs

Foliar lesions typically appear as circular, yellow or reddish-brown lesions on the upper surface of the leaves (figure 11.8). Late in the season, these tend to become smaller, darker and more angular in shape, as they are typically confined between the veins. Directly opposite the lesions, on the lower leaf surface, a white "cottony" growth of the fungus can be seen (figure 11.9). If the weather has been dry recently, this mass may appear flat and pressed to the surface, but it will appear fluffy or downy on mornings following rain or dew.

When infected, young shoot tips or cluster stems often curl and may become covered with the downy white growth of the organism following wet or humid nights (figure 11.10). Berries on infected cluster stems

Figure 11.8 ▪ Circular "oil spot" lesions of downy mildew most commonly seen early in the summer.

Figure 11.9 ▪ Symptoms and signs of downy mildew. Sporulation of downy mildew on lower leaf surface *(right)*; symptoms on upper leaf surface *(left)*.

become brown and eventually shrivel. Berries that are infected directly while very young may become covered with the fluffy white fungal growth and die soon thereafter. Cluster infections that occur more than 2 weeks after bloom cause berries to remain hard ("leather berries"), often appearing mottled gray-green (white varieties) or prematurely red (red varieties) in color (figure 11.11).

Disease Cycle and Conditions for Development

Rainfall and high humidity are the most important environmental factors promoting downy mildew epidemics. The downy mildew organism overwinters as dormant spores (oospores) within infected leaves on the vineyard floor or in the surface layer of the soil. In general, these spores first become active about 2 or 3 weeks before bloom, and the season's first (primary) infections occur when new spores produced from them are splashed onto susceptible tissues during rainstorms. For this reason, the earliest infections are often observed near the ground, on volunteer seedlings or trunk sucker growth. Temperatures must be 52°F (11°C) or higher for primary infections to occur.

Downy mildew epidemics are the result of repeated cycles of secondary infections, spread by spore-containing bodies (sporangia) that comprise most of the white downy growth seen on infected tissues. For this to occur, sporangia require at least 4 hours of extremely high (greater than 95%) relative humidity at night for their formation, followed by rain or heavy dew to allow them

Figure 11.10 ▪ Downy mildew on grape cluster infected before bloom.

Figure 11.11 ▪ Downy mildew on red-fruited grape berries infected after fruit set.

Figure 11.12 ▪ Chardonnay vine nearly defoliated by severe downy mildew infestation.

to germinate and cause new infections. Sporangia formed during a wet night will die the next day if exposed to bright sunlight before given an opportunity to germinate, but can persist and "wait" several days for rain under overcast conditions. With an incubation period of as little as 4 or 5 days, and the dissemination of sporangia within and among vineyards by wind, spread can be "explosive" after a series of warm, foggy nights followed by thundershowers. Highly susceptible, unprotected vines can be almost defoliated under these conditions (figure 11.12).

In contrast, disease development can come to a sudden halt when the necessary sequence of events does not occur. The ideal temperature for sporangium formation (at night) and infection (rain) is approximately 77°F (25°C), but disease development can occur within a range of about 50 to 86°F (10 to 30°C), although it is less intense and slower at the edges of the temperature range.

Period of Susceptibility

Young leaves are highly susceptible to infection while they are still expanding, but they become significantly more resistant soon thereafter. New foliar infections can occur throughout the season as long as new leaves are being produced, and older leaves can still become diseased if infection pressure is high. Young clusters and berries also are highly susceptible, and cluster stems can be seriously damaged before bloom if conditions are favorable for infection. Cluster stems and berries become resistant by approximately 4 weeks after fruit set.

Management

Any vineyard or vine management practice that improves air circulation and speeds drying will aid in the management of downy mildew. However, properly-timed fungicide applications also are necessary with susceptible varieties. Applications should start about 2 to 3 weeks before bloom, depending on the weather, and management efforts should be particularly conscientious until 2 to 4 weeks after fruit set. This is when direct fruit losses can occur and it is the critical time to keep an epidemic from getting started. The necessary intensity of mid-summer spray programs is dictated by variety, weather, and the presence or absence of disease in the vineyard. It often is possible to curtail downy mildew fungicide sprays during extended dry periods, but the disease can build up very rapidly on unprotected tissues should warm, wet weather return.

Some fungicides used for control of downy mildew act only in a protective mode; they must be applied *before* an infection period in order to be effective. Other fungicides exhibit both protective and post-infection qualities. Fungicides that have post-infection activity are capable of suppressing or killing the target fungus if applied within one to several days after the infection period has occurred. Understanding the individual activities of fungicides will improve the efficiency of their use.

Phomopsis Cane and Leaf Spot

The common name for the disease caused by the fungus *Phomopsis viticola* is Phomopsis cane and leaf spot. However, in eastern North America economic losses from this pathogen are more often the result of infections of the fruit and rachises (cluster stems). Although sporadic, fruit and rachis infections can be severe under heavy disease pressure, and such infections have caused significant losses in some vineyards during years with frequent rains in the spring and early summer. Many common *V. vinifera*, native, and hybrid varieties are susceptible. However, serious losses are most frequent on certain native varieties, probably due to their drooping growth habits and the less intensive spray regimes typically used in their culture. Niagara appears to be a particularly susceptible variety. Among *V. vinifera* plantings, infections are most frequent on those that are spur-pruned.

Symptoms and Signs

Lesions on shoots and leaves are the most common symptom of the disease. Infections on shoots result in black, elongated lesions that are most numerous on the first three to six basal internodes (figure 11.13). Cracks or fissures may develop within these lesions, eventually crusting over to produce a rough, blackened appearance by the end of the season. The stem is weakened around these infection sites, occasionally leading to stem breakage in high winds before shoots become attached to the trellis wires. The greater impact of cane infections, however, appears to be the establishment of the fungus within woody portions of the vines, from which it can infect fruit and rachises in subsequent years. During the dormant season, infected canes may become bleached, and numerous tiny, black fungal fruiting bodies (pycnidia) erupt through the surface to produce a speckled appearance.

Elongated, black lesions similar to those on young shoots develop on infected petioles (leaf stems), and these may cause affected leaves to turn yellow and drop if severe. Numerous small, light green lesions with irregular margins develop on infected leaf blades, eventually turning brown to black with yellow margins (figure 11.14). Occasionally, the infected tissue may drop out, causing a "pin-pricked" or "shot-hole" appearance. Heavily infected leaves often become misshapen and crinkled. The impact of such foliar infections is questionable, although they do serve to indicate the presence of fungal inoculum within the vine.

Figure 11.14 • Small black leaf lesions caused by *Phomopsis viticola.*

Lesions on the rachises are sunken and black (figure 11.15), causing the rachis to become brittle; clusters may break at these points under the weight of the maturing crop or during harvest operations, leading to a reduced yield. If lesions girdle the rachis, berries below the infection site shrivel and may fall to the ground. A rapidly expanding brown rot develops on diseased berries during the pre-harvest period. The rotten zone is often first apparent where the berry is attached to its supporting stem (pedicel), but this is not always the case. After the berry has become completely rotted, numerous black fruiting bodies (pycnidia) of the fungus erupt through

Figure 11.13 • Elongated Phomopsis lesions on three most basal internodes of heavily infected shoot.

Figure 11.15 • Black, girdling rachis infections of Phomopsis on Concord cluster with pronounced internode infections.

the skin, giving it a rough texture (figure 11.16). Eventually, the diseased berries shrivel into black raisin-like mummies visually indistinguishable from those caused by black rot, although Phomopsis mummies usually remain somewhat rubbery and soft, whereas black rot mummies are hard.

Fruit rot caused by Phomopsis is frequently confused with black rot, and adequate management of either disease requires accurate identification of the causal organism. In addition to the difference in mummy texture, Phomopsis and black rot may often be differentiated by the timing of symptom appearance, the pattern of pycnidia production, and the attachment of infected fruit to their pedicels. Black rot symptoms typically appear during the first few weeks after bloom and new symptoms are unusual after veraison, whereas the first berry symptoms of Phomopsis generally do not appear until the period between veraison and harvest. Black rot pycnidia may develop in concentric rings as the rot advances through the berry, whereas this pattern does not occur with Phomopsis, and pycnidia do not form until the entire fruit has turned brown. Berries affected with black rot typically remain firmly attached to their pedicels, even after they mummify. In contrast, berries of many grape varieties infected by Phomopsis are easily detached from their pedicels. The detached vascular strands (brush) remain in these berries, leaving dry stem scars after they fall.

Figure 11.16 ▪ Phomopsis fruit rot; disease appears to have expanded from the cluster stem via berry pedicels.

Disease Cycle and Conditions for Development

Rainy weather during the early growing season promotes Phomopsis development. In the late winter and early spring, spores are produced within the black pycnidia formed in previous infection sites. These may occur not only on canes that are retained for fruiting wood, but because shoot infections are most frequent on basal internodes, they are even more common when all of the fruiting wood is composed of spurs. For the same reason, infected pruning stubs also can be important sources of the fungus, and mechanical hedging systems, which retain large numbers of basal cane segments, can be particularly problematic for management of this disease. The fungus can continue to produce pycnidia and their spores for several years after infection occurs; in fact, new research indicates that pycnidia within older wood produce significantly more spores than do those in one-year-old canes. Thus, a serious disease outbreak in one year may provide inoculum for new infections for several subsequent years.

Phomopsis spores ooze from the pycnidia in a sticky mass during humid or wet weather and are then splashed by rain onto newly developing shoots, rachises, and fruit. Because of this dispersal mechanism, significant numbers of spores move only a short distance within the vine and it is not unusual to see the disease occur in "pockets" associated with a nearby inoculum source. Similarly, disease is frequently more severe on tissues directly beneath such sources than on those growing well above them. Infection can occur during wet periods across a wide range of temperatures. Experimental studies show that this occurs most rapidly at approximately 59° to 68°F (15° to 20°C), although severe disease can develop after prolonged rains at much cooler temperatures early in the growing season. Based on New York and Michigan research, most spores are released from their overwintering pycnidia between bud break and bloom, and in most years, few if any are detectable within 2 to 3 weeks after bloom. In Michigan, however, some spore release was observed as late as bunch closure when unusually dry conditions restricted this process earlier in the season. Secondary disease spread during the summer appears to be rare, since additional spores are seldom produced from new shoot or leaf lesions during the current growing season. However, spores are produced from diseased berries after they rot completely, and could potentially infect surrounding fruit if the initial berry infections appear well before harvest, although this is not typical. Some shoot and leaf lesions can appear 3 to 4 weeks after infection, whereas other foliar infections may not appear until the leaves senesce and some shoot infections appear only after pycnidia form on the canes during the dormant season. Rachis infections also may appear within a month after infection, although they typically are most pronounced during the pre-harvest period. Fruit infections usually are not apparent until after veraison and typically become most visible during the last 1 to 3 weeks before harvest.

Period of Susceptibility

On shoots, the young growing tip is the most susceptible to Phomopsis infections, but tissue becomes relatively resistant just two or three internode positions back from it. Thus, susceptible tissue is continually produced while the shoot elongates, and the timing of infection is determined only by the availability of inoculum and favorable weather. Rachises are highly susceptible when clusters first appear, about the time of 2- to 3-inch (5- to 8-cm) shoot growth. Rachis infections during the early period of cluster development are the most damaging and may cause serious girdling with subsequent berry loss. Rachises become relatively resistant to infection after fruit set. Berry infection can occur through the pedicel (figure 11.16) or several weeks during the prebloom period or directly through the fruit epidermis after set. Typically, such infections remain latent (dormant) until the preharvest period, when they can become active and rot the fruit. Some studies indicate that berries become resistant to the initial establishment of infection after the pea-sized berry stage, whereas others show them to remain susceptible to new infections throughout the season. However, because few Phomopsis spores appear to be available beyond midsummer, it seems likely that the prebloom and early postbloom periods are when most infections occur.

Management

Sanitation is the first line of defense against this disease and, as with black rot, it is a critical management component for growers who wish to avoid or minimize standard fungicide sprays. Selectively pruning out infected canes and older wood during dormancy will significantly reduce disease pressure for the coming season. Similarly, protecting new shoots from infection will limit establishment of the fungus in tissues that might otherwise harbor it through subsequent years. Standard protectant fungicides should be applied according to need (weather, inoculum level in the vines, varietal susceptibility) during the early part of the growing season. The first few weeks after cluster emergence is the most important time to control rachis infections. Sprays during this period also provide some control of fruit infections; however, additional sprays to protect against fruit infections may be necessary if the weather is wet during the prebloom through early postbloom period. Some traditional protectant fungicides (captan, mancozeb, ziram) provide good control, but many materials commonly used against other diseases during this period have variable levels of efficacy against Phomopsis; check current, local recommendations. Because all Phomopsis inoculum originates within the wood of the grapevines, it is theoretically possible to reduce disease pressure by thoroughly spraying the vines with eradicative fungicides shortly before bud break. However, sprays of lime sulfur and other materials have provided variable (and, generally, only modest) benefits in controlled trials, although this tactic is still being investigated.

Botrytis Bunch Rot

Several different organisms can cause bunch rots of grapes. The fungus *Botrytis cinerea* is by far the most common in regions with cool to moderate temperatures during the preharvest period, but becomes progressively less important in the more southernly regions of bunch grape production, where other causal organisms tend to predominate. Botrytis can be a particular problem on certain tight-clustered, white varieties of *V. vinifera* and its hybrids such as Chardonnay, Riesling, Sauvignon Blanc, Semillon, Seyval, and Vignoles. Although not a common problem on most red varieties, tight-clustered clones of Pinot Noir are highly susceptible, and Baco Noir can be problematic as well. In addition to direct crop reduction, economic losses result from a decreased quality of the wines made from infected fruit. The Botrytis fungus can also infect leaves and young clusters if the weather is frequently wet during the early growing season.

Symptoms and Signs

Following long periods of rain in the spring, large, irregularly-shaped brown lesions develop on young leaves, typically including the edge of the leaf blade. During humid or wet weather, such lesions are typically covered with the gray, moldy growth (spores) characteristic of Botrytis (figure 11.17). The significance of foliar infections is debatable with respect to vine health, although they may be important sources of spores for the initiation of fruit infections if still active during bloom. If early-season rains are especially persistent, the fungus also may cause entire clusters to rot and fall off before bloom.

The first diseased berries usually become visible at or shortly after veraison, although this may occur even earlier in very wet years. The incidence of these earliest infections is typically low, and even when present they may be difficult to detect without looking carefully in the vineyard. Infected berries of white varieties turn brown, whereas those of red-fruited varieties become reddish. Infected berries tend to shrivel up in dry weather and may even fall from the clusters of varieties or clones with loose cluster architecture. However, infected berries remain plump if conditions are moist and soon become covered with the characteristic fuzzy, gray spores, leading to the term "gray mold" that is sometimes applied to this disease. Once established, infections can spread rapidly from berry to berry during the pre-harvest period, particularly in tight-clustered clones or varieties and under wet conditions (figure 11.18).

Figure 11.17 ▪ Botrytis leaf infection.

Figure 11.18 ▪ Early stage of Botrytis bunch rot in ripening Chardonnay cluster as disease begins to spread via berry-to-berry contact.

Disease Cycle and Conditions for Development

Although it can be very destructive, the Botrytis fungus is a relatively "weak" pathogen, unable to infect healthy, robust tissues. Rather, it preferentially attacks new, succulent growth (young leaves and clusters), dying or senescing tissues (old blossom parts, ripening fruit), or tissues that have been injured, such as berries cracked from excess water or powdery mildew infection or fruit damaged by grape berry moth and other insects. Botrytis can grow at temperatures of approximately 34° to 86°F (1° to 30°C), with an optimum range of 59° to 77°F (15 to 25°C). It thrives under damp conditions with little to no air movement. The above factors are key to understanding its biology and control.

The fungus overwinters in debris on the vineyard floor and on the vine. Old rachises that remain attached to the vine after mechanical harvesting appear to be particularly common overwintering locations. Spores are produced in the spring and are distributed to susceptible tissues by wind. Infection can occur under very high humidity or in the presence of dew, but is most likely when tissues are wet from rains. In addition to the spores produced from overwintered debris, those produced from prebloom infections of young leaves and clusters can serve as inoculum for fruit infections during and after bloom.

Some fruit infections are initiated at or shortly after bloom. Germinating spores colonize aging blossom parts and the fungus then grows into the newly-developing berry, where it becomes latent (dormant). Many latent infections remain inactive even through harvest, but some can become active around the time of veraison, as the berries begin to ripen. The factors that promote the activation of latent infections are poorly understood, but some evidence suggests that they include high humidity, abundant water in the root zone, and high nitrogen availability. New infections can also occur at this time, especially from contaminated blossom debris that became trapped against berries within the expanding clusters, or through wounds. As ripening berries become increasingly susceptible to infection during the pre-harvest period, the fungus can spread rapidly through an affected cluster via berry-to-berry contact (figure 11.18). This phenomenon is much more common in clusters where berries are tightly pressed against each other than in those where the berries are loose. As berries rot, spores produced from them are distributed to nearby healthy clusters by wind, rain and insects, where they can initiate new infections which produce even more spores for additional spread.

Period of Susceptibility

Leaves and supporting structures of the cluster (rachis, pedicels) are susceptible while young and succulent. Blossom parts appear to be most susceptible as they age and wither. Although latent infections can be established in fruit at any time during their development, these occur most frequently when the fungus grows into the berry from some attached "foothold" in which it previously became established, such as infected blossom parts and debris. Berries first become significantly susceptible to direct infection by spores (or through wounds) at veraison, and become progressively more susceptible as they continue to ripen.

Management

Perhaps more than any other common disease, control of Botrytis on susceptible varieties requires an integrated approach that utilizes both chemical and cultural components. When planting susceptible varieties, the choice of clones with loose cluster architectures, should they be available and acceptable for the intended use or market, will greatly aid subsequent management efforts. Similarly, the benefits of choosing a planting site with good air flow cannot be overstated. Training and canopy management systems that provide good fruit exposure and that consequently maximize air movement and spray coverage in the cluster zone are very valuable (see chapter 6, page 124). Selective thinning of the foliage around the fruit clusters also has proved useful for this purpose; check local recommendations for specific details. Balanced nitrogen nutrition is yet another cultural component of an integrated Botrytis management program. Control of powdery mildew and grape berry moth, both of whose injuries provide points of entry for the Botrytis fungus, will restrict these avenues of attack.

Timely fungicide applications can be very beneficial in some years, although these are unlikely to prove satisfactory under heavy disease pressure if cultural management practices are ignored. There are four developmental stages or periods at which fungicide applications can be beneficial, although the need (or lack thereof) for each is heavily influenced by individual vineyard and weather conditions around each of the four periods. Those periods are:

(1) **Late-bloom.** Sprays are intended to protect against the initiation of latent infections in the young berries and the infection of withering blossom parts, which may become trapped as debris within the developing clusters as they are shed.

(2) **Pre-bunch closure.** This is the last opportunity to cover the entire surface of the expanding berries with a protective fungicide residue.

(3) **Veraison.** Sprays are intended to protect against the start of an epidemic as latent infections become active and berries enter their period of peak susceptibility to direct infection.

(4) **One to three weeks pre-harvest** (check fungicide label restrictions). Sprays are intended to prevent the spread of disease to highly susceptible fruit, particularly if Botrytis is active in the vineyard, extended rains are forecast, or harvest delays are expected.

Many fungicides used for control of other diseases have little activity against Botrytis; however, Botrytis-specific fungicides are available. Due to the specificity of the mode of action of these fungicides, there is a risk that Botrytis will develop resistance if any one material (or class of compounds) is used excessively and exclusively. Therefore, both efficacy and the need to rotate materials for resistance management should be considered when choosing the fungicidal component of a Botrytis management program.

Anthracnose

Anthracnose, also called bird's-eye rot, is caused by the fungus *Elsinoë ampelina*. Under commercial conditions, it causes serious disease on only a small group of highly-susceptible varieties. Vidal blanc is particularly susceptible, but problems also have been reported on Cayuga White, Chardonnay, Vignoles (fruit only), and Villard blanc as well as several eastern table grape varieties. As Anthracnose was once the most damaging grape disease in Europe, it is possible that some of the *V. vinifera* varieties commonly grown in eastern North America are more susceptible than commonly recognized, but that it is routinely controlled by the intensive spray programs applied against other diseases. Anthracnose occurs only sporadically in the Northeast, upper Midwest, and Ontario but is much more common in warmer production areas to the south of these regions. In all regions, it is most likely to develop in seasons with frequent rainfall. Vineyards that are closely bounded by woods that support wild grapevines (alternative inoculum source) may be at increased risk.

Symptoms and Signs

Lesions on young shoots and leaves are rounded and brown or tan colored, with a dark purplish margin. Although individual lesions are small and discrete, they may grow together to cause larger necrotic areas. Infected leaf tissue eventually falls out to produce a "shot hole" appearance, and severely diseased leaves become crinkled and distorted. Infections on the rachis and pedicel appear similar, but these may coalesce to cause girdling lesions and the subsequent loss of some or all of the berries on an affected cluster. Berry lesions are slightly sunken and tan to ash gray, eventually becoming surrounded by a distinct dark border to produce the so-called "bird's eye" appearance (figure 11.19).

Disease Cycle and Conditions for Development

The fungus overwinters primarily on infected canes, although it may also survive in infected berries on the vineyard floor. In spring, spores are produced from these overwintering sites and are dispersed primarily by splashing raindrops to young, susceptible tissues. Infection requires at least 12 hours of continuous wetness and can occur across a wide range of temperatures, although

Fritz Westover, Texas AgriLife Extension Service

Figure 11.19 ▪ Symptoms of anthracnose on berries.

temperatures of 75° to 79°F (24° to 26°C) are optimal. Additional spores, which also are splash-dispersed, are produced through new infections, and these can rapidly spread the disease through multiple repeating cycles of infection under warm, rainy conditions.

Period of Susceptibility

Shoots and leaves are most susceptible when young and succulent. Cluster and fruit stems (rachises and pedicels) also are most susceptible when young. Berries reportedly remain susceptible until veraison.

Management

An application of lime sulfur at the end of the dormant season can reduce much of the overwintering fungal inoculum, and this has proven to be an extremely valuable component of anthracnose management in vineyards where the disease is problematic. Some fungicides also offer good protection when applied on a regular basis during rainy portions of the growing season. Sprays should be aimed at protecting young, succulent tissues from infection.

Bitter Rot and Ripe Rot

Bitter rot and ripe rot are diseases of mature fruit caused by different fungi. Bitter rot is caused by *Melanconium fuligineum* (*Greeneria uvicola*) whereas ripe rot is caused by two fungi that are closely related to each other, *Colletotrichum gloeosporioides* (*Glomerella cingulata*), and *C. acutatum*. Both bitter rot and ripe rot develop in similar fashions and are subject to similar control programs. Traditionally, these diseases have been rare in the Northeast, upper Midwest, and Ontario, but their incidence appears to be increasing, and significant losses have occurred on unprotected fruit in some vineyards following frequent and/or prolonged warm summer rains. These diseases can be common and serious under wet conditions in the warmer viticultural regions of eastern North America. In Virginia for example, bitter rot is more common than ripe rot, although both can be serious in wet seasons.

Symptoms

Usually, the bitter rot fungus first invades the berry from the pedicel after veraison. As the rot advances, the infected regions of light colored berries turn brown and show concentric rings of small, dark fungal fruiting bodies (acervuli) (figure 11.20). The advancing rot and acervuli are less apparent on red/black berries, but may give them a rough appearance. Once it is completely rotted, the berry becomes covered with acervuli, which appear as numerous raised pustules that rupture through the cuticle. Within a few days, diseased berries soften and may drop; others shrivel into firmly attached mummies

Figure 11.20 ▪ Raised pustules of bitter rot fruiting bodies (acervuli).

that resemble those caused by black rot, Phomopsis, and ripe rot. The disease name was derived from the bitter taste of infected berries, and wines made from as little as 10% affected fruit may be unpalatable.

It is important that growers and winemakers be able to distinguish between bitter rot and black rot. Because black rot infects only green berries, the fungicide sprays for controlling it are generally stopped at or before veraison. However, stopping sprays at this time could be disastrous if bitter rot is present, because it spreads from fruit to fruit after the berries have changed color. In addition to the timing of symptom appearance (black rot infections should be apparent at or before veraison), the two diseases also can be distinguished by (1) the appearance of the fungal fruiting bodies on infected fruit (those of black rot are round and uniform in size, whereas those of bitter rot are irregular and variable in size); and (2) the tendency of fruit infected with bitter rot to leave hands sooty black if handled when wet (whereas those infected with black rot will leaves hands clean).

Symptoms of ripe rot do not develop until after veraison and are most common on ripe fruit at harvest. Infected berries initially develop circular, reddish brown lesions on their skin which eventually expand to affect the entire berry. Under humid conditions, small "dots" of slimy, salmon-colored spores may develop across the rotten berry. Infected fruit shrivel and mummify, but usually remain attached to the rachis on bunch grape varieties, although diseased muscadine grapes may fall to the ground. No foliar symptoms are produced.

Disease Cycle and Conditions for Development

The bitter rot fungus colonizes dead tissues of the grapevine (fallen leaves and berries, damaged shoots, necrotic bark), where it overwinters and produces spores the following spring. After flowering, some spores are moved by splashing raindrops onto the pedicels (stems) of the developing berries, where they germinate and cause a latent (dormant) infection. When the berries mature, the fungus resumes growth and advances into them, causing the fruit to rot. The acervuli that cover the diseased berries contain abundant spores, which spread the disease as they are splashed onto healthy berries during subsequent rains. Infection occurs through any type of injury, including rain cracking, insect damage, or bird injury. Temperatures of approximately 82° to 86°F (28° to 30°C) are optimal for infection.

The ripe rot fungus overwinters in mummified fruit, infected pedicels, and dead bark and cankers. Spores are produced from these sites in the spring and are dispersed by splashing and blowing rain. Fruit may be infected at any stage of their development, but the fruit remain symptomless (infections remain latent) until the berries begin to ripen. During warm rainy periods (77° to 86°F, or 25° to 30°C, is optimum), the salmon-colored spores produced upon diseased fruit can spread the disease to additional berries, which become increasingly susceptible to infection as they ripen. Frequent rains during the pre-harvest period can result in severe crop loss if the disease is not managed.

Management

Cultural practices, such as pruning out dead spurs, removing overwintered mummies, and removing weak or dead cordons, are important to help reduce the inoculum in the vineyard. Both diseases are frequently controlled in the early- to mid-summer by fungicide sprays targeted against other diseases, such as downy mildew and black rot. However, not all fungicides applied for control of these latter diseases will provide control of bitter rot and ripe rot; check current recommendations to determine which products may be appropriate. Additional, specific sprays to control bitter rot and ripe rot may be needed in the late season if the weather is warm and wet. In southerly regions where the diseases are consistent problems, it is typically necessary to apply protectant fungicides on a 2-week schedule from bloom until harvest. Because fruit are especially vulnerable in their final stages of ripening, pre-harvest sprays are particularly important where these diseases are active. French-American hybrids are generally more resistant to bitter rot than are varieties of *V. vinifera*. There is little information on the relative susceptibility of specific varieties of *V. vinifera*, *V. labrusca*, or French-American hybrids to ripe rot.

Eutypa Dieback

Eutypa (you·**type**·uh) dieback, caused by the fungus *Eutypa lata*, is a chronic canker disease that affects grapevines throughout the world. The fungus infects numerous woody hosts other than grapevines, and is widely distributed throughout the viticultural regions of eastern North America. Most *V. vinifera*, native, and hybrid varieties are susceptible. Typically, symptoms of Eutypa do not appear in vineyards until they become mature, but the disease can gradually spread throughout the planting thereafter. Damage is often restricted to a single trunk, arm, or cordon on an individual vine, but cumulative yield losses of infected vines can be significant in any particular year, and especially over time. Eutypa dieback is not the only canker disease that affects grapevines in the East, but it is the easiest to recognize and is the only one to have received significant study in the region.

Symptoms

The symptoms most characteristic of Eutypa dieback are best observed in the spring. Young shoots on infected vines are severely stunted, with small, yellowed, cupped leaves. Such leaves also may develop small dead spots and become tattered along the margins. These symptoms are most apparent during the prebloom period, when affected shoots stand out in stark contrast to healthy shoots, which are elongated and green (figure 11.21). However, affected shoots become increasingly difficult to detect as the season progresses, because they become obscured by the developing canopy that is provided by healthy shoots that arise from unaffected arms or cordon spurs proximal to the canker.

Upon careful examination, a canker can usually be found around an old, large pruning wound "downstream"

Figure 11.21 ▪ **Characteristic foliar symptoms of Eutypa dieback.**

Contrast with healthy shoots on proximal portion of same cordon and on adjacent vine to right.

(toward the roots, or proximal) on the trunk or cordon that bears the affected shoots. These cankers are not sunken or otherwise conspicuous, but removing the bark will reveal a zone of dark, dead wood expanding in an elliptical pattern in both directions from the wound. A cross-sectional cut through the cankered region reveals a wedge or pie-shaped zone of dead wood with its "point" in the center of the cordon or trunk. This particular symptom also is produced by other canker-forming fungi, such as *Botryosphaeria* spp., that invade pruning wounds. Relatively recent surveys of older vineyards (15 to 25 years old) in Maryland, Virginia, New York, and New England revealed that species of the *Botryosphaeria* fungus are as commonly associated with cankers as is *Eutypa*. Therefore, a grower should surmise that *Eutypa* is the causal organism only when cankers occur in conjunction with the above-mentioned foliar symptoms.

The Eutypa fungus growing within cankered wood produces a toxin, which is responsible for the characteristic foliar symptoms and frequently affects all shoots distal to (beyond) the point of infection. Although shoots beyond the cankers may not show symptoms every year in the early stages of the disease, the cankers expand slowly over time and eventually kill the affected trunks, arms, or cordons.

Disease Cycle and Conditions for Development

The Eutypa fungus persists within cankers on infected grapevines and other woody hosts, both native and cultivated. Spores can be liberated during rains throughout the year. Data from Michigan and New York research indicate that peak discharge occurs from January through March, often as a result of the moisture provided by melting snow. Similar data from more southern viticultural regions in the East are lacking, but sufficient spores are available to cause numerous infections during the relatively warm winter pruning season in California and Australia. Spores can be carried long distances in wind currents. Infection occurs when they land on fresh pruning cuts and germinate in a film of moisture. The susceptibility of these wounds is reported to decrease rapidly during the first two weeks after they are made, with infection unlikely after four weeks. The period of susceptibility is longer at cold temperatures than under warmer springtime conditions.

The probability of infection increases substantially with increasing diameter of the pruning wound, hence most cankers are observed around cuts that were made through wood that was at least 2 years old. The disease progresses slowly, and symptoms usually are not apparent until 3 or 4 years after infection. Several additional years are usually required until the affected arm or trunk is killed. Numerous spores produced from established cankers spread the disease to additional vines within the planting; disease spread from these wind-blown spores is also suspected to occur over long distances (miles). Although woody hosts outside the vineyard may therefore serve as additional sources of fungal inoculum, the density of spores from diseased vines within the vineyard is likely to be much higher and more significant if diseased wood is allowed to remain within or near the planting.

Management

Eutypa can be a debilitating disease in older vineyards if it is allowed to progress, and is difficult to "turn around" once it becomes well established. These factors, together with the widespread distribution of the causal fungus and the impracticality of controlling it through a regular spray program, make a conscientious cultural management program critical for the long-term productivity of the vineyard. Inspect the vineyard for the tell-tale shoot symptoms during the prebloom period each spring, when healthy shoots are 12 to 18 inches (30 to 45 cm) long. Infected arms or trunks can be removed at this time or they can be flagged for later removal. To ensure that all of the fungus is removed, make pruning cuts at least 12 inches below the canker when pruning out infected wood. The Eutypa fungus can grow in dead wood and continue producing spores for many years. Pruned, infected wood (two-year-old and older) should therefore be burned (or buried), not retained on the vineyard floor or in a pile at the edge of the vineyard. There is no need to take this same precaution with normal one-year cane prunings, since they will not contain the Eutypa fungus. Preventing Eutypa from establishing a firm "foothold" in the vineyard is the best way to prevent future problems.

Because cankers are established primarily in large pruning cuts (through wood 2 or more years old), it is desirable to avoid such cuts until spring, when spore numbers may be lower and wound healing faster. Alternatively, it may be possible to practice "double-pruning" where large cuts to trunks or cordons must be made. In double-pruning, the first cut is made during dormant pruning operations, leaving a 6-inch (15 cm) "stub" beyond the desired point of removal. This is flagged and subsequently removed (along with any incipient infections that entered through the first wound) by cutting at the desired point after growth begins in the spring, when the probability of infection is reduced.

Whether canker diseases are caused by *Eutypa, Botryosphaeria,* or other wood-infecting pathogens, it may be possible to reduce infection of large (nickel-size and larger) pruning wounds with topical application of fungicides, detergent, or wound-dressings. Such treatments are laborious, but may be appropriate when reworking an established vineyard to a new variety or training system. Check local sources for current recommendations and regulations. Multiple-trunk training systems and regular trunk renewals also significantly limit the potential damage from individual, slowly-expanding cankers.

Nonspecific Fruit Rots

Ripening grapes are subject to decay by a number of secondary pathogens, including yeasts, other fungi, and bacteria. Some of these produce objectionable by-products such as acetic acid (a primary component of vinegar), leading to the term "sour rot". Other pathogens may impart undesirable flavors during fermentation or in the bottle. Fortunately, losses from such causes are sporadic. These opportunistic microorganisms often are of consequence only after berries have been wounded by hail, birds, insects, or a primary pathogen such as the powdery mildew fungus or *Botrytis*. As with most microorganisms that negatively affect grapevines, the growth of these secondary organisms is promoted by high humidity, abundant rainfall, and poor ventilation.

Control

Control is based upon good general vine management. Prevent or minimize damage from birds and insects, especially grape berry moth. Provide good control of powdery mildew, Botrytis, and other primary diseases. Choose planting sites and canopy management practices that provide good ventilation around the ripening berries. Regular applications of broad-spectrum fungicides targeted against other diseases may provide some additional control, although the sheer diversity of possible secondary rot organisms almost guarantees that no single product will be effective against all of them.

BACTERIAL AND PHYTOPLASMAL DISEASES

Crown gall and Pierce's disease are the two bacterial diseases of greatest importance on grapevines in eastern North America. Additionally, North American Grapevine Yellows, caused by phytoplasmas, can be destructive to some varieties in certain locations.

Crown Gall

Crown gall, caused by the bacterium *Agrobacterium vitis*, is a problem primarily on *V. vinifera* varieties that have been subjected to freeze injury or other trauma. Virtually all varieties of this species appear to be susceptible. The disease is much less common on native and hybrid varieties, although a few have shown damage in specific sites following severe freezing episodes. Gall tissues interfere with the plant's vascular system and, when severe, can seriously compromise the general health of the vine. Although a similar disease occurs across a wide range of broad-leaved plants, the organism that attacks grapevines is specific to this host only.

Symptoms

The major symptom of crown gall is the tumor-like galls that form on the trunks of affected vines. Although galls are most frequent near the soil line and the graft-union of grafted vines, they can extend well above that point. Bark in the affected regions splits as the bumpy or pimple-like gall tissue proliferates and expands beyond the woody portion of the trunk (figure 11.22). Current-season galls appear in the summer as white, fleshy callus growth, but become brown, dry and corky as the season progresses. New gall tissue often forms at the edge of older galls, where these exist.

Disease Cycle and Conditions for Development

The crown gall bacterium is present in the wood of nearly all commercial grapevines and nursery stock. When the vine is injured, the bacteria transfer a portion of their

Tom Burr, Cornell University

Figure 11.22 ▪ Bumpy, tumor-like galls on the trunk of grapevine affected by crown gall.

genetic material into the plant cells in and around the wounded tissue. The host cells are then "reprogrammed" to produce a large mass of undifferentiated, callus-like tissue, leading to the formation of galls or "plant tumors." This growth may remain relatively superficial or it may continue to expand until the trunk is girdled and the vine dies. The severity of gall formation is determined by the grape variety, the bacterial strain involved, and the severity of the injury that initiated the disease. Although freeze damage is by far the most common form of injury leading to the development of crown gall, mechanical injury or even grafting and the conditions of graft union callusing may also elicit the disease on occasion.

Management

Crown gall is managed primarily by matching appropriate varieties to specific planting sites and by other cultural practices to minimize the risk of low temperature injury. Propagating vines from rootstocks and scion budwood that are free of the bacterium is the ultimate control practice, but very little of such material is currently available. There is also some potential for biological control of the crown gall organism. Multiple-trunking and trunk renewal systems help to limit losses when the disease does occur.

Pierce's Disease

Pierce's disease (PD) is a destructive bacterial disease that affects all bunch grapes in regions of the US that experience warm winter temperatures, including Virginia and states to its south. This disease is the single greatest limiting factor to the expansion of bunch grape production in much of the southeastern US, where warm winters and abundant native host plants are particularly favorable for its persistence and spread. PD also occurs westward through parts of Texas and into California. PD is caused by a bacterium (*Xylella fastidiosa*), which is transmitted principally by several different species of leafhoppers (small, winged insects).

Symptoms

Symptoms of PD vary with the season and variety but may include delayed bud break; stunted shoot growth; marginal "burning" or dying of leaf tissues; wilting or premature coloring of fruit; uneven maturity of canes; and gradual dying of the root system, all resulting in degeneration of the vine (figure 11.23). Symptoms

Anna Wallingford, Virginia Tech

a. Young vine collapse with marginal "burning" or drying of leaves resembling advanced drought stress symptoms.

Anna Wallingford, Virginia Tech

b. Close-up of affected leaves of red-fruited variety illustrating zone of necrotic (brown), reddened, chlorotic and healthy (green) tissue.

Figure 11.23 ▪ Symptoms of Pierce's disease.

tend to be most severe in vines that are stressed, as by drought, and intensify in late-summer as fruit begin to mature. Infection occurs when leafhoppers that have acquired the causal bacterium from infected grapevines or native host plants subsequently feed on a susceptible vine. After the PD organism has been transmitted to the vine, it reproduces repeatedly and forms a large aggregate mass of bacteria. The bacterial masses, as well as gums produced by the vine as a defensive response to infection, block the xylem, or water-conducting tissues of the vine. The resulting symptoms are largely due to this blockage and resemble many of the effects of drought stress. Infected vines may die within a year of infection, or they may persist for five or more years. Because the PD bacterium is supressed or killed by low winter temperatures, the disease is decidedly more severe in regions that experience mild winters. Varieties differ in their susceptibility to PD: Chardonnay and Pinot noir are particularly susceptible, whereas Riesling is generally more tolerant. However, no varieties of *V. vinifera* or *V. labrusca* are totally immune.

Management

Insecticidal control of vectors has provided some level of disease control in California, Texas and the southeastern US, especially with the advent of systemic, persistent insecticides that are specific for leafhoppers and closely allied insects. Particle film technology (kaolin, Surround™) has also reduced the frequency of vector feeding events, thus reducing infection incidence; however, this material must be applied repeatedly in areas with frequent summer rains. The most practical measure for avoiding PD is to not plant susceptible vines in regions where the disease is endemic. This includes much of the coastal plain areas of the Carolinas and the southeastern portion of Virginia, including portions of the Delmarva Peninsula (see chapter 2, page 14). If in doubt, check with your state authorities. Mildly affected vines may recover if subsequently exposed to freezing temperatures during the winter. However, the specific conditions necessary for this recovery are poorly understood. Constant inspection of the vineyard and removal of newly symptomatic vines has also helped reduce spread of the disease.

North American Grapevine Yellows

North American grapevine yellows (NAGY) is a destructive disease of grapevines with variable incidence in eastern North America. North American grapevine yellows is a variant of several forms of grapevine yellows diseases that occur globally. Other variants of the disease include *Flavescence dorée* and *bois noir* in Europe, and Australian grapevine yellows in Australia. The various grapevine yellows diseases differ in causal pathogens and with respect to the severity of disease symptoms. North American grapevine yellows was first observed in New York State; however, the incidence and severity to date appear to be greater in the more southerly states of Virginia and, to a lesser extent, Pennsylvania and Maryland. Affected vines typically die within two or three years of symptom onset. In Virginia, some Chardonnay vineyards have experienced annual losses of 5% or more of the original planting, with attrition rates approaching 30% over a 5- to 10-year period. Affected vines are often most abundant near the vineyard edge, especially edges that are closely bounded upwind by mixed stands of hardwood trees and wild grapevines.

Grapevine yellows diseases are caused by phytoplasmas, single-celled organisms similar to bacteria, but lacking a rigid cell wall. At least two different phytoplasmas have been implicated in NAGY expression. Leafhoppers and possibly some related insects transmit phytoplasmas from wild grapevines or other hosts to cultivated vines, and possibly from vine to vine within the vineyard. To date, wild grapevines (*V. cordifolia* and *V. riparia*) have been the most consistently-detected alternative hosts, and an abundance of these species near cultivated vineyards is associated with increased incidence of NAGY. Chardonnay and Riesling are highly susceptible, but other varieties have also expressed symptoms in vineyard sites prone to disease development (see chapter 2, page 14).

Symptoms

Symptoms of NAGY are often confined to one or several shoots of the vine in the first year, followed by a comprehensive involvement of the vine in the second or third season. Specific symptoms include withering of clusters at or after bloom, rolling and yellowing of leaves, shoot tip dieback, and failure of shoot stems to uniformly develop the brown bark of canes (periderm). Shoot stems may have a blue-gray cast and the stems often droop due to lack of internal lignification (figure 11.24). North American Grapevine Yellows can be visually distinguished from other diseases that express some of these symptoms by the presence of *all* such symptoms on affected vines or shoots. The abortion of fruit clusters, for example, may occur with early- and late-season bunch stem necrosis; however bunch stem necrosis does not occur concomitantly with the failure of periderm to mature. Similarly, grapevine leafroll virus (described below), may cause leaf rolling, but does not typically cause cluster abortion or failure of periderm maturation.

a. Vine in photo shows poor wood maturation, downward rolling leaves, leaf yellowing, and cluster abortion in August *(arrow)*. A representative shoot from a nearby healthy vine was cut and placed on the affected vine to contrast the appearance of affected and healthy shoots.

b. Close-up of an affected cluster in June.

Figure 11.24 ▪ Symptoms of North American Grapevine Yellows on Chardonnay.

Management

The current means of NAGY management is to avoid planting susceptible varieties such as Chardonnay and Riesling in areas where local experience suggests that a high attrition rate due to this disease is apt to occur. While NAGY is known to occur in several eastern US states, the greatest vineyard losses to date have occurred in Virginia. Based on surveys in Virginia, the vineyards *most* at risk of NAGY are Chardonnay plantings in close proximity to wild grapevines or *Prunus* species (especially black cherry and ornamental cherry), and in the western piedmont and Shenandoah Valley regions (see chapter 2, page 14). Removal of wild grapevines and *Prunus* spp. from the vineyard environs may help reduce the incidence of NAGY in cultivated vines; however, this approach has not been rigorously researched. Efforts to prune out affected trunks or cordons in the first season of disease observation have generally failed to prevent the systemic spread of symptoms to the entire vine, at least with Chardonnay. In Virginia, where NAGY has been observed for over 15 years, it is recommended that vines showing clear NAGY symptoms be removed during the first season of symptom expression to lessen the possibility of these vines serving as an additional source of inoculum. Foliar or soil-applied, systemic insecticides, such as used for Pierce's disease management, might offer some mitigation of NAGY phytoplasma transmission. In parts of western Europe, *Flavescence dorée* can be managed with a single insecticide application that targets the single vector, *Scaphoideus titanus*, which is known to have only one generation per year and completes its life cycle only on grapevines. Unfortunately, research in Virginia has revealed multiple insect (leafhopper) vectors, with most having multiple generations per year. Thus, susceptible vines are essentially at risk throughout the growing season. Attempting to manage the disease with repeated insecticide sprays would be extremely expensive and environmentally insensitive.

VIRAL DISEASES

There are a number of viruses that infect grapevines worldwide and pose at least a potential threat to eastern North American grape producers. Although only a few have caused economic losses in eastern regions to date, continued expansion of the wine grape industry provides the opportunity for additional problems to be introduced. Viruses are seldom spread to uninfected vineyards by natural means. Rather, they typically are introduced via infected planting or propagating material such as vines, rootstocks, or budwood. Therefore, the easiest and most effective method to avoid new virus problems is to use uninfected plant material when establishing new vineyards, or when replanting or reworking existing sites. Such material is available through formal certification programs that test and maintain clean "mother vines" used for subsequent propagation. A number of reputable nurserymen also independently identify healthy propagating material that they use in their operations. Additional precautions should be taken before planting in order to avoid two viruses that are endemic in the East. There are no sprays or other therapeutic treatments for curing a vine once it has become infected with a virus.

Leafroll

Leafroll symptoms are most obvious on red-fruited *V. vinifera* varieties. In the late-summer and fall, leaf tissues between the veins turn deep red, while the areas along the main veins remain green (figure 11.25). Leaf yellowing may occur on some white varieties. The margins on infected leaves curl or roll downwards, giving the disease its name. Symptoms are first apparent on leaves at the base of the shoots. Infected *V. vinifera* vines may be slightly stunted, and although infection does not kill or severely debilitate the vine, the disease causes chronic damage. Yield losses of 20% in each remaining year of the vine's life have been reported. Leafroll infection also delays ripening and reduces sugar content and color of the fruit. Limited data suggest that the disease may decrease winter cold hardiness as well. Recent data from New York show that leafroll infections are common on the major native varieties in that region, although such vines are symptomless, and the economic impact of such infections is questionable. Confirmation of suspected leafroll infection can be provided by a laboratory test conducted by one of several commercial or university-affiliated laboratories. Accurate testing usually requires that samples be collected in a prescribed manner, which should be established with the laboratory in advance.

Anton Baudoin

Figure 11.25 ▪ Characteristic leafroll virus symptoms on red-fruited *V. vinifera* variety.

Control

In the East, leafroll is thought to be spread almost entirely by the use of infected propagating material. However, mealybugs have been implicated in the vine-to-vine spread of leafroll disease in California. Although this mode of transmission has yet to be demonstrated in eastern vineyards, recent research has shown the presence of effective mealybug vectors in some heavily infected eastern plantings. The likelihood of introducing leafroll virus into vineyards can be greatly reduced by using certified plant material. Particular care should be taken with rootstocks, which are derived from native grape species that remain symptomless when infected but can transmit the virus to *V. vinifera* scions after grafting. Again, the best insurance against establishing infected vines is to purchase them only from a certified clean stock program. In contrast, random collections of budwood from affected vineyards ensure the continued spread of the disease. Although leafroll decreases both the yield and the quality of fruit from infected vines, the cost of removing and replacing such vines may outweigh the benefits. On the other hand, growers and vintners who wish to maximize the quality of their crop and resultant wines may wish to rogue out confirmed cases of leafroll, particularly with red-fruited varieties.

Tomato Ringspot and Tobacco Ringspot

Tomato ringspot virus is endemic to eastern North America, where it infects a wide variety of fruit crops and common broad-leaved weeds (such as dandelions, sheep sorrel, common chickweed, red clover and narrow-leaved plantain). Because it is seed-transmitted, tomato ringspot virus can be introduced into new sites by the movement of seeds from these plants. Once established in a vineyard, it can be spread to healthy grapevines from infected weeds or other vines through the feeding activities of the dagger nematode, a microscopic roundworm that also is common in eastern soils.

Tomato ringspot virus can cause a serious decline on certain own-rooted hybrid varieties, most commonly Baco noir, DeChaunac, Chelois, and Vidal blanc. Although *V. vinifera* varieties are genetically susceptible to infection, most are tolerant of the virus (that is, they show few symptoms despite being infected), and a number of common rootstocks also show good field resistance. Consequently, tomato ringspot is seldom a serious problem in *V. vinifera* plantings. Varieties of *V. labrusca* are resistant.

Like tomato ringspot, tobacco ringspot virus is endemic to the Northeast. Most aspects of the biology and transmission of the two viruses are very similar (weed hosts, seed transmission, nematode vector), and their symptoms are indistinguishable. Tobacco ringspot is potentially more severe on *V. vinifera* varieties, although confirmed reports of damage are few. Interspecific hybrids are reportedly less susceptible, although disease has been detected on Villard blanc and Chambourcin vines. *Vitis labrusca* varieties are resistant. The same control program is employed against both tomato and tobacco ringspot virus.

Symptoms

Symptoms differ significantly in the different viticultural regions of the East. Despite the names, actual "ringspot" patterns on the foliage of infected vines are rare. In colder, northern production areas, disease caused by tomato ringspot virus is typically evident during the second year after infection. New growth may be sparse, since many buds of infected plants are killed by low winter temperatures. Shoots are stunted due to their significantly shorter internodes, and leaves are small and distorted. Berries on infected fruit clusters are sparse and develop unevenly. Affected vines may be killed outright within 3 or 4 years, or develop new growth from only a few buds at the base of the trunk. Frequently, infected vines are clustered into elliptical or roughly circular patches in the vineyard, with dead and dying vines in the center and others in less severe stages of decline radiating out from this epicenter. In the Mid-Atlantic region, some or all of the clusters on Vidal blanc vines infected with tomato ringspot typically show sparse berry set and/or a severe reduction in fruit size (figure 11.26). The decline symptoms associated with shoot stunting and winter injury are much less common in this region than they are further north. As with leafroll, suspected infections by the two ringspot viruses can be confirmed by specialized commercial laboratories.

Control

As with all virus diseases, care should be taken to ensure that new vines are produced from clean propagating material. Several common rootstocks are either resistant to or have shown good field tolerance of the virus, including 5C, 5BB, SO4, and C-3309. Although the economics of hybrid grape production may not typically favor the planting of grafted vines, this option should be used if replanting into a block that suffered from tomato ringspot virus or if a new planting of a highly susceptible variety is planned for a site that is known to be

Figure 11.26 ▪ Tomato ringspot symptoms on Vidal blanc.

infested with dagger nematodes capable of transmitting the viruses. Because the viruses persist in broadleaf weed host "reservoirs", it is desirable to maintain new planting sites as free from such weeds as possible during the year before the vineyard is established.

Replant situations. Occasionally a vineyard infested with tomato or tobacco ringspot virus must be reestablished due to poor yields or actual vine loss, or the grower simply wishes to change varieties. Once the decision has been made to renovate a vineyard, one of two courses of action might be considered:

(1) The ideal approach would entail removing all grapevines and the trellis. Kill remaining vines in the late-summer after harvest with glyphosate herbicide used at a rate specified on its label for woody weeds. Two applications may be necessary. Pull the vines the following winter, removing as much of the larger root system as possible. Make any desired soil amendments (such as liming) and thoroughly cultivate the site. Conduct a nematode assay and follow the recommendations for nematode control provided in the pre-plant vineyard renovation sidebar (see chapter 4, page 71). Afterwards, plant the site to a grass cover crop, such as fescue, and control weeds through mowing for another one to two years to allow former grapevine roots to decompose. Reestablish the vineyard using certified stock grafted to a suitable rootstock (see chapter 3, page 37). If Vidal blanc, Baco, DeChaunac or other virus-sensitive hybrid varieties are to be re-established, they too should be grafted.

(2) An alternative approach, but one which has some inherent limitations is as follows: Kill and pull vines as described above, but leave the trellis intact. Control weeds under the trellis and in the alleys. Lime if needed. Cultivate row middles if a grass cover crop is not present and establish a cover crop. Keep the site free of broadleaf weeds and replant in one to two years with grafted vines. Plant the vines between former vine locations, rather than using the same spacing in the row. This will alter the vine spacing/post interval, but should result in better establishment of the new vines. One fundamental limitation of this approach is a phenomenon termed "replant malady", which is characterized by poor growth of vines that are planted close in time and space to formerly present vines. The cultivation and offsetting of new rows, as described in option #1, minimizes the expression of replant problems, but is of course more time-consuming. There is also some evidence that the use of composts that are rich in microbial flora and fauna may promote the establishment of replanted vines.

CONCLUSION

One might conclude that there exists an overwhelming array of diseases waiting to damage or destroy cultivated vines and their crops in eastern North America. Most of these diseases, fortunately, are avoidable or can be managed with cultural and fungicidal options. Some, such as Pierce's disease, require careful consideration about vineyard site selection. Chemical disease management options vary from year to year due to changes in product registration. Grape producers are strongly encouraged to maintain close communication with their state and university specialists who can provide state-specific management guidelines.

References

HTTP://WWW.IPPC.ORST.EDU/CICP/FRUIT/GRAPE.HTML

HTTP://WWW.NYSIPM.CORNELL.EDU/FACTSHEETS/GRAPES/

Delp, C. J. 1954. Effect of temperature and humidity on the grape powdery mildew fungus. *Phytopathology* 44:615–626.

Isaacs, R., Schilder, A., Zabadal, T., and Weigle, T. 2003. *A Pocket Guide for Grape IPM Scouting in the North Central and Eastern U.S.* Michigan State University Extension Bulletin E-2889.

Pearson, R. C. and Goheen, A. C., eds. 1988. *Compendium of Grape Disease.* APS Press, St. Paul. 93 pp.

—12—

Major Insect and Mite Pests of Grapes

Author
Douglas G. Pfeiffer, Virginia Tech

A diverse group of arthropods feeds on grapes and grapevines. Most occur only sporadically; a few may be extremely abundant. It is important to be able to identify insects found on grape in order to plan effective management strategies. Since the pests themselves may not be seen, it is equally important to be able to recognize damage characteristics of major pests. In cases where adult feeding differs from that of the larvae, the most apparent damage may not be the most important. Although relatively minor, such symptoms may serve as an indicator of the presence of the insect. Timely recognition of a pest's presence is extremely important in certain cases where damage may occur rapidly, as with grape berry moth, or where timing of control measures is critical, as with grape root borer.

Pests may be divided into two general groups: pests causing direct injury and those causing indirect injury. Direct injury occurs on the harvested part of the crop plant—that is, the fruit. Generally, only a low pest population causing direct injury is required to cause economic damage. Examples of pests causing direct injury include grape berry moth, redbanded leafroller, and grape curculio. Indirect injury affects non-harvested parts of the plant such as leaves, canes, trunk, and roots. Grape leafhopper, grape plume moth, and European red mite are examples of pests that can cause indirect injury (grape plume moth is generally a leaf feeder, though young clusters may also be fed upon if webbed together with leaves). Vines usually tolerate larger populations of indirect feeders. Some notable exceptions to this generalization occur. The grape root borer larvae, hidden beneath the soil surface, can be devastating.

Insect and mite grape pests are discussed in this guide in the order of those injuring berries, buds, and clusters, foliage, shoot stems, canes, trunk and, finally, roots. Identification keys for pests and their damage symptoms are included in appendix 4 (page 307). The biological information given here is vital for effective pest management. Specific information on chemical control measures is not provided because that information can change from year to year. For current recommendations, consult cooperative extension materials appropriate for your state or the Pest Management Regulatory Agency for Canada (for example, in Virginia, publications 456-017 and 456-018). Many states post grape pest management recommendations on-line (or provide links for ordering hard copies). See appendix 5 (page 312) for a list of state publications.

Pest populations will often be present at levels sufficiently low or localized that special control measures will not be required. Research in grape integrated pest management is currently being conducted by entomologists in several eastern states to determine "economic thresholds," or population densities that justify the cost of control measures. After identifying a grape pest using this manual, grape growers may wish to consult university or extension specialists in their state in order to determine the appropriate response to the level of infestation.

The following insects and mites occur with some regularity in eastern vineyards. By convention, common names are followed by scientific genus and species names where a specific pest is identified. Scientific names may be followed by the naming authority, the person who originally named the insect or mite. Authority names in parentheses are the original describers of the pest, but the pest genus was subsequently reclassified to its current taxonomic classification by someone else.

PESTS THAT AFFECT BERRIES, BUDS AND CLUSTERS

Grape Berry Moth — *Paralobesia viteana* (Clemens)

The grape berry moth is one of the most severe grape pests in eastern North America. Populations have been increasing in many vineyards in recent years. Economic damage is primarily from larvae feeding on the berries, although minor (noneconomic) injury is also done to foliage. This species overwinters as pupae in grayish silken cocoons in fallen leaves. Survival of overwintering pupae is higher in damp leaf litter. Adult moths have irregular markings of different shades of brown with a blue-gray band across the wings (figure 12.1). Earlier workers reported that adults emerge when grape foliage is expanded, near bloom time. More recent work in the mid-Atlantic region has shown a prebloom generation as well, with mature larvae present by bloom. This has not been noted in Michigan.

The larvae are creamy in color at first, turning green or gray-green and eventually nearly purple when mature, and nearly 0.4 inch (10 millimeters) long (figure 12.2). The first-generation larvae feed on blossoms or small berries, leaving silken strands where they have crawled, resulting in webbed clusters. One larva may injure several berries. Initially they feed externally on the small

Rufus Isaacs, Michigan State University

Figure 12.1 ▪ Grape berry moth, adult.

Figure 12.2 ▪ Grape berry moth, larva.

berry; but as fruit reach 0.2 to 0.3 inch (6 to 8 millimeters) in diameter, larvae begin to feed internally, making a small hole about 0.04 inch (1 millimeter) in diameter. On red grape varieties, a red spot develops at the point of entry of the larva soon after it enters a berry; winding dark tunnels are evident through the surface of the berry (on dark-colored varieties, this is most apparent before berries start to color). Larvae tunnel into the fruit, feeding on pulp and seeds. This causes fruit to shrivel and fall and, in severe cases, may lead to loss of most of the crop. Berries may be webbed together and sticky with juice. Damage can lead to fruit rot that affects other berries in the cluster. After feeding for 3 to 4 weeks, larvae drop to the ground to pupate. The pupal stage lasts 10 to 14 days, except for the final generation, where the pupa enters diapause and overwinters in leaves under the trellis. Two generations annually have been recorded in season. Pennsylvania sites only 10 miles apart have been shown to vary in the number of generations. Work in Virginia indicates three to four generations, with a possible fifth in some seasons.

Recent research in Pennsylvania has centered on the termination of diapause. The first adults appear in the spring at 266 degree-days above a developmental threshold of 47°F (148 degree-days above 8.4°C). Development of a complete generation requires 900 degree-days (50°F) (500 degree-days above 10°C). This can be divided into about 763 for development from egg to adult and about 133 for mating and preovipositional behavior (about 424 and 74 degree-days Celsius, respectively). However, timing of spring emergence is variable in this species. The Pennsylvania researchers found that after cold exposure, adults emerged 20 to 26 days after a return to warm temperature. But if pupae were not exposed to cold, individuals emerged within a month before or after the cold-treated insects. Therefore, in springs following mild winters, it may be difficult to accurately predict adult emergence. At dusk, females lay an average of twenty eggs singly on grape stems, blossom clusters, or berries. The eggs hatch after about 4 to 8 days.

The second flight occurs near bloom time, continuing for several weeks. Third- and fourth-generation eggs are laid from late July into September on berries. Larvae of these generations are more numerous than those of the first generation. The final generations can attain extremely high levels and cause the most loss, and pose a threat of contamination that may be detected by processors. Rot infections may follow penetrations made by grape berry moth larvae.

Populations of grape berry moth may be monitored with a commercially available pheromone trap. A noneconomic species that feeds on sumac—*Episimus argutanus* (Clemens), or sumac moth—is also attracted to grape berry moth pheromone traps (figure 12.3). The sumac moth is light brown with fine darker brown striations across the wings (lacking the blue-gray band found in grape berry moth). Trap numbers cannot be used to predict specific levels of injury, merely to indicate presence and periods of activity. The relationship between grape berry moth trap captures and fruit damage is strongly influenced by vineyard surroundings. In vineyards surrounded by open terrain (cornfield, pasture, and so on), trap numbers may be high but actual injury low. In vineyards surrounded by woodland (with wild grapevines), trap captures may be low but accompanied by high levels of injury. In a Michigan study, pheromone traps were placed at varying heights up to 30 feet (9 meters), and grape berry moth trap captures were found to increase with height.

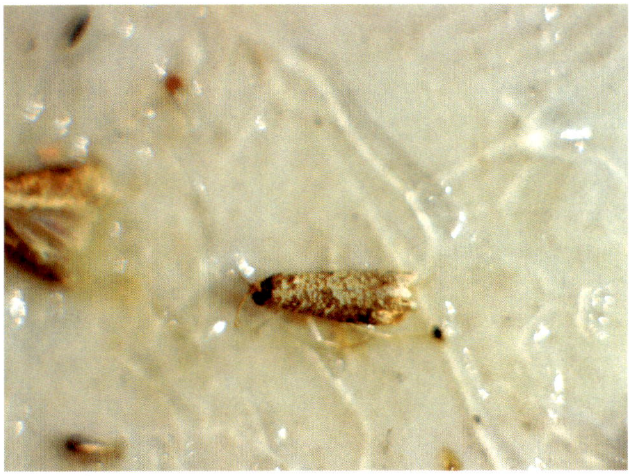

Figure 12.3 ▪ *Episimus* (Sumac moth), adult.

Pennsylvania research has addressed the effects of photoperiod on the initiation of diapause. Initiation of diapause is stimulated by a critical photoperiod occurring during the egg stage. At latitude 42°N (the latitude of the study in Northeast, Pennsylvania), if eggs are subjected to 15 hours of light, the resulting pupae will produce adults with no diapause. When subjected to 14 hours 44 minutes of light, 25% of pupae entered diapause. At 14 hours 30 minutes, 80% entered diapause; and at 14 hours light, 99% went into diapause. Control measures applied just before most individuals are entering diapause may prevent heavy injury from the last generation, but the details must be worked out for each region. At 38°N (Charlottesville, Virginia), maximum day length at summer solstice (June 21) is 14 hours 48 minutes. Critical day lengths associated with given proportions of the population entering diapause may be shifted accordingly in geographic races.

Researchers in New York have established a risk-assessment system for grape berry moth. Factors associated with high risk are: history of high grape berry moth injury, wooded vineyard margins, and prolonged snow cover. Low-risk factors include open vineyard settings and low historical incidence. While this scheme was created for juice grapes and table and wine grapes were considered to be always at high risk, the presence of these risk factors also calls for increased management in wine grape vineyards. High populations and damage have been seen in Virginia after consecutive mild winters. Substantial winter mortality occurs after several days of very cold temperatures (–6°F to +5°F), when pupae are not buried by snow.

There is often a marked edge effect with grape berry moth, with higher levels of berry infestation in rows near deciduous woods, especially where wild grapevines are found. If the rows are parallel with the woods, sprays can be directed at edge rows. This edge effect can be compounded in mating disruption vineyards, since there may be the normal immigration from wild hosts, and pheromone levels in the air tend to decline at vineyard edges. Pay special attention to vineyard margins when scouting for mobile pests.

In addition to chemical control, mating disruption and cultural control are useful for managing grape berry moth. In high-risk vineyards, no single control tactic provides adequate control; a multi-pronged approach is best, incorporating chemical control, mating disruption, cultural control, and biological control. Where this pest has been troublesome, leaves should be raked and burned in the fall. The soil beneath rows may be cultivated to bury overwintering pupae. Soil from row centers should be piled beneath vines about a month before harvest. In the spring, about 2 weeks or more before bloom, this ridge is pulled back into the row center and disked or plowed. This management strategy would not be an option in vineyards that use permanent cover crops in row middles. An egg parasite, *Trichogramma minutum* Riley, provides some biological control. However, grape berry moth does not appear to be an optimal host for this wasp, and resulting adult parasites have poor vigor and exhibit developmental abnormalities. Inundative release of insectary-produced *Trichogramma* may be a solution.

Grape Flea Beetle — *Altica chalybea* Illiger

This metallic blue-green beetle is almost 0.2 inch (5 millimeters) long (figure 12.4). Adult grape flea beetles overwinter in debris in and near the vineyard. They become active early in spring and lay eggs in cracks in the bark, at bases of buds, between bud scales, and on leaves. Eggs are light yellow and are laid in masses; they hatch in a few days and larvae feed on grape leaves for 3 to 4 weeks. Larvae are brown with black spots and reach a length of 0.4 inches (10 millimeters). Larval feeding damage consists of characteristic chain-like feeding marks on leaves. However, the damage by adult grape flea beetles is more important. The beetles eat holes into the sides of buds and gouge out the contents as the buds swell; the injury appears similar to that of climbing cutworm (figure 12.6), and they also feed on the unfolding leaves.

Figure 12.4 ▪ Grape flea beetle, adult.

Mature larvae drop to the ground and pupate in an earthen chamber. Adults emerge 1 to 2 weeks later, in July and August. These adults feed for the rest of the summer but cause little damage. In the fall they seek protected places in which to overwinter. In addition to wild and cultivated grape varieties, grape flea beetles feed on Virginia creeper (*Parthenocissus quinquefolia*).

The grape flea beetle is more common in neglected vineyards, but some commercial growers consider this species their main insect pest, especially in vineyard rows near deciduous forest. Sprays and fall vineyard cleanup for grape berry moth aid in controlling flea beetles; but where a history of damage is noted, sprays may be needed for this early-season pest. Flea beetle feeding injury may be mistaken for that caused by climbing cutworms (see next section). Injury by the latter is more likely to be ragged in appearance, though there is overlap in appearance. Proper identification is increasingly important with the loss of organophosphate insecticides; more selective products—for example, methoxyfenzide (Intrepid) or the loss of organophosphate insecticides; more selective products—for example, methoxyfenzide (Intrepid) or spinosad (SpinTor, Entrust)—are not as effective for beetle control as they are for moth larvae.

Climbing Cutworms

Cutworms are larval stages of several species of moths in the family Noctuidae that are usually gray or brown with either dark or light markings. The species differ in how they overwinter (egg, larva, or pupa) and in the number of generations per year (one to four).

The caterpillars vary in appearance according to species, often smooth brown or gray (figure 12.5), and with longitudinal stripes. In daytime they remain concealed in weeds or soil in a curled position. At night they climb onto the vine to feed on primary grape buds and cause injury similar to that of grape flea beetles (figure 12.6). Damage is usually localized within the vineyard, and the period of susceptibility is generally short. Only caterpillars present during time of full bud swell to 3-inch (about 8-centimeter) shoot growth cause significant injury. However, during springs when cool weather causes a protracted period of bud swell, extensive injury may occur. The economic impact extends beyond simple loss of primary buds. The subsequent development of secondary buds can cause uneven berry ripening, aggravating efforts to achieve uniform maturity at harvest. Good weed control within the row is especially helpful in limiting cutworm populations. We have also observed increased cutworm populations in vineyards that use minimal or no under-trellis herbicides, and in vineyards where mulch has been applied around vines. Growers who choose to follow those management options should be particularly vigilant for cutworm activity in the following spring.

Figure 12.5 ▪ Climbing cutworm, larva.

Figure 12.6 ▪ Climbing cutworm injury to swollen bud.

Yellowjackets, Other Wasps

Wasps, especially yellowjackets (*Vespula* spp.), paper wasps, and hornets, may break open the skins of grape berries in order to reach the sweet contents. In the early part of the growing season, these wasps are mainly predatory. Toward the end of the season, the wasps' foraging behavior changes as the sugar content of the fruit increases. Although honey bees are sometimes seen on broken fruit consuming the juice, they are not responsible for the initial injury because their mandibles are not strong enough to puncture the fruit skin. Yellowjackets may sometimes be seen foraging around harvest lugs at harvest time. While yellowjackets may be somewhat less aggressive when fully fed or preoccupied, caution should be exercised if yellowjackets are feeding at clusters to avoid hand stings. Control measures can be directed against the overwintering yellowjacket queens by establishing bait stations containing an attractant, sometimes with a pesticide, in early spring. Destruction of nearby nests is helpful but difficult, because the underground nests may be hard to locate. Night is more suitable for control of nests because wasps are more docile.

Banded Grape Bug — *Taedia scrupeus* Say

This insect overwinters in the egg stage. Eggs hatch when there are 2 to 5 inches (5 to 13 centimeters) of shoot growth. Nymphs are from ⅛ to ½ inch (3 to 13 mil-but then move to developing clusters when they become available. The number of clusters per vine is reduced, as is berry weight. Research in New York has shown that injury is most severe in the edges of vineyards near woods. The adult stage is reached after about 3 weeks of development. Adults are predatory and are not injurious. They deposit eggs in crevices in 2-year-old wood. This insect is mainly a problem in the northern part of eastern US, especially Michigan and the Lake Erie region of New York (injury has not been noticeable in the Finger Lakes region). A threshold suggested in New York is one nymph per ten shoots. If banded grape bug nymphs are common, an insecticide should be applied between the time of 6 to 12 inches (15 to 30 centimeters) of shoot growth and prebloom.

Gallmakers

Galls of various shapes occur on grapevines as a result of attack by small gall midges (family Cecidomyiidae). Galls occur on leaves, tendrils, and blossom buds and can occasionally cause considerable injury (figure 12.7). The species causing this symptom, grape tumid gall, was formerly considered to be a complex of species. The most prominent are the grapevine tomato gall, grape apple gall, and grape tube gall. No practical chemical control for these galls is known, nor is economic injury an issue. Removing the galls by hand and destroying them may reduce future populations if the galls are affecting trunk or cordon establishment.

Figure 12.7 ▪ Grape tumid gall.

Rose Chafer — *Macrodactylus subspinosus* (Fabricius)

This general feeder is related to the Japanese beetle and has a similar life cycle. Rose chafer was reported as a grape pest as early as 1810, later extending its host range to include a wide assortment of plants. Grape remains among the most severely injured crops. Larvae overwinter in soil, resuming development in the spring. Adults are tan, long-legged beetles about 0.5 inch (12 millimeters) long and emerge near the time of grape bloom. Mating and egg-laying occur continuously for about 2 weeks with each female depositing 24 to 36 eggs. The average life span of the adult is about 3 weeks, during which time they feed on blossoms, newly set fruit, and leaves. In about 2 weeks, eggs hatch, and larvae begin feeding on grass roots. This pest is more common in areas with light, sandy soils. There is one generation per year.

Control is seldom needed for this pest in most vineyards, but vigilance should be maintained early in the season in case of high numbers. In severe cases or in vineyards with a history of rose chafer, blossom buds can be completely destroyed, resulting in little or no grape production. Population levels vary from year to year. Insecticide sprays immediately following bloom for grape berry moth will also help manage rose chafers.

Grape Curculio — *Craponius inaequalis* (Say)

This species is a small, black weevil about 0.1 inch (2.5 millimeters) long. Unlike the two weevil pests discussed later in this chapter (grape cane girdler and grape cane gallmaker), grape curculios feed directly on the grape berries. Adults overwinter in sheltered places and become active about the time of Concord grape bloom. Feeding on leaves results in small groups of short, curved lines. After about 2 weeks, oviposition begins. The female chews a cavity beneath the skin of the grape and inserts an egg. The egg hatches in about 6 days, and the larva, a legless grub, feeds on the flesh and seeds. After about 3 weeks, the larva drops to the ground and pupates. New adults emerge 3 to 4 weeks later, and these feed on foliage for the rest of the summer. There is one generation per year. Populations are generally of low economic significance.

Redbanded Leafroller — *Argyrotaenia velutinana* (Walker)

This leafroller is a pest of several crops, including grape. Pupae overwinter in debris on the ground. Adults emerge at about 3 to 5 inches (8 to 13 centimeters) shoot growth, and females deposit egg clusters on canes. First-generation larvae feed on new foliage and fruit clusters. Larvae are pale green, including the head capsule, and about 0.6 inch (16 millimeters) long (figure 12.8). After the pupal stage, moths appear in midsummer. Two more generations occur, the third being the most numerous and damaging. Webbing and larvae may be seen in clusters or leaves. Larvae will eat holes into the sides of fruit but do not enter the berries (distinguishing it from grape berry moth).

Figure 12.8 ▪ Redbanded leafroller, larva.

Organophosphate insecticides generally work well against this species, as does methoxyfenozide, an insect growth regulator used against grape berry moth. A sprayable pheromone formulation is available for mating disruption of this species [the main component should be (*Z*)-11-tetradecenyl acetate].

Multicolored Asian Lady Beetle — *Harmonia axyridis* (Pallas)

This insect is an important predator in most of the eastern states, having become the most common lady beetle in many sites. It is a fairly large species, about ¼ inch (6 millimeters) long. Adults are variable in the degree of

Figure 12.9 ▪ Multicolored Asian lady beetles.

redness (ranging from light pink to orange to dark red) and also in the degree of development of the ten dark spots on each wing cover, ranging from almost invisible to almost merging (figure 12.9). The pronotum (the section just behind the head) is white with a black M- or W-shaped mark. Larvae are yellow and black and are recognizable in having forked spines along the back. The multicolored Asian lady beetle has aroused concern in many areas where it has become established by its habit of seeking protected places in which to overwinter, often entering human homes by the thousands. Another cause for concern is the fact that beetles may enter clusters close to harvest and be collected into picking lugs or bins, especially with mechanical harvesting. When beetles are alarmed or crushed, they release a foul-tasting compound. This has been shown to impart an off-flavor to wine if the beetles are crushed with the grapes. This has been reported as a problem mainly on the Niagara Peninsula, the Finger Lakes region, and southeastern shore of Lake Erie. The economic threshold may be as little as fifteen beetles per lug (about 25 pounds of fruit).

PESTS THAT AFFECT FOLIAGE

Japanese Beetle— *Popillia japonica* (Newman)

This familiar beetle may be a severe pest of grapevines during the summer, feeding mainly on foliage but rarely on berries. The large, white, C-shaped grubs overwinter several inches below the soil surface (in southern parts of the range, grubs may not descend as far below the root zone). In spring they move closer to the surface and feed on grass roots; larvae begin pupating in May. In late June (early June in early parts of the range), adults emerge and begin feeding on a wide range of plants. Adults are shiny green with copper-colored front wings (elytra, or wing covers) (figure 12.10). Beetles skeletonize leaves of some varieties or eat completely through the leaves of others. Fruit are rarely attacked. Beetles are gregarious; they may be present in great numbers on a few vines, and feeding is concentrated in the upper part of the canopy. Peak adult activity is in July but may continue into September. Eggs are laid in the soil, hatching in about 2 weeks. Females make several trips to the soil, laying one to four eggs each time, and prefer moist grassy areas. Young larvae feed on grass roots until cold weather, when they descend deeper into the soil.

There are reports that Japanese beetle is most severe in the mid-Atlantic states, where there are large acreages of larval habitat (pasture) adjacent to preferred adult food (vineyards); this combination is favorable for Japanese beetle populations.

In a feeding study with Japanese beetle on the French hybrid Seyval Blanc in the upper Shenandoah Valley of Virginia, four feeding treatments were compared: a natural unprotected treatment; a controlled treatment, where beetle feeding was prevented; and two caged treatments where high numbers of beetles were contained on vines (1) from the beginning of beetle activity to veraison and (2) from veraison to harvest. The natural infestation did not result in any significant reduction in fruit quality, yield, or vine growth relative to protected control vines, despite greater leaf area loss than in the control (6.5% versus 3% leaf area loss, respectively). Intensive feeding by Japanese beetle after veraison caused the most severe effects on fruit quality. These vines had 11% leaf area loss when averaged over the whole vine (initial visual impact of feeding may be misleading because feeding is more intense on the upper part of the canopy). This loss occurred in less than half the time, relative to natural feeding—about 3 weeks compared with 6 weeks, respectively. Veraison marks the onset of the process of berry ripening; and during the post-veraison period, berries become the sink for photosynthates (see chapter 16, page 282). While established vines can tolerate significant Japanese beetle feeding, young vines can be totally defoliated and should be more rigorously protected, especially when grown in grow tubes. Therefore, the two most critical periods for Japanese beetle control are when vines are young and post-veraison with fruiting vines.

Figure 12.10 ▪ Japanese beetle, adults.

Several parasites have been introduced in attempts to suppress Japanese beetle. *Tiphia* wasps are ectoparasitoids on Japanese beetle grubs and use chemical cues to find hosts beneath the soil. A tachinid fly, *Istocheta aldrichi* Mesnil, parasitizes adult beetles. Ants may also be a source of natural mortality for Japanese beetle eggs. Japanese beetle larvae are subject to attack by a bacterium, *Paenibacillus popillae* (Dutky) (formerly *Bacillus popillae*) (milky spore disease). This biological control agent can be used in grassy areas with large larval populations, but it is ineffective against adults entering the vineyard. Adults are capable of flying great distances and will come into the vineyard from untreated areas. Because biological control is not reliable, insecticide applications will sometimes be necessary. Commercially available attractant traps are generally not adequate to protect vineyards but can be useful in monitoring population fluctuations. Traps should be placed some distance away from vines they are intended to protect, because the lures are extremely powerful and will increase beetle abundance in the immediate area.

Green June Beetle — *Cotinus nitida* (Linnaeus)

This relative of the Japanese beetle has a similar life history and causes similar damage. This is species is most common from Virginia southward. Adult green June

Figure 12.11 ▪ Green June beetle, adult.

beetles feed on the foliage of many shrubs and trees and will attack most tree fruits and berries. The adult is about 1 inch long and ½ inch wide (25 millimeters long and 13 millimeters wide), and flat on the top. Beetles are dull, velvety green above with deep yellow to bronze margins, and metallic green below (figure 12.11). Grubs are grayish white and considerably larger than Japanese beetle grubs, and less C-shaped than other white grubs (though when disturbed will coil tightly). There is one generation each year. Grubs overwinter up to a foot or more below the soil surface. They gradually make their way close to the surface during the spring and feed mainly on decaying plant matter and, to a lesser degree, on grape roots. Larvae may leave their protected sites and crawl on their backs to establish a new site elsewhere. The grubs pupate by May, and adults emerge in early July and August. Eggs are laid in soil with decaying vegetation. Beetles feed on petioles, leaves, and fruit; a single adult can cause significant damage. Adults are often found in groups and take large chunks from the fruit. Fruit injury is more common than with the Japanese beetle and is more likely to occur when populations are large. Most injury is seen in late July and August and, unlike injury from the Japanese beetle, can occur on both unripe and ripe fruit. Traps are available for Japanese beetle and are somewhat effective for green June beetle, but are used only to indicate the initial adult emergence. Adults may be monitored by quietly moving along the vineyard row, gently shaking the canopy, and observing how many fly off. Direct fruit counts are the most effective way of assessing damage. Clusters should be examined by the method outlined for other insects. Since feeding may be "clumped" or unevenly distributed, care should be taken in looking at a representative sample before making a spray decision. A treatment is justified if feeding exceeds 1%.

European Red Mite — *Panonychus ulmi* (Koch)

European red mite overwinters as tiny red eggs around cane nodes. Eggs hatch in the spring, and mites move to foliage where several generations occur annually. Populations are normally highest in midsummer; however, early-season infestations occur sporadically and can retard shoot development. Concord is among the most susceptible varieties. Among *vinifera* grapes, Riesling is quite susceptible. Mites are small and dark red, with hairs protruding from white dorsal tubercles (figure 12.12). During the growing season, eggs are laid on undersides of leaves and near leaf veins on the tops of leaves. After eggs hatch, the first active stage is the six-legged larval stage. The subsequent stages are eight-legged protonymph, deutonymph, and adult stages. Toward the end of the deutonymph stage, the female mite becomes quiescent and exudes a pheromone that attracts a male mite. The male remains close to the female until she emerges as an adult. Mating then takes place. Mated females lay eggs that give rise to females; unmated females lay male eggs. Immature and adult mites feed by piercing leaf epidermal cells and extracting cell contents, causing chlorosis and bronzing. Fruit sugar reductions of up to 1.5% have been recorded in Pennsylvania. However, economic

Figure 12.12 ▪ European red mites.

thresholds of two to five motile mites per leaf have been shown to be too conservative; infestation levels of up to 3,500 mite-days had minimal impact on leaf function (mite-days reflect both mite numbers and time. One mite per leaf for 5 days results in 5 mite-days, as does a population of five mites per leaf for 1 day). A suggested, provisional economic threshold is 10 mites per leaf on thin-leaved varieties, such as *vinifera*, and 20 mites per leaf on thick-leaved varieties, such as Concord; but as these levels are approached, observe the condition of the foliage and the presence of predatory insects and mites.

European red mite is the main spider mite species affecting grape in the mid-Atlantic. Serious infestations have occurred in some Virginia vineyards over the past several years, and European red mite has been an important pest in nearby states. In some states (for example, Michigan) and in western North America, two-spotted spider mite predominates. Populations are normally held in check by several predatory insects and mites.

Two-Spotted Spider Mite — *Tetranychus urticae* Koch

This spider mite is the most agriculturally important mite species in the world. However, they are not as common on grapevines as is the European red mite; although where they do occur, they are very damaging. Injury is from removal of chlorophyll from leaves by mites feeding from the underside of the leaves. Symptoms are seen as yellow stippling of the upper surface of the leaves where chlorophyll has been extracted. Delayed fruit ripening and reductions in quality occur when significant feeding occurs after veraison.

This species spends the winter as an orange-colored adult in the ground cover. In the spring, adults move into the vines and lay round, clear eggs on the foliage. Summer adults are yellow-white with two dark markings. Since populations must disperse upward into the canopy each season, two-spotted spider mite infestations usually occur midseason to late-season. If populations remain low, these mites may be included in the total for European red mites during monitoring. If they become common, however, the economic threshold should be halved. When using chemical control, observe label comments that sometimes call for higher use rates against two-spotted spider mites than European red mites, since two-spotted spider mites may be harder to kill. Management options for two-spotted spider mite include the use of several registered miticides, light summer horticultural oil applications, and/or release of mite predators such as *Neoseiulus fallacis* (Garman) or *Phytoseiulus persimilis* Athias-Henriot.

Grape Phylloxera — *Daktulosphaira vitifoliae* (Fitch)

This aphid-like insect is native to the eastern US but does not pose a serious problem in most eastern US vineyards. It is progressively more severe in northeastern and north-central states. There are two forms of this insect, one on the roots (radicicola) and the other on the foliage (gallicola) (figure 12.13). Root forms, more common on *vinifera* roots, are more damaging than foliar forms. Foliar forms are more common on non-*vinifera* *Vitis* species, such as the wild grapes in eastern North America. Phylloxera genetic variability and vine condition also influence galling ability. Wild vines (*V. riparia*) are highly susceptible to foliar forms. The interspecific hybrids Seyval, Chambourcin, and Villard Blanc are also highly susceptible.

The foliar form overwinters as eggs on canes. Eggs hatch when new leaves have appeared in the spring. Nymphs move to upper leaf surfaces and begin feeding. Their feeding induces the formation of galls on young leaves, shoots, and tendrils. After about 15 days, the

Figure 12.13 ▪ Grape phylloxera galls.

adult stage is reached within the galls. Young are produced parthenogenetically in the absence of mating, and six to seven generations occur on the leaves. Crawlers exit the gall, moving from older leaves to leaves at shoot tips. Occasionally, some individuals may drop to the ground and move to the roots. In the fall, females lay from three to eight eggs that hatch and mature into insects of both sexes, including winged forms. These adults mate, and females lay a single egg that overwinters. Injury by gallicoles results from reduced photosynthesis, rather than removal of photosynthates. Infested leaves may have photosynthesis suppressed by up to 50%, but neighboring leaves may be affected as well, although to a lesser extent. Fruit production on heavily infested canes may be reduced; the effect is greatest when galls occur in the 2 weeks following bloom.

In the root form, phylloxera overwinters on the surfaces of root galls. Two types of growth enlargements occur: tuberosities (localized growths of phloem parenchyma cells) and nodosities (enlarged growths that cause the root to bend at the feeding site). Tuberosities are the more damaging type and are less common on roots of American grape species. Nodosities may develop on American roots to a limited degree but have less impact. When roots start growing in the spring, these asexual insects lay eggs and produce several generations. Sexual males and females are eventually produced, mate, and give rise to overwintering eggs. Fungal infection in roots is more severe in the presence of phylloxera.

This pest illustrates one of the classic cases of host plant resistance. Native American grape rootstocks are resistant to the root form of grape phylloxera, while European (*V. vinifera*) varieties are susceptible (however, some American species—for example, *V. riparia*—are susceptible to foliar forms). When the insect was introduced into Europe in about 1860, the French wine industry was nearly destroyed. However, the *vinifera* vines were saved by grafting them onto imported American rootstocks. Resistant rootstocks are still important in limiting damage by the grape phylloxera. Nymphal establishment and adult reproductive rates are lower on resistant rootstocks than on susceptible types. In the early 1980s, a biotype appeared in California with the ability to colonize some resistant rootstocks, particularly the *vinifera* hybrid rootstock AXR-1. Consequently, many vineyards that had been planted to AXR-1 rootstock declined and were replanted with rootstocks that comprised only native-American species.

Grape Leafhoppers — *Erythroneura* spp.

Several related species comprise this group, with minor differences in their life cycles. Unmated adults overwinter in plant debris in and near the vineyard. Adults are about ⅛ inch (3 millimeters) long, pale yellow with red, yellow, or black markings. In the first warm days of spring, the adults become active, feeding on any green foliage before grape leaves appear. In May, leafhoppers go to vines and imbibe plant juices but little damage is done at this time. After feeding for 2 to 3 weeks, females oviposit just below the surface of leaf tissue. Eggs hatch in about 2 weeks, depending upon temperature. Females lay an average of one hundred eggs each. Nymphs are pale and wingless, with red eyes. They usually remain on the undersides of leaves and are quite active, often running sideways. There is a 3- to 5-week development period, and two to three generations per season.

Both nymphs and adults suck sap from leaves, resulting in pale stippling (figure 12.14). With heavy infestations, leaves become yellow or brown, and drop. Numbers are highest late in the summer, especially on end vines and border rows, and in seasons of above-average temperatures. Quality (sugar content) and quantity of fruit may be greatly reduced. However, high pest popu-

Figure 12.14 ▪ Grape leafhopper injury.

lations are required to reduce fruit quality. Researchers in New York have suggested a conservative, provisional action threshold of five nymphs per leaf before August 1 and 10 nymphs per leaf thereafter (if no stippling is apparent on the tops of leaves, it is unnecessary to count leafhopper nymphs on the leaf undersides). Surrounding vegetation can have an important impact on grape leafhopper densities as well as on an important egg parasite, *Anagrus epos* Girault. Because grape leafhopper overwinters in the adult stage, the presence of other leafhopper species on other hosts that overwinter in the egg stage may allow more stable populations of parasites.

Potato Leafhopper — *Empoasca fabae* Harris

Potato leafhopper feeds on many crops in addition to potato, including grape, apple, soybean, and alfalfa. Adults and nymphs are both active and light green in color, adults reaching a length of about ⅛ inch (3 millimeters). There is a row of six white spots on the front edge of the pronotum (the section just behind the head). Nymphs run sideways when alarmed and will often run to the other side of a leaf while being observed. Unlike grape leafhopper, potato leafhopper feeds in vascular tissue of the young leaves. A toxic saliva causes phloem and xylem tissue to collapse, which leads to a cupping, marginal yellowing, and necrosis—a collection of symptoms called "hopperburn." Adult potato leahoppers spend the winter in the Gulf Coast states, extending north into the Tidewater section of Virginia in pine habitats. In the spring, there is a northward dispersal, followed by two to four generations during the season. Injured shoots often prematurely stop growing. Nymphs are present until September and adults into October, though often in low numbers because of a lack of suitable foliage at that time of year.

Leafhopper-Borne Grape Diseases

Sharpshooters and Pierce's Disease

Pierce's disease (see chapter 11, page 216) is a bacterial disease of grapevines spread by several species of sharpshooters. Sharpshooters comprise a group of xylem-feeding leafhoppers, not the same as grape leafhoppers (*Erythroneura* spp.). Pierce's disease is a severe problem in the southeastern US, extending well into Virginia. It is not currently a problem in the northeastern US or Canada. The causal agent is a bacterium (*Xylella fastidiosa* Wells et al.) that occurs in many plants and is common in many grasses and sedges that are the usual hosts of many of the sharpshooter vectors. While many plant species can harbor *X. fastidiosa* populations, not all support bacterial inoculum at levels high enough to allow transmission to a new host, and not all host races can infect grape.

Pierce's disease has been present in California for many years, where the most efficient vector is the blue-green sharpshooter, *Graphocephala atropunctata* (Signoret). Other vector species include the green sharpshooter (*Draeculacephala minerva* Ball) and the red-headed sharpshooter (*Carneocephala fulgida* Nottingham). Recently, another more important vector has moved into southern California, the glassy-winged sharpshooter, *Homalodisca coagulata* (Say). While the blue-green sharpshooter feeds more distally on shoots, and the disease may be partially controlled by vigorous pruning, the glassy-winged sharpshooter feeds more basally, leading to more systemic infections. It will feed on woody stems and on wood during the winter. The glassy-winged sharpshooter is originally from the southeastern US. The distribution in the mid-Atlantic states is unclear. However, Virginia seems to be outside the main historical range of the species. It was not reported in the leafhopper species list for Virginia by Stearns (1927). Turner and Pollard (1959) reported it only as far north as Fayetteville, North Carolina. In Georgia and South Carolina (where it is also the vector of phony peach disease, a disease caused by a different strain of the same bacterium), it is collected only east of the Fall Line, the geological division between the Piedmont and the Coastal Plain.

Vectors involved in transmission of Pierce's disease in the northern part of its range in the southeastern states are unclear. One known vector that is commonly seen in vineyards outside of the deep south is *Oncometopia orbona* (Fabricius), the broadheaded sharpshooter (figure 12.15), which overwinters in woods and moves into vineyards in the spring. Another known vector, *Graphocephala versuta* (Say), has been collected from northern and central Virginia.

The ecology of the disease in complicated. In California, Pierce's disease mostly invades from nearby weedy hosts, so infections are more common at the edges of

Figure 12.15 ▪ *Oncometopia orbona*, the broadheaded sharpshooter.

vineyards. In the southeastern states, it is more likely to move from vine to vine. Insecticidal control of vectors has not been very effective. Recent work in Georgia has shown that imidacloprid may prolong vineyard life somewhat but with limited effect. In addition to imidacloprid, bifenthrin, fenpropathrin, cyfluthrin, thiamethoxam, acetamiprid, and thiacloprid have shown promise in experimental trials; however, not all of these insecticides are registered on grape.

Leafhoppers and North American Grapevine Yellows

North American grapevine yellows is a disease caused by phytoplasmas (see chapter 11, page 216). Symptoms include necrotic shoot tips, tendrils and inflorescences, downward-curling leaves, and vine death. It is similar to another yellows-type disease in Europe, *Flavescence dorée*, although North American grapevine yellows spreads within a vineyard more slowly than *Flavescence dorée*, which is known to be transmitted by various leafhoppers and planthoppers. A major vector of *Flavescence dorée* in Europe that was introduced from North America is the leafhopper *Scaphoideus titanus* Ball, the most common deltacephaline leafhopper associated with grapevines in New York. Research in Virginia has illustrated several potential vectors of North American grapevine yellows, including the leafhoppers *Agallia constricta* (Van Duzee), *Exitianus exitiosus* Uhler, *Macrosteles quadrilineatus* Forbes, *S. titanus*, and *Deltocephalus flavicosta* (Stal).

Grape Mealybug — *Pseudococcus maritimus* (Ehrhorn)

The grape mealybug is white with a flattened, oval shape. Filaments protrude along the perimeter of the body, the longest protruding from the rear. First instar nymphs overwinter in a white cottony ovisac produced by the female in the fall. They become active in April or May, disperse over the vine, and begin to feed at bases of shoots or pedicels of grape clusters. Numbers are usually not high enough for damage to be caused at this point. Adults appear in late June, and ovisacs containing eggs are deposited beneath loose bark. Young nymphs appear a few days later and may get into fruit clusters. Adults again appear in late August. Egg-laying continues until cold weather, but eggs that do not hatch before winter do not survive.

Damage is primarily due to excretion of honeydew, a thick sugar solution, by feeding mealybugs. Honeydew deposited onto berries may support growth of a sooty mold or attract ant populations that may eventually feed on fruit as it ripens, especially if there is any fruit cracking or other wound entries (from grape berry moth or other pests). High numbers of mealybugs in fruit clusters may be objectionable. Populations are most likely to develop on vigorous vines with heavy foliage that supplies greater shade and nutrition. Some mealybug species have recently been implicated in the transmission of grapevine leafroll virus in California (grape mealybug, obscure mealybug, longtailed mealybug, and citrus mealybugs); the status of mealybugs in virus transmission in the East is unclear.

Grape Plume Moth — *Pterophorus periscelidactylus* Fitch

While this pest is not a major problem in commercial vineyards, it may be one of the most frequently seen spring pests in eastern vineyards and has the potential to cause significant injury, especially on young shoots and fruit clusters when infestation levels are high. Eggs are laid in groups of two to ten on vines and hatch at or near budbreak. Caterpillars are light green with white hairs, and are about 0.8 inch (20 millimeters) long. They feed on the upper leaf surfaces, tying the edges together with silk. Pupation occurs in early June to mid-June. The adult appears in late June. The moth is light brown with lighter markings; possesses narrow, plumed wings; and is about 0.6 inch (14 millimeters) long.

Eightspotted Forester — *Alypia octomaculata* (Fabricius)

Like the grape plume moth, this species is not a major pest in commercial vineyards but may become a significant pest in certain vineyards, especially those where low-spray methods are practiced. The pupa is the overwintering stage. Adults emerge and oviposit on grape shoots and leaves in May and June. Moths are black with white and yellow markings (two yellow spots on each forewing, two white spots on each hindwing). The caterpillars feed on foliage, leaving petioles and larger veins. The markings of the larvae are distinctive, consisting of orange, yellow, black, and white stripes (figure 12.16). Although commercial vineyards are not damaged severely, small areas within a vineyard may have concentrated infestations and defoliation. Damage rarely occurs in 2 consecutive years. When larvae are full-grown, they drop to the ground to pupate in soil, in trash on the ground, or in crevices in old wood. There are probably two generations annually.

Figure 12.16 ▪ Eightspotted forester, larva.

Grapevine Looper — *Eulythis diversilineata* (Hübner)

The grapevine looper is not currently considered an important pest of grape, although it has risen to pest status in the past. Eggs overwinter on canes in crescent-shaped rows of 8 to 12. Eggs hatch in June. Larvae feed on foliage for 6 to 8 weeks. Food plants include grape and Virginia creeper. Larvae are of the measuring-worm type, with prolegs only on the last two segments of the abdomen. Color of larvae is highly variable, ranging from yellow-green to red to nearly black. Larvae reach a length of about 1.6 inches (4 centimeters). The pupal stage lasts about 10 days and is spent in a loose web spun before pupation on a leaf or berry cluster. Adults emerge in midsummer and lay overwintering eggs. Adults only live 2 or 3 days, and there is one generation annually. A related species—the greater grapevine looper, *Eulythis gracilineata* Guenee—has been more commonly collected in North Carolina in some years. Effective control of loopers may be provided by *Bacillus thuringiensis* products applied when larvae are small.

Sphinx or Hawk Moths (Family Sphingidae)

Several species of sphinx or hawk moths feed on grape. Three of the more common are the Virginia creeper sphinx, *Darapsa myron* (Cramer); the achemon sphinx, *Eumorpha achemon* (Drury); and the white-lined sphinx, *Hyles lineata* (Fabricius).

Achemon sphinx pupae overwinter in the soil and adults emerge in early May. Eggs are laid on foliage and hatch in 6 to 9 days. Larvae are hornworms, and a horn-like "tail" protrudes upward from the end of the abdomen (this horn drops off older larvae). Larvae feed voraciously on leaves for about 25 days, then descend to the soil to pupate. Second-generation moths appear in early July. The second-generation larvae are more numerous than in the first generation and do their greatest damage in August.

The life cycles of the other species of sphinx moths are similar, although the number of generations may vary. While rarely an economic threat, young vines should be closely monitored and protected, if necessary, to avoid losing the small leaf area on such vines to this pest.

Spotted Pelidnota — *Pelidnota punctata* (Linnaeus)

This beetle feeds on grape foliage and fruits but is not an important problem in maintained vineyards. The adult is a tan beetle with black spots, 0.8 to 1 inch (2 to 2.5 centimeters) long. Larvae feed in decaying stumps and logs, so grapes attacked by the adults are often near wooded areas.

Grape Erineum Mite — *Colomerus vitis* (Pagenstecher)

This mite is very small and difficult to see without magnification. It overwinters beneath loose bark of 1-year-old canes. In spring, mites move to leaves where they cause a felt-like "erineum" gall on the lower leaf surface, made up of abnormally curled plant hairs (figure 12.17). The patches are white at first, though these turn reddish eventually. On the upper surface, above the erineum, the leaf surface appears swollen (figure 12.17); these spots eventually turn necrotic. Mites feed and reproduce in this patch of hair-like growth. It has been difficult to correlate erineum mite populations with yield losses. In addition to the erineum strain, there is a leaf curl strain that is not known to occur in eastern North America.

Figure 12.17 ▪ Erineum mites.
Lower leaf surface *(left)* and upper leaf surface *(right)*.

Grape Leaffolder — *Desmia funeralis* (Hübner)

Pupae of this moth overwinter in folded, fallen leaves. Adults are dark brown with white spots on the wings. Both sexes have two white spots on the forewings; on the hind wings, males have one irregularly shaped white spot, and females have two white spots. Moths appear in late May or June. Females oviposit on canes and leaves, especially in the angle of the leaf vein and surface. After hatching, larvae begin feeding on and folding the leaves; leaves are rolled in thin-leaved varieties. Caterpillars are light green with dark markings near their heads and reach lengths of 0.8 inch (2 centimeters). Feeding continues for about a month, and the resulting pupal stage lasts about 10 days. A second generation is present in August and September. The second generation is more damaging. Each larva will make two or more rolls during its development. Infestations are often localized within a vineyard and usually occur first along borders. In some seasons, grape leaffolder may be commonly seen in some eastern vineyards, although infestations are generally of minor importance.

Grapeleaf Skeletonizer — *Harrisina americana* (Guerin)

This species occurs widely in eastern North America. Pupae overwinter in cocoons among debris on the ground. In late spring, adults emerge and eggs are deposited in yellow clusters on lower leaf surfaces. After hatching, larvae feed on upper leaf surfaces, often feeding side-by-side in a row. Young larvae skeletonize leaves; older larvae feed on entire leaves. Larvae are yellow with four black tubercles on each body segment. Adults are small, black, narrow-winged moths. There are two generations annually. Infestations are generally of minor importance.

Grapevine Aphid — *Aphis illinoisensis* Shimer

The winter host of this aphid is *Viburnum* (black haw). Winter is passed in the egg stage, mainly around buds. Eggs hatch in early spring over a 2- to 3-week period. A few parthenogenetic (reproducing without males) generations occur before winged migrants leave for grapevines. Colonies of dark brown aphids develop on young shoots and leaves. When populations are high, some may feed on fruit clusters, causing some berries to drop. In fall, winged individuals are again produced; these return to viburnum and produce males and females. The sexes mate, and overwintering eggs are produced.

This aphid species is usually not important enough to necessitate specific treatments. Grapevines are of sufficient vigor to tolerate some attack, and the aphids are usually controlled by predators (for example, ladybird beetles and lacewing larvae). Dry weather is conducive to buildups of aphid populations.

Grape Colaspis — *Colaspis brunnea* (Fabricius)

This beetle passes the winter in the larval stage in the soil. Development is completed in the spring; adults emerge in June and are present through August. Adult beetles are light brown, about 0.2 inch (5 millimeters) long, with longitudinal rows of evenly spaced punctures on the wing covers. Feeding marks are elongated oval scars. Damage is usually dispersed, occurs after foliage is developed, and is considered of minor importance. Besides grape foliage, grape colaspis adults feed on strawberry and a variety of vegetable crops. Eggs are deposited in the soil during summer and fall. Highly creviced soils, such as clay loam, are preferred. Larvae feed on roots of grasses, clover, soybean, and rice until cold weather but are not known to feed on grape. There is one generation per year.

PESTS AFFECTING SHOOTS, CANES, AND TRUNK

Grape Cane Girdler — *Ampeloglypter ater* LeConte

This is a small, black weevil about ⅛ inch (3 millimeters) long. Its damage may look alarming but is actually not too important unless the affected shoots are intended to be retained as trunks or cordons. Adults overwinter in debris on the ground. In early June, usually before bloom, the female encircles a shoot (figure 12.18) with a series of punctures made with her mouth parts. Eggs are deposited in these holes. She then girdles the cane a few inches higher. The grubs feed in the cane pith, and both injured portions may break off (usually at the more distil girdle first). The damage is usually inconsequential because the girdled areas are usually beyond the clusters, and lateral regrowth will compensate for lost leaf area. If shoot damage affects trunk development, chemical and cultural control options may be exercised, but timing of chemical control is well before one observes the shoot breakage; see local pest management recommendations.

Figure 12.18 ▪ Grape cane girdler oviposition injury.

Larvae pupate in July, and adults appear in late July and August. If injured portions are to be pruned as a means of control, this should be done below the lower girdle before adults emerge; and the prunings should be destroyed.

Grape Cane Gallmaker — *Ampeloglypter sesostris* (LeConte)

This weevil is closely related to the grape cane girdler and is similar in many respects in size and life cycle. The species is red-brown rather than black. The main difference is in the biology of this species: the female punctures the shoot just above one of the lower joints and places an egg in the puncture. There are additional punctures above this in which no eggs are laid. As the larva feeds in the pith, a red gall or swelling develops in the area, about 1 to 2 inches (2.5 to 5 centimeters) long, about twice the diameter of the cane. The cane may break at this point when green; but after it hardens, it will usually survive to produce a crop the next season. The shoot still grows, so damage is usually not important even though this species may be a common pest. Infested shoots may be destroyed as soon as the swellings are visible. Recent research in Pennsylvania has shown no effect on berry quality or vine vigor. As with other incidental pests, the impact is potentially greater in young vineyards due to the paucity of leaf area on young, developing vines.

Ambrosia Beetles — *Xylosandrus germanus* (Blandford), *Xyleborinus saxeseni* (Ratzeburg)

This group of beetles is apparently increasing in importance in southeastern states and in recent years some have caused concern in North Carolina and Virginia by tunneling into grapevines. *Xylosandrus germanus* was introduced to North America from Japan, and *X. sax-*

eseni from Europe. Ambrosia beetles, unlike other bark beetles which make galleries beneath the bark, burrow straight into the wood. *X. germanus* tunnels penetrate into the trunk for about 0.4 inch (1 centimeter), then divide into three or more arms. *X. saxeseni* tunnels likewise proceed straight into the trunk but then become a long chamber parallel with the grain of the wood. Beetles inoculate the tunnel walls with a fungus on which the larvae feed during their development. Ambrosia beetles do not normally attack dead or healthy wood but prefer hosts weakened by disease, drought, and other stresses. However, some species will more readily attack a healthy host, and most species are able to attack such host plants during population outbreaks.

There are generally two generations per year, although this may vary from one to three depending on temperatures. Adults overwinter in their galleries. Emerging adults fly with the warming of spring, peaking in mid-April to late April. The vines seem able to repel many of the attacks by copious sap flow from wounds, pushing out the invading beetles.

Adult females are about 0.08 inch (2 millimeters) long, brown, and have truncate posterior ends; males are generally smaller. Males are a relatively small proportion of the population, never leave their native tunnels, and are seldom collected. Tunnels are about 0.04 to 0.08 inch (1 to 2 millimeters) in diameter and are often marked by sap running down the trunk, sometimes congealing into a clear to dark gel. The economic importance of ambrosia beetles is not clear at this time. The fungus inoculated by the beetles as a food source for the larvae is not known to be pathogenic to the host plant; however, other fungi, such as *Fusarium* spp., are sometimes introduced incidentally; and wounds made by the boring insects might provide infection sites for systemic fungal or bacterial pathogens (for example, crown gall pathogen).

Grape Cane Borer/Apple Twig Borer — *Amphicerus bicaudatus* (Say)

This beetle in the family Bostrichidae feeds on a variety of trees, including apple, pear, peach, plum, forest and shade trees, and ornamental shrubs. The adult beetle is about 0.4 inch (1 centimeter) long, brown with a cylindrical body. The head is directed downward, and a blunt abdomen bears a pair of horn-like projections. Adults overwinter in burrows in wood. They become active early in spring, boring into axils of grape and other plants. Eggs are deposited in May in dying or recently killed wood. Neither young, healthy canes nor dead, dry wood is suitable for larval development. Larvae develop during the summer; and, after a pupal stage, adults appear in the fall.

The adult feeding is the source of injury to grapevines, because larvae feed in dying wood. Young shoots may suddenly break off at the injury site or die. Near the base of the injured section is a small hole, with a burrow leading into the main stem. The beetle may be found in this burrow. All new growth may be killed in a severe infestation, with potentially greater impact in young vineyards. Dying trees and prunings should be removed from the vicinity of the vineyard.

European Fruit Lecanium — *Parthenolecanium corni* (Bouche)

These soft scale insects attack a variety of shade and fruit trees in addition to grape. They are brown, oval, and about 0.3 inch (8 millimeters) long. Immature scales overwinter on canes. Maturity is reached in spring, and the females produce eggs that fill the cavities beneath the scales. In June, light-colored nymphs appear and crawl onto foliage where they temporarily establish on the undersides of leaves. Nymphs crawl back onto canes in late summer and overwinter on one-year-old wood. These scales rarely reach damaging levels. Fortunately, soft scales such as this species are normally controlled by a range of natural enemies. Chemical treatment applied when crawlers are present will suppress scale populations. Oil sprays are effective and can be applied during the dormant period. Recent work has evaluated plant-based, soybean oil as an effective alternative to petroleum oils.

Grape Scale — *Diaspidiotus uvae* (Comstock)

Grape scale, an armored scale, is related to the more widely known San Jose scale and is similar in many respects; but grape scale is economically important only on grapevine. It has also been reported on sycamore, hickory, and Virginia creeper. There are one to two generations per year. Grape scale is widely distributed in the eastern US, from Florida through New Jersey.

Immature scales overwinter on the grapevine trunks and resume development in the spring. In May or mid-June, crawlers disperse from the maternal scales. After wandering for 24 to 48 hours, the crawlers settle, begin to feed, and secrete their scale coverings. There are two nymphal instars, which remain under their scale coverings. The females never move from their sites; but, when the males are mature in the spring, they leave their scales, fly to the females, and mate.

Adult scale colonies may create whitish patches on vine trunks. Vigorous vines are less severely affected. Grape scales seldom reach damaging numbers in commercial vineyards, although in the early years of the 20th century, important damage was reported. This species may be seen most frequently on backyard vines.

Periodical Cicada (*Magicicada* spp.)

The most important periodical cicadas adults appear in large numbers every 17 years (13-year forms are seen in parts of the South). However, some areas may have more than one overlapping brood. The intervening years are spent as nymphs feeding below ground on the sap of tree roots. Nymphs do not establish well on grape roots. The body of the adult is black with red or orange markings, about 1.2 inches (3 centimeters) long (figure 12.19). Clear wings are held tent-like over the body. Periodical cicadas are smaller and darker in color than annual cicadas.

Adult periodical cicadas appear in May, generally in very high numbers. They may invade vineyards from wooded areas. Females oviposit principally in green shoots, but occasionally in canes, laying in excess of five hundred eggs in groups of about fifteen. A series of slits 3 to 4 inches (7 to 10 centimeters) long is made in a shoot a little thicker than a pencil. The shoot may break in the current or following year. Damage can be significant during outbreaks and will impact young vines more so than older, established vines. For this reason, it is prudent to learn when to expect adult emergence years in your area and to avoid planting vines in the 2-year period prior to an emergence. Alternatively, one can net young vines to avoid the ovipositioning injury.

Figure 12.19 ▪ Periodical cicada, adult.

PESTS AFFECTING ROOTS

Grape Root Borer — *Vitacea polistiformis* (Harris)

More of a problem in the southeastern states, grape root borer is near the northern edge of its distribution in Virginia, although it does extend into several southeastern counties of Pennsylvania. A recent survey revealed that grape root borer extends farther to the north than previously thought. Substantial research has been conducted on this pest in Georgia, where it has severely affected muscadine grapes. In southern Mississippi, vines younger than 12 years old had little infestation by grape root borer, vines 12 to 15 years old exhibited moderate injury, and older vines often had severe injury.

This species overwinters as larvae in two different stages of development. The life cycle takes 2 or possibly 3 years to complete, and almost all of this is spent as larvae feeding on grape roots. They bore into the roots and crown below the soil surface, reducing the productivity of the vine. Vines eventually die; there may also be increased susceptibility to cold injury.

Moths are wasplike in appearance, have dark brown bodies with yellow or orange bands on the abdomens (figure 12.20), and are daytime fliers. After flying for several days, females begin ovipositing on grape foliage, canes, and weeds. Each female lays an average of 300 eggs. About 2 weeks after hatching, first instar larvae drop to the ground and tunnel to roots. Young larvae are spread throughout the root zone, while older larvae are found on larger roots close to the trunk. Ninety percent of the pupae are within 14 inches (35 centimeters) of the trunk, and the mean depth in the soil is 3.5 to 4 inches

Chris Bergh, Virginia Tech

Figure 12.20 ▪ Grape root borer adult.

Figure 12.21 ▪ Grape root borer, larvae.

(9 to 10 centimeters). A lack of plant vigor is usually the first sign of the presence of this pest.

Full-grown larvae are about 1 inch (25 millimeters) long, white, with brown heads (figure 12.21). They leave the roots and pupate in cocoons near the soil surface beginning in June. Adults emerge 35 to 40 days later, beginning about mid-July. The greatest numbers are present in the last 2 weeks of July or early August, continuing through September. Besides declining vine vigor, another indication of infestation is the presence of cast pupal skins (exuviae) protruding from the soil near the base of the trunk in late July and August (figure 12.22).

The greatest natural mortality occurs between egg hatch and larval establishment on roots. Only 1.5% to 2.7% survive the first stage because of predation, parasitism, and desiccation; but, once larvae are established in roots, mortality is very low. Infested vines are usually encountered randomly across a vineyard. Larvae do not travel very far in the soil, usually remaining on the roots of a single vine.

Control of this pest is difficult. Contact insecticides are ineffective against subterranean larvae. Soil barrier treatments with the organophosphate chlorpyrifos have been shown to be effective, and this has been the only insecticidal control available recently. However, the long preharvest interval of chlorpyrifos makes it difficult to use in southern grape regions because grape root borer oviposition and hatching extend well into this 1-month preharvest interval. Although soil injection of fumigants once showed promise, the most effective materials have now been banned. An effective cultural control method involves mounding soil beneath vines after borers have pupated and then leveling the ridges in the fall or spring. When adults leave the cocoons, they are unable to dig to the surface. Timing is important because if mounding is done too early, the larvae merely tunnel up into the ridge before pupating. This is not only time-consuming but also impractical for shallow-rooted *Vitis* spp. Proper weed control appears to be important in borer management because of increased larval mortality at the exposed soil surface. Thus, interest in under-trellis cover cropping (see chapter 13, page 262) might aggravate management of grape root borer.

Entomopathogenic nematodes have recently been considered as control agents for grape root borer. Research in Ohio showed that *Heterorhabditis bacteriophora* (Poinar) caused more than 85% mortality of grape root borer in the laboratory; another species, *Steinernema carpocap-*

Chris Bergh, Virginia Tech

Figure 12.22 ▪ Grape root borer, cast pupal skins (exuviae).

sae (Weiser), induced 69% mortality. Percent mortality caused in the laboratory does not necessarily correlate with field results for a variety of reasons, including soil type, which influences efficacy dramatically. When vines are infested, nitrogen fertilization may help ameliorate the effects of feeding.

The commercially available pheromone blend used for monitoring grape root borer is (*E*,*Z*)-2,13-octadecadienyl acetate (ODDA) and (*Z*,*Z*)-3,13-ODDA (99:1 ratio).

Grape Rootworm — *Fidia viticida* Walsh

Larvae of this beetle overwinter in the soil among grape roots, from 0.4 to 0.8 inch (1 to 2 centimeters) to more than 20 inches (50 centimeters) deep. In spring, feeding on roots is resumed. Larvae pupate in cells close to the surface, usually 16 to 24 inches (40 to 60 centimeters) from vine bases, in late May and June (about the time of grape bloom). Adults appear about 2 weeks later. Beetles are about ¼ inch (6 millimeters) long, chestnut brown with yellow-white hairs (figure 12.23). Adults feed on foliage for a month or more, making chain-like feeding marks (similar to those made by larval grape flea beetles) (figure 12.24). Females deposit creamy-white egg clusters on canes or under loose bark, averaging about one hundred eggs per female, with 20 to 30 eggs per cluster.

Eggs hatch in 1 to 2 weeks; and larvae drop to the ground, enter the soil, and feed on grape roots until cold weather. Larvae consume smaller roots and eat pits into larger ones. This damage is much more important than the adult foliar injury. Vines become unthrifty, yield is reduced, and vines may even die after several years of infestation. Root damage by grape rootworm will be compounded by planting in poor soil. The life cycle takes 1 to 2 years for completion. A related species, *Fidia longipes* (Melsheimer), the southern grape rootworm, has been found to be more common than *F. viticida* in North Carolina. Its distribution in other states is not known.

Figure 12.23 ▪ Grape rootworm, adult.

Figure 12.24 ▪ Grape rootworm adult feeding injury

Chemical control is directed toward adults soon after emergence, before oviposition. First and second grape berry moth sprays may control grape rootworm. Pupae may be destroyed by intensive shallow cultivation of soil until adults emerge in late June.

CONCLUSION

Grapes are host to a wide range of insect and mite pests, but fortunately, most vineyard operators will not have to deal with all of these potential pests every year. A few species are important in almost all vineyards every year, but most pose problems only in certain environmental settings. Grape berry moth is a key pest, and one that will be a concern in most years across a wide geographic area. Growers should begin monitoring for grape berry moth using pheromone traps as well as rachis and cluster examinations in susceptible areas, as indicated in the accompanying Pest Management Schedule (appendix 6, page 313). The most important leaf feeding pests are Japanese beetle and European red mite. Japanese beetle creates conspicuous feeding injury, although established vines can tolerate quite a bit of feeding before grape yield or quality are affected (young vines should be stringently protected). European red mite is the main spider mite that affects grapevines in the East, and can severely damage foliage during outbreaks. Spider mites are normally held in check by natural enemies, and these beneficial predators can be harmed by indiscriminate pesticide application.

Over time, vineyard operators will gain an appreciation of the common pest problems of their own vineyards. This will allow a more focused approach to pest management, specifically with respect to critical monitoring periods. For example, some vineyards historically have particular problems with climbing cutworm just before (bud swell) and during budbreak. Under those conditions, growers should pay particular attention to scouting for cutworm injury, as indicated in the Pest Management Schedule (appendix 6). Be aware that changing environmental conditions (such as droughts, extreme winter cold, increased average winter temperatures) may either lessen or increase the abundance of particular insect pests. Pay particular attention to common insect and mite pests but monitor vines closely for the unusual pest occurrence as well. Consult local Cooperative Extension or other resources for annual updates to both cultural and chemical pest management recommendations.

References

All, J. N., J. D. Dutcher, M. C. Saunders and U. E. Brady. 1985. Prevention strategies for grape root borer (Lepidoptera: Sesiidae) infestations in Concord grape vineyards. *Journal of Economic Entomology*. 78: 666–670.

Golino, D. A., S. T. Sim, R. Gill and A. Rowhani. 2002. California mealybugs can spread grapevine leafroll disease. *California Agriculture*. 56(6): 196–201.

Martinson, T. E., C. J. Hoffman, T. J. Dennehy, J. S. Kamas and T. Weigle. 1999. Risk assessment of grape berry moth and guidelines for management of the eastern grape leafhopper. *New York Food and Life Science Bulletin*. 138: 1–10.

Pfeiffer, D. G., T. J. Boucher, M. W. Lachance and J. C. Killian. 1990. Entomological research in Virginia (U.S.A.) vineyards. p. 45–61. In N. J. Bostanian, L. T. Wilson and T. J. Dennehy (eds.). *Monitoring and Integrated Management of Arthropod Pests of Small Fruit Crops. Intercept*, Andover, U.K. 301 p.

Snow, W. J. [sic], D. T. Johnson and J. R. Meyer. 1991. The seasonal occurrence of the grape root borer, (Lepidoptera: Sesiidae) in the eastern United States. *Journal of Entomological Science*. 26: 157–168.

—13—

Vineyard Weed Management

Author
Jeffrey F. Derr, Virginia Tech

There are three basic types of plants in a vineyard —grapevines, weeds, and ground covers. Weeds are defined as plants growing where they are not desired. It is possible that a plant species is considered a weed when growing within a grapevine row while the same species is not considered a weed when growing in the area between the rows. Ideally, a ground cover suppresses weeds both within and between grapevine rows, requires little mowing, does not excessively compete with grapevines, and does not harbor insect or disease pests. Finding such a living mulch has been difficult; therefore, the vineyardist needs to manage vegetation within each row and between the rows to minimize adverse effects on grape yield and quality.

WHY WORRY ABOUT WEEDS?

Weeds can compete with grapevines for water, nutrients, and light, with competition for water being of greatest concern. Competition for water is especially important in dry summers and with young vines. Nutrient competition seems to be especially important for nitrogen and explains why grapevine leaves can exhibit a light-green color when vines compete with weeds. Growers cannot simply solve the problem of this competition by adding fertilizer because this also encourages weed growth. Tall weeds or climbing weeds such as morningglory can grow up into the grapevine canopy and shade the vine's leaves, reducing their photosynthesis rate. This loss of carbohydrate production results in lower grape yields. It can also adversely affect fruit quality. Weed stems, leaves, and seed pods mixed in the vine canopy can become contaminates of the fruit at harvest, especially with machine-harvested crops. Weeds also can interfere with harvest. One example of this is poison ivy, which contains a toxin to which many people are sensitive. Weeds with thorns, such as horsenettle and wild blackberry, are another nuisance to pickers.

Weeds harbor insect pests, disease pathogens, and rodents that can damage grapevines. For example, some broadleaved weeds serve as alternative hosts for nematodes such as *Xiphinema americanum* that can vector tomato ringspot virus to grapevines (see chapter 11, page 216). One way to help manage disease and insect pests is through control of weed species that serve as alternative hosts (See Cultural Control, page 267). Weed growth can reduce air flow through the vineyard, increasing humidity, which could lead to increased disease severity. Weeds can intercept fungicide or insecticide sprays meant for grapes and vine canopies, thereby reducing the effectiveness of these sprays. Some weeds, such as white clover, may be attractive to bees. This could be a concern if insecticides need to be applied to the vineyard. Weeds also provide cover and a food supply for mice, voles, and other rodents.

WHAT TYPES OF WEEDS INFEST VINEYARDS?

Weeds can be divided into taxonomic groups or they can be grouped by life cycle. Understanding the taxonomic plant groups will assist in understanding selectivity of herbicide action and will guide the cultural and chemical control options needed for management.

Taxonomic Plant Groups

The major weed species that commonly occur in vineyards can be classified as dicots, which have two seed leaves (cotyledons) when they germinate, or monocots, which have one seed leaf. Dicots are more commonly referred to as broadleaves; an example is dandelion. Broadleaves generally have wide leaves with reticulate (net-like) leaf venation and flower parts occurring as twos, fours, or fives. Monocots tend to have narrow leaves with parallel leaf veins (parallel venation) and flower parts in threes (for example, 3 sepals and 3 petals). Monocots can be divided into three categories: (1) grasses, such as large crabgrass, with round or flattened stems, (2) sedges, like yellow nutsedge, with triangular stems, and (3) other monocots, such as wild garlic and wild onion. It is important to distinguish grasses from sedges because control programs differ for these two plant groups.

Life Cycle Grouping

Weed species also can be grouped by their life cycle into the following categories: annuals, biennials, and perennials.

Annuals

Annuals live for only one growing season and reproduce strictly by seed. These weed species can be divided into summer annuals and winter annuals. Summer annuals start to germinate in March, April, or May, depending on species and location, and continue to germinate though the summer. They will flower in summer or early fall and generally die with the first killing frost. Summer annuals can be categorized as broadleaves, grasses, or sedges. Annual sedges are not a major problem in grape production. Winter annuals start to germinate in early fall (generally September) and continue to germinate in October and November, depending on species and location. There may be additional germination in late-winter to early-spring (February to April). Winter annuals will flower in spring, and then die off with the onset of hot, dry weather. Winter annual weeds are divided into the categories of broadleaves and grasses. Sedges are warm-season plants that do not grow during the winter.

Summer annuals are of greater concern to grape growers than are winter annuals. Summer annuals are detrimental because they compete for water and nutrients with actively-growing grapevines during the growing season. Winter annuals do not pose this competition concern while grapes are dormant, in fact, they may be beneficial for reducing soil erosion during the winter months. Winter annual weed species, such as horseweed, that grow tall and thus interfere with grapevine growth after budbreak, or species that climb up grapevines, like narrowleaf vetch, however, need to be controlled. Certain species, such as common groundsel, overlap the summer and winter annuals categories.

Biennials

Biennials live for two years and also reproduce strictly by seed. They form a vegetative rosette during the first year and then flower in the second year. The biennial weeds of concern are all broadleaves. A biennial species may sometimes behave as a winter annual or a short-lived perennial.

Perennials

Perennials live for many years and generally are the hardest species to control. There are two types of perennials—simple and creeping. Simple perennials, like common dandelion, spread only by seed. Creeping perennials, such as yellow nutsedge, spread by seed and also vegetatively through rhizomes, stolons, bulbs, tubers, or true roots. It is more difficult to eradicate creeping perennials than simple perennials. Some perennials die back to the ground in winter and regrow in spring from rhizomes, tubers or other underground structures, such as yellow nutsedge. Others, like quackgrass, are green throughout the year. Perennial species can also be classified as herbaceous (lacking a woody stem), such as common dan-

Common Weed Species, Grouped by Life Cycles

SUMMER ANNUALS

Summer Annual Grasses

- barnyardgrass
- large crabgrass
- goosegrass
- giant foxtail
- yellow foxtail
- fall panicum
- witchgrass

Summer Annual Broadleaves

- wild buckwheat
- cocklebur
- hairy galinsoga
- jimsonweed
- common groundsel
- ladysthumb
- common lambsquarters
- tall morningglory
- redroot pigweed
- smooth pigweed
- purslane
- common ragweed
- Pennsylvania smartweed

WINTER ANNUALS

Winter Annual Grasses

- downy brome
- annual bluegrass
- cheat
- annual ryegrass

Winter Annual Broadleaves

- hoary alyssum
- hairy bittercress
- common chickweed
- redstem filaree
- common groundsel
- henbit
- horseweed
- prickly lettuce
- mayweed
- pennycress
- Virginia pepperweed
- shepherdspurse,
- narrowleaf vetch

BIENNIALS

Biennials (broadleaves)

- burdock
- wild carrot
- white cockle
- common eveningprimrose
- yellow goatsbeard
- common mallow
- common mullein
- wild parsnip
- sweetclover species
- tansy ragwort
- teasel
- plumeless thistle

PERENNIALS

Perennial Grasses

– **Clump-Type Perennial Grasses**

- tall fescue
- orchardgrass

– **Creeping Perennial Grasses**

- bermudagrass
- johnsongrass
- wirestem muhly
- quackgrass
- timothy

Perennial Sedges

- yellow nutsedge

Other Monocots

- wild garlic
- wild onion

Spore Producers

- field horsetail

Perennial Broadleaves

– **Simple Broadleaf Perennials**

- chicory
- broadleaf dock
- curly dock
- dandelion
- broadleaf plantain
- buckhorn plantain
- pokeweed

– **Creeping Broadleaf Perennials**

- aster species
- wild blackberry
- field bindweed
- hedge bindweed
- Virginia creeper
- dewberry
- hemp dogbane
- goldenrod species
- Japanese honeysuckle
- horsenettle
- poison ivy
- St. Johnswort
- common milkweed
- mugwort
- stinging nettle
- red sorrel
- tansy ragwort
- Canada thistle
- toadflax
- trumpetcreeper
- yellow woodsorrel

delion, or woody, like poison ivy. Perennials can be categorized taxonomically as broadleaves, grasses, sedges, or other monocots. Field horsetail, which is in a separate taxonomic group, is a non-flowering primitive plant that spreads by spores or vegetatively by rhizomes.

COVER CROPS

Cover crops are living ground covers that are planted either before or after vineyard establishment. Cover crops may be planted annually, maintained perennially, or used on a semi-permanent basis, depending on species used and the vineyardist's goals. Cover crops can be planted between rows, in the row, or both between and in the row.

There are various benefits to cover crops, the most important of which is reducing soil erosion. Cover crop foliage reduces the force of rain drops hitting the soil surface, which helps reduce soil compaction. Cover crops also slow the surface movement of water, thus reducing the ability of surface water to transport soil particles. Cover crops improve soil structure and water penetration through formation of pores from root growth, increased organic matter level, and increased aggregation of soil particles.

Cover crops can be used to regulate vine growth. Nitrogen-fixing cover crops increase soil nitrogen levels, potentially increasing grapevine growth. Cover crops can also be used to compete with grapevines for water and nutrients thereby reducing grapevine growth. Depending on species, cover crops can provide a habitat and food for beneficial insects and mites. Cover crops also permit operation of equipment in the vineyard sooner after heavy rainfall than in the absence of cover crops.

Annual Cover Crops

Annual cover crops are typically seeded in late-summer or fall and then tilled into the soil or killed with herbicides the following spring. This schedule works best on relatively flat sites where soil erosion is not a risk. In addition to reducing soil erosion, cover crops established in late-summer can serve to reduce vine vegetative growth, perhaps improving the vine's adaptation to winter cold. A no-till seed drill could be used to seed the cover crop if soil erosion is a concern. Tillage could be restricted to within the row and mowing used to maintain the areas between rows. An annual cover crop should be killed in the spring before it exceeds 2 feet in height, otherwise it might compete undesirably with vine growth. The cover crop can be shredded and incorporated into the soil to increase soil organic matter. Two examples of annual cover crops are rye and annual ryegrass. Legumes such as hairy vetch, subterranean clover, and red and white clovers have been investigated as cover crops because they increase nitrogen levels in the soil; however, growers need to evaluate existing vine vigor to determine whether or not added nitrogen would be of benefit to canopy management goals (see chapter 6, page 124).

Perennial Cover Crops

Many eastern North American vineyards are maintained with perennial cover crops between the rows. This usually entails planting a grass species, because broadleaf plants can lead to a buildup of nematodes. Cool-season grasses tolerate frost and will go dormant during transient droughts. Cool-season perennial grasses maintain green foliage in the winter, which is beneficial for reducing soil erosion and for permitting machinery operation during the dormant period. Perennial grasses are competitive with grapevines and thus can be used to reduce vine growth. Fine fescues, such as hard fescue and creeping red fescue, are less competitive than other perennial grasses. Fine fescues grow shorter, thus requiring less mowing, but are slower to establish and do not tolerate hot, dry conditions as well as other perennial grasses. Kentucky bluegrass, perennial ryegrass, tall fescue, and orchardgrass are more competitive grasses for row middles.

All of these grasses are cool-season perennials, meaning that they grow best in spring and fall. These grasses are best planted in the fall, generally in September in the mid-Atlantic, and as early as August further north. Weed infestation of the newly seeded cover crops is less problematic during fall than it is during spring and summer. Aggressive summer weeds such as large crabgrass can overtake and thus prevent the establishment of a spring-seeded cover crop. Spring seeding may not allow for sufficient establishment prior to the onset of hot, dry weather. A legume such as white clover could be added to provide nitrogen for grass growth, but growers should evaluate the need for added nitrogen before pursuing this option.

WEED MANAGEMENT STRATEGY

A general strategy for vineyard floor management is to maintain some kind of ground cover in the row middles, while minimizing weeds in the row, at least during the growing season. As indicated above, the benefits of ground covers between the rows include reduced soil erosion, reduced potential for nutrients or pesticides to move offsite, competition with vines to reduce excessive vegetative growth, and the ability to operate machinery soon after rains. Commonly used ground covers between the rows include Kentucky bluegrass, tall fescue, creeping red fescue, hard fescue, orchardgrass, and perennial ryegrass. Because ground covers as well as weeds compete with grapevines for water and other resources, areas within the row are generally kept free of weeds, especially in young plantings. Bare ground will also aid in frost protection. Thus, a 3- to 4-foot wide weed-free strip is generally maintained during at least the first 3 years after planting (figure 13.1).

Vine growth increases in proportion to the width of the weed-free strip. For example, an 8-foot wide strip results in more vine growth than does a 4-foot strip. However, the increased width of the weed-free strip must be balanced against the increased risk of soil erosion, as more soil is exposed to the elements with increased band width. The width of the killed strip can be reduced or eliminated in older, bearing vineyards if one desires a reduction in vine growth from grass competition (figure 13.2).

Figure 13.1 ▪ Young vineyard illustrates weed-free area under trellis with perennial grass cover crop between rows.

W.G. Giese, Surry Community College

Figure 13.2 ▪ Established Cabernet Sauvignon vineyard with grass cover crop extended under trellis.

In regions that are subject to winter cold injury, the base of grapevines is hilled-up with a 4- to 6-inch mound of soil in the fall to protect the graft union and lower portion of the scion from cold injury. The hilling operation also controls weeds, provides a soil ridge that can reduce downhill movement of water, and prevents depressions around the vine where herbicides could concentrate and cause vine injury. This hilling is accomplished by moving soil from the area next to the row using a grape hoe, plow, disk, or other tool. This soil mound is removed from the base of the vine at least every other year to discourage scion rooting.

PRIOR TO PLANTING

Perennial weeds, especially perennial broadleaf weeds, need to be eliminated from sites where grapes are to be planted. Perennial weeds can be controlled through preplant applications of a nonselective herbicide such as glyphosate or through repeated tillage. Smother crops, such as sudangrass, could be grown in fields to be planted in future years to grapes (see chapter 4, page 71). Crops such as field corn, which are tolerant of selective broadleaf herbicides, could be grown for several years prior to planting grapes. Check herbicide labels for any carryover restrictions prior to planting grapes. There is no easy way to remove perennial broadleaf weeds *after* planting unless protective grow tubes or vine shelters are used in conjunction with nonselective herbicides.

After perennial weeds are controlled, a ground cover can be established. Glyphosate can be applied in strips in the fall to control the ground cover and grapes can then be planted into the killed strips (see chapter 4, page 71).

NEW PLANTINGS

Competition from weeds is most detrimental in the first few years after planting because this will delay vine establishment and full cropping potential. It is therefore desirable to maintain a weed-free strip within the row and, because competition with the ground cover in the alleyways will restrict grapevine root development, a wider weed-free strip, within reason, is beneficial.

WEED CONTROL OPTIONS

Weeds can be controlled in vineyards through cultural methods, through herbicide application, or through a combination of these two strategies.

Cultural Control

Hoeing and hand-weeding are labor-intensive, but acceptable, options for weed control, especially in small vineyards. Rows will need to be hoed approximately every 2 to 4 weeks during the growing season. Mechanical hoes and cultivators are available and can be used to eliminate under-trellis weeds (figure 13.3). Considerable care is required, however, to avoid trunk and root injury.

Figure 13.3 ▪ Mechanical grape hoe used for under-trellis cultivation.

Additional hand-weeding may be necessary to control weeds growing close to the grapevines.

Cultivation with a grape hoe, rototiller, or other equipment can be used for weed control in the row. If the land is relatively flat and soil erosion is not a concern, cultivation with disks or rotovators is possible. When cultivating, use caution to avoid damage to vine trunks and roots. Cultivated soil is subject to erosion and both soil structure and organic matter content are decreased with cultivation, which are drawbacks to mechanical cultivation. Mixing of soil also brings weed seeds to the surface where they can germinate, so frequent cultivation is needed to maintain a weed-free zone.

Mowing is used to suppress vegetation in the areas between rows but is generally not useful for in-row weed control. However, research in New York suggests that mowing has limited impact in reducing competition from cover crops in the row middles. Growers with small acreage may be tempted to "mow" weeds in the row with hand-held, string weed-trimmers or "weed whackers."™ These devices have, unfortunately, caused considerable injury to vines by girdling the trunk and causing long-term vine injury. Their use is therefore not recommended in vineyards.

Mulching has been used in grape production, although it is not widely used. Mulches can reduce soil crusting, conserve soil moisture, and suppress weed growth. Use of mulches is limited by cost, availability, and difficulty in spreading. Organic mulches can serve as a haven for rodents. Mulching suppresses weed growth, which reduces competition for soil moisture and nutrients, increases rainfall penetration, conserves soil moisture, provides nutrients as the mulch decays, and reduces soil erosion. In vineyards where vine vigor is low, mulching can improve vigor. A thick layer of mulch on poorly drained soils should be avoided because this could retain excess soil moisture in the root zone, leading to root diseases. Mulch can also be a fire hazard when material such as hay or straw is used in dry weather.

Most organic materials such as bark, straw, hay, cobs, sawdust, composted grape pomace, or wood chips can be used as mulching materials. The mulch may be applied to the entire vineyard floor or confined to a 4- or 5-foot wide band beneath the trellis. Adding mulch to row middles will reduce soil erosion. The desired depth of the mulch can vary, with 2 to 4 inches a general guideline. The necessary mulch depth depends on the quantity available, density of the material used, and the cost. Greater depths of loose material like straw could be used. If the mulch material has not gone through a composting process, soil nitrogen immobilization can occur as soil microbes degrade the mulch. This in turn could lead to a nitrogen deficiency observed in the grapevines.

Mulches control annual weeds but generally do not control perennial ones. Also, as a mulch breaks down over time, it may provide a suitable habitat for weed seed germination.

Mulches can be a source of weed seed so use mulch of high quality. Once the initial mulch has been established, annual additions are necessary to maintain the desired depth. In general, a ton of straw is required for an inch of mulch spread over one acre. Other materials may require varying amounts to form an inch of mulch, depending on their density and bulk. In addition to introducing weed seeds, mulches have tended to aggravate certain insect problems in some vineyards. Climbing cutworm (see chapter 12, page 241) pressure, for example, is often increased by use of mulches around the grapevine trunks. The mulch may provide a haven or refuge during the day for these insects. Mulches (and under-trellis vegetation) may also increase the survival rate and subsequent injury caused by grape root borer larvae.

Black plastic and landscape fabrics provide greater weed suppression than do organic mulches but are not commonly used in vineyards due to their relative costs. Black plastic provides excellent annual weed control and will suppress perennial weed growth. Black plastic does not allow for water penetration or gas exchange, although drip irrigation could be installed under solid black plastic to irrigate the crop. Landscape fabrics do allow for water or gas movement between the soil and air but allow greater weed growth compared to solid black plastic. Weed shoots can penetrate through openings in landscape fabrics. Weed roots can penetrate down through such openings from above, especially if a mulch is used above the fabric. Black plastic and landscape fabrics also could provide protection for rodents, and create a disposal problem if one desires to remove them from the vineyard. Mowers can damage black plastic or fabrics if edges are caught by the blades.

Propane flame weeders can be used to control small, emerged weeds, but the bark of young vines as well as the grapevine foliage needs to be protected during flame cultivation and vines should be growing under good soil moisture. Repeated flaming is needed during the growing season to control weeds because underground parts of perennial weeds and buried weed seed are unaffected by flaming.

Chemical Control

Herbicides are commonly used for weed control in vineyards because they are relatively inexpensive and they provide a high level of control. Herbicides must be properly applied, otherwise injury to grapevines can occur. Herbicides can be divided into those that are applied prior to weed emergence (preemergence herbicides) or after weed emergence (postemergence herbicides).

Preemergence Herbicides

Preemergence herbicides are applied prior to weed germination and provide residual control for approximately 2 to 6 months, depending upon the chemical used, application rate and timing, and weather conditions. The more rain that a site receives, the shorter a preemergence herbicide will last. Most preemergence herbicides will not control weeds after they emerge. Emerged weeds can be controlled through addition of a postemergence herbicide or a preemergence herbicide could be applied to bare soil after cultivation.

Preemergence herbicides are applied primarily to control annual weeds, and have little or no effect on perennial weeds. The choice of preemergence herbicides for a given vineyard depends on the age of the planting. Certain herbicides, such as oryzalin, can be applied after planting, while for other chemicals like simazine or diuron, the vineyard must be established for at least 3 years. Herbicide choice is also based on the weed spectrum present at the vineyard. Some herbicides, like oryzalin or napropamide, are more effective for control of annual grassy weeds, while diuron, flumioxazin, oxyfluorfen, and simazine are more effective on annual broadleaf weeds. Yellow nutsedge is not controlled by most preemergence herbicides, although suppression can be seen with dichobenil and norflurazon. Combinations of preemergence herbicides can be used to broaden the spectrum of weed control.

Scout the vineyard in late spring to identify winter annual weeds and in late summer for identification of summer weeds. Check herbicide labels and cooperative extension recommendations for herbicide selection, as some states and even counties have more stringent pesticide use restrictions than do others.

Preemergence herbicides should be applied in early spring for control of summer annual weeds and in late summer if winter weed control is desired. All preemergence herbicides must be activated in the soil, either by rainfall or irrigation. If a preemergence herbicide remains on the soil surface without activation for several weeks, the herbicide could be lost due to volatilization and photodecomposition. Then, when rains do occur, there may not be sufficient herbicide to control germinating weeds.

Postemergence Herbicides

Postemergence herbicides are applied after weed emergence because these chemicals are absorbed by weed foliage. Most postemergence herbicides have little to no soil activity, so a preemergence herbicide could be added to extend the period of weed control.

Postemergence herbicides can be separated into selective and nonselective types. Selective herbicides cause little to no crop injury but only control certain weed species. Nonselective herbicides will damage all plants including grapevines if the spray contacts foliage or immature bark. Directed sprays or shielded equipment can be used to minimize potential for crop injury.

One group of selective herbicides for use in vineyards is the postemergence grass herbicides clethodim, fluazifop, and sethoxydim. These three herbicides only control grasses and generally have no effect on sedges, other monocots like wild onion, or any broadleaf weed. It therefore is important to tell the difference between grasses and sedges when utilizing this group of herbicides.

Postemergence herbicide can also be divided into contact and systemic types. Contact herbicides have a rapid effect on weed foliage but do not translocate to roots. Contact herbicides are thus more effective on annual weeds because perennial weeds will regrow from the unaffected root system. Contact herbicides can also be used for burn-down of grapevine suckers if they are not needed for trunk renewal. Systemic herbicides are slower acting, but will translocate to and control root systems of weeds. Systemic herbicides are therefore a better choice for perennial weed control. Nonselective, translocated herbicides pose a substantial risk of crop damage due to the potential for systemic injury. Injury from contact, nonselective herbicides will be limited to only the foliage directly contacted by the spray.

Glyphosate is the primary nonselective systemic herbicide used in grape production. Overall, glyphosate is the most effective herbicide for controlling perennial weeds but poses the greatest risk of crop injury. This risk is increased as the season progresses because mature leaves translocate the glyphosate to other parts of the vine, including the roots. Use shielded sprays or wiper applications to reduce the potential for contact with vine leaves or stems. By contrast, paraquat is strictly a contact herbicide, and thus does not control the underground portions of perennial weeds. Repeat applications would be needed to control regrowth from roots or rhizomes; eventually the underground reserves of the weed will be depleted, in a similar manner to repeated close mowing or flaming of an upright-growing perennial weed. Another postemergence herbicide, glufosinate has principally contact activity, with a limited degree of systemic action. Glufosinate poses less risk of grapevine injury than glyphosate, but provides less control of perennial weeds. On the other hand, due to its systemic action, glufosinate generally provides greater perennial weed control than paraquat does.

Apply postemergence herbicides when weeds are actively growing under good soil moisture conditions. Poor weed control can result if postemergence herbicides are applied to weeds under drought stress.

Check the herbicide label to see if an adjuvant is needed when applying a postemergence herbicide. Surfactants improve leaf coverage by spreading spray droplets and are needed when applying contact herbicides such as paraquat. Adjuvants are generally not needed with preemergence herbicides.

There are many options for effective weed management programs. Several such options include: early season applications of effective preemergence herbicides, or combinations of preemergence herbicides, with the addition of postemergence herbicides if weed growth is present at the time of application; split applications of effective pre-and postemergence herbicides if permitted by the labels; two or more applications of postemergence herbicides only where contact with desirable grapevine suckers and foliage can be avoided; and combinations of mechanical and chemical methods of control.

Herbicide Injury

Grapevines can be injured by herbicides by several modes. Preemergence herbicides can injure vines if the herbicide rate was too high. Certain postemergence herbicides can injure vines if they contact the grapevine foliage. This can be particularly serious with systemic herbicides such as glyphosate. Grapevines can also be injured by drift of herbicides applied to nearby fields, highways and utility right-of-ways.

Follow vine age restrictions on the label to minimize the potential for preemergence herbicide injury. Use lower rates on light, sandy, or gravelly soils that are low in organic matter, and higher rates on heavier clay soils higher in organic matter. Calibrate the sprayer to ensure accurate application. Do not base preemergence herbicide rates on the amount of product per gallon but rather on the amount of product per unit area of land to be treated.

Herbicides applied to crops grown adjacent to the vineyard can injure grapevines from spray drift or vapor drift. Chemicals such as 2,4-D and related growth regulator herbicides can cause severe systemic injury to grapevines from spray drift and vapor drift. Avoid applications under windy conditions, use low spray pressures and add a drift control agent if applications will be made near actively growing grapevines. Where possible, make applications to pastures and other crops when vines are dormant. Use *amine* formulations of 2,4-D instead of *ester* forms when treating areas near actively growing grapevines. Ester formulations are more susceptible to vapor drift than are the amine forms. With the increased use of glyphosate-tolerant crops, there is also increased potential for injury from drift of this herbicide. Consider use of trees as a wind break to limit drift of herbicides from nearby fields and discuss the sensitivity of grapevines to herbicide drift with your neighbors.

Herbicide Resistance

If the same herbicide is repeatedly used in a vineyard, two adverse situations can develop. Weed species shifts can occur as the herbicide-tolerant weed species will increase in density. Also, weed species that were initially susceptible to that chemical could develop resistance. Examples of weeds that have developed resistance to herbicides include common lambsquarters and smooth pigweed that have developed resistance to triazine herbicides such as simazine, and horseweed developing resistance to glyphosate. To reduce the potential for herbicide resistance development, rotate or tank-mix herbicides with different modes of action. Utilize cultural controls where possible.

CONCLUSION

Vegetation management both within and between vineyard rows is an important management issue for grape growers. Weeds compete with grapevines for water and other resources and may ultimately cause vine size decline and reductions in crop yield. Weeds also interfere with harvest, reduce crop quality, and harbor insect, disease, and rodent pests. Cover crops are maintained for erosion control, drivability in vineyards soon after a rain, and improved soil structure, and can be used to regulate vine growth. Growers should develop a comprehensive vineyard floor management plan that considers weed management in the row middles and within the row, as well as desirability of mulches or green manures to increase soil organic matter.

References

Dami, I., B. Bordelon, D. C. Ferree, M. Brown, M. A. Ellis, R. N. Williams, and D. Doohan. *Midwest Grape Production Guide.* Ohio State University. Bulletin 919-05. HTTP://OHIOLINE.OSU.EDU/B919/0001.HTML

Ingels, C., R. Bugg, G. McGourty, P. Christensen, 1998. *Cover Cropping In Vineyards—A Growers' Handbook.* UC ANR Publications #3338, Oakland, California. 162 pages.

Uva, R. H., J. C. Neal, and J. M. DiTomaso. 1997. *Weeds of the Northeast*, Cornell University Press, P. O. Box 6525, Ithaca, NY 14851-6525.

Zabadal, T. J. 1997. *Vineyard Establishment II: Planting and Early Care of Vineyards*, Michigan State University Extension Bulletin E-2645 (HTTP://WEB1.MSUE.MSU.EDU/MSUE/IMP/MODFR/26459701.HTML)

Weed Science Society of America. Encyclopedia of North American Weeds DVD, 3rd edition (HTTP://WSSA.NET/WSSA/STORE/INDEX.HTM)

Weed identification and management websites

Michigan State (weed seedlings) – HTTP://WEB1.MSUE.MSU.EDU/MSUE/IAC/E1363/E1363.HTM

Rutgers University – HTTP://WWW.RCE.RUTGERS.EDU/WEEDS/

University of Massachusetts – HTTP://WWW.UMASSGREENINFO.ORG/FACT_SHEETS/WEED_HERBARIUM/COMMON_NAME_LIST.HTM

Virginia Tech weed identification – HTTP://WWW.PPWS.VT.EDU/WEEDINDEX.HTM

Virginia Cooperative Extension pesticide (including herbicide) recommendations for grapes and other crops – HTTP://WWW.EXT.VT.EDU/PUBS/PMG/

—14—

Wildlife Deterrence

Author
Tony K. Wolf, Virginia Tech

Various vertebrate animals, including birds, deer, raccoons, foxes, and bear are attracted to grapes and grapevines. The consumption of berries by animals provides one means for widespread distribution of seeds; one line of thought is that the color of pigmented berries is a natural adaptation to animal attraction and subsequent dispersal of seeds. This ecological vignette provides little comfort to the grower who is experiencing crop loss from wildlife depredation. Deterrence measures are considered here for the major vertebrate pests. "Deterrence" is emphasized, as "management," or "control," might not be practical or legal, depending on the species in question.

BIRDS

Many species of birds are fond of ripe grapes and will quickly cause appreciable crop loss if not discouraged. This is especially true of flocking birds such as starlings. Robins, mockingbirds, and finches are also frequently observed feeding on grapes, and vineyards that are located near wild turkey habitat occasionally suffer feeding losses from these large birds. Birds are daytime feeders and can be identified if you are in the vineyard when birds are present. Otherwise, the clues to bird feeding are peck marks in individual berries, remnants of berry skins retained on the rachis, and a selective feeding on individual berries of the cluster, with the rachis left intact (figure 14.1). Birds tend to consume the darkly pigmented berries first, leaving the greener, unripe berries for a later day. Vines under or close to roosting areas such as a treeline or overhead powerlines are often the most vulnerable to bird depredation. Dark-fruited, small-berried wine grape varieties are particularly susceptible, as are all seedless table grape varieties. The more common vineyard conditions that promote bird depredation are shown in table 14.1.

Options to reduce bird feeding are diverse and include recorded distress calls played on audio equipment in the vineyard, various reflective materials intended to frighten, propane cannon with loud reports, various balloons and kites suspended above the vineyard to simulate bird predators, and bird netting. All of these devices have their limitations and most birds will quickly habituate to static scare tactics, at which point the feeding will resume. The efficacy of the visual deterrents is improved if used in combination with audible deterrents and if the devices are moved every three to seven days within the vineyard. Shooting birds is often considered by the grape producer but, with few exceptions, killing birds is strictly prohibited under the provisions of the Migratory Bird Treaty Act of 1918. Rock pigeons, English sparrows, and European starlings are classified as introduced, non-native spe-

a. Grape cluster with bird pecking injury.

Imed Dami, Ohio State University

b. Grape clusters after extensive bird feeding.

Figure 14.1 ▪ Bird damage.

Affected berries are subject to rot and attract bees and fruit flies.

Table 14.1 ▪ Conditions that encourage bird depredation of grapes. The more conditions that apply, the greater the likelihood that the vineyard will suffer bird depredation

Conditions that encourage bird depredation of grapes	Check if condition exists at your vineyard
Adjacent tree lines, bush, or woodlots	
Presence of overhead power lines or other roosts	
Source of water nearby	
Position in flight path of migrating birds	
Best source of food in vicinity	
Absence of nearby growers protecting crops	
Early-ripening, very sweet; darkly colored grapes	
No regular human activity around vineyard	

Adapted from Fraser et al. (1998).

cies and, as such, are granted no protection under existing federal (US) regulations. Individual states, however, may regulate conditions under which these species may or may not be hunted or shot. Land owners may be allowed to "harass" protected, depredating bird species by firing blank shot-shells or by use of other non-lethal means; however, endangered or threatened species are excluded from that permission. Given the complexities of federal and state wildlife laws, it is well worth checking with state wildlife agency staff before using shooting as an option.

Netting, although expensive and laborious to apply and remove, is the choice when effective, environmentally benign control is desired (figure 14.2, page 274). Tractor-mounted equipment is available to both play out and retrieve the over-the-row netting. Because lateral shoots often will grow through the applied netting, some growers use opposing panels of netting that "sandwich" the fruit zone of VSP-trained vines, leaving the top of the canopy open (figure 14.3, page 274). The opposing panels must be secured to each other through the canopy with ties or clips.

Netting with ⅝- to ¾-inch extruded plastic mesh, which is something of an industry standard dimension, will significantly reduce bird feeding; however, determined birds will find openings or tears in the netting or will feed through the netting where the net lies close to the fruit clusters. On Long Island, where bird pressure on vineyards is intense, loss of 10% of the crop even with standard mesh netting is not uncommon. Recent netting trials on Long Island have shown that finer mesh netting can further reduce fruit loss to birds, but the material costs increase as well (table 14.2). In one trial with Merlot, the incidence of bird damage to clusters was reduced from 5.2% with standard netting, to 0.1% with Gintec™ netting and 1.9% with Vineside®.

Table 14.2. Relative costs, properties and application labor associated with three types of vineyard row netting on Long Island, New York, in 2005

Type of net	Material cost/acre[a]	Mesh dimensions (inches)	Degree of shade	Application method	Man-hours/acre to apply[b]
Vineside®	$1450	0.6 x 0.6	Minor	Side net	4
Gintec	$1620	0.16 x 0.16	11%	Side net	4
Standard	$918	0.63 x 0.75	Minor	Over the row	3

Adapted from Wise, (2006).

[a] Based on 8-foot row width and 2005 product costs (US dollars), Long Island, NY.
[b] Based on laying out two rolls of net per row middle per pass.

Alice Wise, Cornell University

Figure 14.2 ▪ Bird netting applied over-the-row to individual rows.

Alice Wise, Cornell University

Figure 14.3 ▪ Bird netting applied only to sides of vertically shoot-positioned canopies.

DEER AND BEAR

A fondness for grapes extends to the mega-fauna, white-tailed deer and black bear in eastern North America. White-tailed deer are remarkably adaptable and can be found in rural as well as suburban settings. Deer depredation may be identified by sighting the deer in the vineyard or by their pattern of feeding. Deer lack upper incisors and feed by tearing off leaves, shoots, and ripening grapes. This feeding produces jagged edges that can distinguish deer browsing from that caused by other animals. Look for rachises that are torn or shredded and shoot tips and leaves that have been stripped. Deer may be deterred from vineyard feeding by various scare tactics, repellents, fencing, or by regulated shooting. Each method has its limitations. Whatever method or methods are used, they should be implemented well before the damage becomes intolerable. It is difficult to overemphasize the point that once deer have "learned" about the source of food, it will be exceedingly difficult to discourage them.

Black bear are less common vineyard problems, but are sometimes troublesome in areas that bound national forests and parks (sanctuaries); bears have even been observed in suburban areas. Bear deterrence may involve vineyard dogs or woven or electric fencing, as described below, but each has some limitations with determined animals. Electric fencing is probably the most effective bear deterrence. Most of the balance of this section will pertain to deer deterrence, although the fencing section has some bearing (sic) on bear deterrence.

Scare Devices

Acoustically scaring deer (or bears) is, at best, a delaying tactic. Acoustic scare devices include propane cannon, frightening acoustic recordings (including radio), pyrotechnics, and physically patrolling the vineyard with people or dogs. Noise emitters should be moved about every few days, but deer and bear quickly learn that sounds do not pose a threat, and often habituate to the sounds within hours. Their principal disadvantages are inefficacy, a power requirement, and the fact that they may be a nuisance to vineyard owners or neighbors.

Repellents

Deer repellents may be taste- or odor-active, or both. Odor-active repellents deter deer by scent alone. Some products include animal-based proteins or higher fatty acids that produce odors that deer avoid. Depending upon formulation and label statements, the odor repellents may be sprayed on or around vines or mounted on the trellis (e.g., animal tankage bags). Research has shown that odor-active materials generally out-perform taste-based products. Repellents, must, however, be registered as pesticides by the Environmental Protection Agency as well as individual states if applied to grapes in the US. Very few commercial repellents are labeled for food crops. The legal restrictions, combined with relatively poor performance, limit the utility of repellents in vineyard wildlife deterrence. If attempting to repel deer with repellents, consider the following:

- Apply prior to anticipated damage. This might be predicted by past experience. Do not allow a feeding pattern to become established.
- Use on a non-crop area (e.g., vineyard perimeter) if material is not registered for use on grapevines.
- Feeding pressure will be greatest when alternative food sources are scarce. Repellents may work well when other food is available but may perform very poorly if little else is available for deer. This may partially explain year-to-year variation in repellent effectiveness or mixed results among different vineyards.
- Monitor repellent effectiveness and reapply or alternate with other tactics if necessary.
- Be prepared to implement alternative strategies such as fencing if repellents prove ineffective.

Besides sprayable repellents, growers have attempted to discourage deer with human hair, animal "tankage" and soap bars placed around the vineyard. These are, at best, hopeful measures.

Fencing

Fencing is the most effective means of excluding deer from vineyards. Although the initial costs can be very high, the near-perfect protection afforded makes fencing economical, especially when considered over the 20+ year service of a well constructed fence. Fencing may be either electrified or non-electric. Non-electric fences are usually of a woven mesh and may be 8 to 12 feet in height. The advent of high tensile (HT) strength fence wire, coupled with high voltage, low impedance electric fence chargers, has provided an economic alternative to the more expensive woven wire fences. Many designs exist but the least complicated may be the most effective and easiest to install and maintain under low- to moderate-pressure situations. The 6-wire vertical design depicted in figure 14.4 is a modified version of the Penn State™ 5-wire design. An optional hot (+) wire located about 4 or 5 inches from the ground will provide some deterrence of raccoons, opossums and other small animals; however, it is essential that the soil under the fence be kept free of weeds that would otherwise reduce the effectiveness of the fence charger if something were to

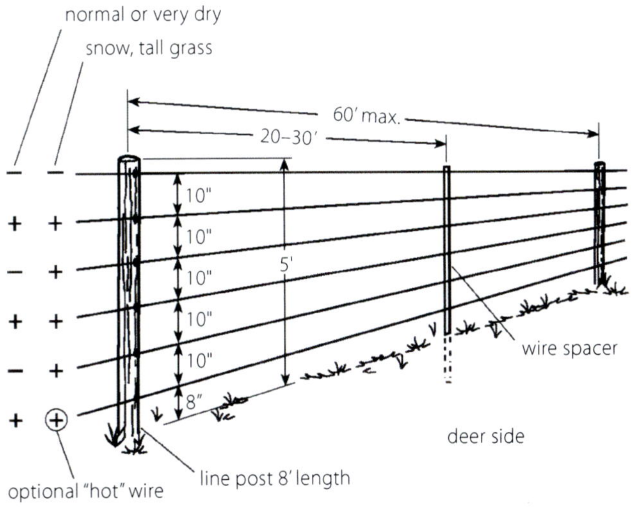

Figure 14.4 ▪ Vertical, 6-wire electric fence.
The changing pattern of hot wires (+) is governed by the soil and ground cover conditions.

contact the positive wire(s). The 6-wire fence is only about 5 feet tall, a height that deer have no limitation in jumping. But approaching deer will first attempt to crawl through or under the fence before jumping. The high energy output of the charger "modifies" deer behavior, training deer to avoid the fence.

Designs for more elaborate slanted and three-dimensional fences are available and may be more effective with moderate to high deer pressure.

Properly charged fences produce an extremely unpleasant, but non-lethal shock. Therefore, electric fences should always be posted to alert people to avoid accidental shock.

Electric fences should be kept charged 365 days per year, and the line voltage should be routinely checked with a dependable voltage meter. Upon questioning, most growers who complain about ineffective electric fence operation confess that the fence was not continuously charged. Erecting the fence prior to the vineyard ever bearing a crop is the most effective approach; the deer are much less tempted to investigate what's on the other side if they are trained to avoid the planting from the outset. Clear at least 10 feet of brush and trees from the outside (deer side) of the fence. This gives uninitiated deer plenty of room to approach the fence, touch it with their moist noses, and get a nasty shock. Keep vegetation, including weeds, clear of the hot wires. This "shorting" significantly increases fence resistance resulting in rapid battery discharge and ineffective shocking energy. Use a preemergence weed herbicide under the fence unless soil erosion is an issue. Some wildlife experts recommend "baiting" the electric fence with dabs of peanut butter smeared underneath aluminum foil tents placed at intervals on the hot wire of a perimeter fence.

Depending upon terrain and how much brush clearing is involved, a battery-operated solar charged, six-wire electric fence can be installed around a five-acre vineyard for $5,000 to $7,500 in material costs ($1.00 to $1.50 per linear foot, excluding labor).

Non-electric woven wire fencing up to 12 feet in height has been used in some high-pressure vineyard situations (figure 14.5). This fence design is similar to that used around airports and other high-risk areas where there is zero tolerance for deer intrusion. The initial expense is high ($5 to $10 per linear foot, materials and installation); however, the maintenance is low and the efficacy is nearly 100%. Check local zoning ordinances or homeowner association covenants before construction to ensure that you will not be at variance with fence style or height restrictions.

Figure 14.5 ▪ Non-electric, woven-wire fencing.

To avoid unnecessary expense and retrofitting, the vineyardist should seek local assistance or advice from wildlife agency staff or state viticulturists to help determine the sophistication needed in a fence prior to installation. In high-pressure situations, it may be most economical to start from the beginning with a 12-foot high woven wire fence. Low pressure situations might, alternatively, be managed with a 6- to 8-wire vertical, electric fence.

Domestic dogs equipped with radio-controlled shock collars, and "fenced" or "penned" within the vineyard by a buried perimeter radio frequency wire, are used with varied success. The advantages over conventional fencing include the absence of an unattractive fence, and costs that can be competitive with, or less than, woven or electric fencing. The disadvantages include animal care, including veterinary bills, potential liability if escaped dogs attack neighbors, winery visitors or livestock, and the occasional lazy dog that lacks a sense of duty.

Regulated Shooting

Under certain conditions, farmers, including grape growers, can obtain permits to selectively destroy deer (and bear) that significantly damage crops. Shooting permits are typically issued by state wildlife agency staff or game wardens after a review of the vineyard problem. Depending on the state program, kill permits may be issued at the time the damage occurs, or the harvest of nuisance deer might be deferred to the regular hunting season. Information on kill permits can usually be found under state listings for department of natural resources, or the state's game and fisheries agency. While lethal measures such as shooting can provide immediate damage abatement, fencing or other long-term strategies should be considered for sites with chronic deer damage. A limitation to the off-season kill permits is finding someone willing to carry out the task. Wildlife agency staff may be of assistance in identifying responsible shooters. Grape growers and their neighbors should also understand that regular hunting seasons and deer bag limits are set by deer biologists as the principal means of regulating the deer carrying capacity of a habitat. While not always balanced, this approach to deer population management remains vital in the effort to limit destruction of agricultural crops. Grape producers can and should encourage responsible hunting on their property if deer are causing damage. Land-owners of large tracts of land can derive added farm income by leasing hunting rights to individuals or hunt clubs.

SMALL MAMMALS

Rabbits and groundhogs will occasionally nip young shoots from grapevines; however, this is a rare occurrence in the author's experience, perhaps because grapes break bud relatively late in the spring, when forbs and other food sources are relatively abundant. Raccoons can be a significant problem in some areas and may be difficult to identify as well as to deter. Raccoons are fond of ripening fruit and are particularly fond of the more aromatic varieties. Healthy raccoons are typically nocturnal in habit and oddly behaving raccoons observed during the day should be avoided due to the possibility of their being rabid. If you do not see the raccoons in the vineyard, you might not suspect this thief. Clues of raccoon feeding include mud on the trunks and cordons, shoots that have been bent or broken over along the cordon, and berry skins deposited under the vines. Over time, the raccoon(s) may leave visible trails in the grass as they enter and leave the vineyard on their foraging trips. Deterrence of raccoons can be difficult. Electric deer fences can be modified with a band of wire mesh along the bottom of the fence (not touching/grounding the fence) to force the raccoons into contact with the electric fence wires if they attempt to penetrate the vineyard. Trapping is another option if you can identify the point of entry into the vineyard. Check with state wildlife or game officials as to the legality of lethal trapping. Because raccoons are a rabies virus vector, trapping should only be conducted by experienced personnel who have obtained pre-exposure vaccinations. In many states, rabies vector species may not legally be live-trapped and transported. Foxes (red and gray) can also forage grapes and may feed nocturnally. Again, fencing is an effective means of management.

CONCLUSIONS

Wildlife deterrence and management of wildlife damage in vineyards is complicated by the sporadic nature of damage, the lack of easily applied control measures and regulatory aspects of wildlife conservation. Growers are encouraged to use preemptive strategies such as fencing for deer and to be prepared to use a range of other measures for other wildlife, including small mammals and birds.

References

Fraser, H.W., K.H. Fisher, and I. Frensch. 1998. Bird control on grape and tender fruit farms. Ontario Ministry of Agriculture, Food and Rural Affairs. Factsheet 98-035. Available electronically at: HTTP://WWW.OMAFRA.GOV.ON.CA/ENGLISH/ENGINEER/FACTS/98-035.HTM

Wise, A. 2006. Personal communication.

Further Reading

Kays, J. S. (2000). *Managing Deer Damage in Maryland.* (Extension Bulletin 354). College Park, MD: UMCP, MCE. 20 pp. Cost $2

— 15 —

Grape Purchase Contracts and Vineyard Leases

Author

Mark L. Chien, The Pennsylvania State University

The focus of much of this production manual deals with materials and management practices used to produce and ripen wine grapes. For the independent grape grower, the approach of harvest brings another question: How will the grapes be sold? Ideally, the grower has, by this point, invested time in market exploration and has one or more wineries lined up for purchase of the crop. Long before the harvest occurs, the vineyard owner, or grower, should prepare by negotiating a legally binding grape purchase contract that serves and protects both the vineyard owner and the winery. In addition, the winery may be required to lease the vineyard in order to comply with state or provincial farm winery regulations. We will explore the provisions of these agreements within this chapter.

GRAPE PURCHASE CONTRACTS

Keep in mind that a commercial vineyard is a business. The use of grape contracts and vineyard leases makes good business sense and should not be interpreted as a sign of distrust between two parties. Instead, contracts clarify and record obligations, thereby helping to avoid conflict between growers and wineries.

In general, a contract is a written, legally enforceable agreement involving two or more parties. Key contractual assumptions typically include the desire by both the winery and the vineyard to produce a high-quality product and a willingness to enter into a good faith agreement. Experience indicates that long-term relationships based on fair pricing result in a more consistent product and are, in general, more rewarding than short-term relationships such as those that are based on year-to-year negotiation. Wineries will benefit from working with the same growers year after year, as the winemakers will become familiar with the vineyard and fruit and will, therefore, be able to make a more predictable product. Wineries need a secure supply of grapes that match the style and price of the winery's product line. Wineries may specify crop levels and minimum standards for pest management and arrange for seasonal vineyard visits by winery representatives. They may also specify in the contract the type, quality, and quantity of grapes they want. This degree of detail helps in planning for storage space and creating a grape purchase and wine marketing budget.

The security of a home and an established value for a perishable crop offer peace of mind for the vineyard owner. Some intrepid growers may feel that, unless grapes are in oversupply, contracts are unnecessary because winery buyers are easy to find. These growers may play the spot market as harvest approaches, in hopes of getting top dollar for their grapes in a short market. Most growers, however, want the certainty of having a home for their fruit and receiving a fair price in a timely manner. As you, the grower, establish a reputation for producing a high-quality product, winemakers will want to buy your grapes and "deals" will be made long before harvest occurs.

A contract is an excellent record of the agreement between you and the winemaker regarding quality parameters, pricing, and quantity. This document will help you to avoid confusion during the busy harvest period and it is something to which both parties can refer if the crop is short or long or if the quality or price of fruit is in dispute. A contract also is a guide for responsible vineyard management, specifying, for example, cultural practices like fertilization and canopy management or the use of pesticides. Most contracts contain a section devoted to the types of vineyard practices a winery expects the vineyard to use. These may be either generic or very specific. Any significant deviation from the assigned methods could render the contract void. Contracts typically designate quantity of fruit by weight. Fruit is picked and weighed on a certified scale and delivered or weighed at the winery.

As winemakers seek greater control over vineyard practices—particularly crop yields—there has been increased interest in acreage contracts, in which the winery buys fruit from a designated section of the vineyard and pays a flat fee, usually with a minimum tonnage clause. Under this agreement, the risk of growing grapes is shared by the winery. Rather than contracting for the value of the crop (dollars per ton), the winery negotiates with superior growers for a flat payment per acre of vineyard. For example, the winery may contract to pay the grower $5,000 to $7,000 per acre, with a lower catastrophic "cap" to this payment in the event of a natural disaster that eliminates the crop. The winery will set a crop yield limit, for example 2 to 3.5 tons per acre, and will likely dictate other aspects of vineyard management. The principal advantage to the grower is less risk due to crop adversity; however, the potential for profit is also "capped." The advantages to the winery are a known cost for grapes and the ability to strictly control many aspects of grape production. Typically, the wine pricing reflects lower yields, with retail bottle price in the ultra-premium category or higher.

Contracts should include provisions for very practical concerns such as the type and size of harvest bins, preferences for mechanical or hand picking, delivery schedules, bin cleaning, and type of transport vehicle used (size and type, refrigerated or unrefrigerated). Another important provision concerns load weights taken at certified truck scales. Certified scale receipts are a legal record of weight and should be retained by the vineyard.

Contracts can be long or short, complex or simple. All contracts should contain basic information and, of course, the signatures of the parties involved. *Force majeure*, transfer of title, assignments, remedies, and dispute resolutions are all legal matters; however, these issues do not govern the success of a contract or the delivery of the sound grapes that make good wine. It is not necessary to have an attorney involved in drafting a contract, but it may be wise to have an attorney review the contract before it is used.

A contract should stipulate an agreed-upon price for each grape variety and clone (if necessary) as well as the quantity of each. It should include a payment schedule for the winery with interest assignments for late payments. A typical grape contract will ask for payment in full within 30 days of delivering grapes to the winery. In some cases a down-payment is requested to secure commitment from the winery. Payment schedules are sometimes used but are not considered highly desirable unless interest can be collected on any unpaid balance. In the case of non-payment, each state has specific regulations governing payment default. In some cases, a lien may be placed against the winery's inventory. Collection agencies represent a drastic last resort.

The terms of a contract should be clearly stated. A one-year contract is common. A long-term contract, on the other hand, is valid until both parties agree to its nullification. An evergreen contract renews on an annual basis until the parties agree to cancel it.

Contracts may either reward a grower for delivering fruit of superior quality, or penalize the grower's payments if the fruit fails to meet minimum quality standards. Fruit rots and mildew, bird damage, unripe fruit, and the presence of large amounts of material other than grapes (MOG) are common factors contributing to price reductions or even rejection of crops. These parameters must be clearly stated and agreed upon in the contract.

At every stage of a contract—negotiation, execution, and payment—communication is the key to success and satisfaction for both parties. From problems in the vineyard to concerns about payment, each side must keep the other informed. This will allow both sides to react and plan ahead. If the grapes develop a disease that will affect wine quality, for example, the winery should know about this so that plans can be made to handle the grapes accordingly. In the same way, if a winery cannot meet its payment obligation, it must inform the vineyard so that the vineyard can find alternative ways to pay the pickers.

Ultimately, a contract is only as good as the character of the two parties who consummate it. Try to find wineries with good reputations in the realm of business partnership as well as the production of high-quality wines. It might be a good idea to commit to an initial short-term (1- or 2-year) contract so that you can determine if your goals and temperament represent a good match for both vineyard and winery. If they do not, it is easy at this point for you and the winery to go your separate ways. If you do find a good match, you may renegotiate a long-term agreement that will add stability, consistency, quality, and satisfaction to your business and to the relationship.

A representative Grape Purchase Agreement might contain the following clauses:

- Description of parties
- Statement of contract intent: Grower will sell and winery will buy the grapes described in the contract.
- Terms of contract:

- Grower's commitment to sell and winery's commitment to purchase. Clause may include language specifying either a short-term (e.g., one-year) or long-term contractual relationship.
- Quantities of grapes. Tonnage of grapes can be specified by variety here or in an appendix. Description of what happens in event of under- or over-estimation of quantity delivered. Clause specifies who has responsibility for weighing crop and by what means (for example, certified scales, winery scales, or other means).
- Price to be paid. Specifies minimum base payment, payment schedule, and method of setting the minimum base payment structure. This clause may also stipulate which party (grower, winery, or independent third party) conducts the juice or must analyses that determine minimum price.
- Modification to price. Factors that might increase or decrease the minimum price per ton will be spelled out in this clause or in an appendix. Factors may include elements such as MOG, rots, mildew, or incorrect variety.
- Transportation. Specifies who will pay for picking, loading, and transport of fruit. Need for refrigeration and special picking bins or containers can also be included.
- Grower responsibilities. Specification of vineyard best management practices or other minimum standards of maintenance. Estimate of crop yield made shortly after fruit set and updated through season. Specification of fruit sampling to predict harvest (how and by whom samples are collected and who does analysis of samples). Winery may also request a written record of vineyard inputs such as fertilizer, irrigation, and pesticides.
- Winery responsibilities. Winery may opt to provide vineyard consulting services and may elect to void contract if grower fails to meet industry standards of vineyard management or is plainly negligent in tending the vineyard.
- Harvest dates. Specifies (ideally) a joint decision between winery and grower based on the ripeness of fruit, threats to crop, and potential for improvement or deterioration of crop quality.
- Payment. Specifies time frame and form (cash, equity shares, other means), and whether a deposit is required.
- Terms. Included or in separate clauses will be an identification of the state or province governing the laws under which the agreement will be binding, whether (typically) the contract can be assigned to other parties (such as if the vineyard is sold mid-season), exemptions due to *force majeure*, and responsibilities in the event of need for legal action (arbitration and legal fees).
- Period or term of agreement. short- or long-term, dates are specified, and language is included that allows either party to terminate the contract with sufficient notice (for example, with six-month or one-year advance notice).
- Signatures and dates.

A sample contract can be found in appendix 7, page 314.

VINEYARD LEASES

Vineyard leases allow a winery to secure a grape supply without having to make a large capital investment in property and development. Often a winery will do business for a number of years with an independent vineyard and, once the quality of grapes is recognized, will offer to lease the vineyard from the owner. A lease differs from a grape purchase contract in that the winery assumes full operational and financial control over the vineyard with the lease. This provides the vineyard owner with a steady stream of income and shifts some burden of financial and agricultural risk to the winery. The winery is able to exert partial or full control over the production of the grapes. In some leases, the grower will continue to be responsible for the farming duties, but in others the winery will assume operational control of the vineyard. Responsibilities assigned to each party should be clearly stated in the lease. Leases can vary in length of term and should contain a clause that will allow for termination upon consent of both parties.

LOCAL REGULATIONS

Each state has an agency of state government that oversees matters related to alcohol beverage production, transfer, and sales. For example, in Pennsylvania, this agency is the Pennsylvania Liquor Control Board (PLCB), and in Virginia, it is the Virginia Department of Alcoholic Beverage Control (ABC). Each state has created its own specific set of regulations to govern winery production and sales and, in some cases, the way in which independent vineyards can do business with wineries. For example, Virginia's laws recently changed and are now somewhat more lenient in regulating vineyard and winery relationships, depending on the class of Virginia Farm Winery license held by the winery (Class A or Class B). The Class A license stipulates that 51% of the crop used by a Virginia Farm Winery comes from vineyards that the winery owns or leases. The Class B license does not require 51% to be grown by the winery; instead, 75% of the fruit must come from Virginia. However, to obtain a Class B license, a winery must first hold a Class A license for at least 7 years. In addition to farm winery licenses, most states also issue commercial winery licenses, which generally do not restrict where the grapes or bulk wine is sourced. A given winery may hold both a farm winery license and a commercial winery license, although the product lines fall under different production and marketing regulations.

Grape prices are subject to market conditions and may vary from year to year. New York State's Department of Agriculture and Markets provides a means of gathering and listing grape prices in advance of harvest, while some other states provide a census of those data after the crush but in advance of the following year's harvest. In either case, the grower has at least a sense as to the fair market price of grapes.

The laws can be confusing, so it is important to consult with your state's beverage control agency as well as the state wine association about compliance with these rules and regulations before the vineyard is planted. If a vineyard plans to ship grapes across state lines, it should be aware of interstate commerce regulations regarding the transport of perishable products across state lines. Inspection and declaration requirements may be involved.

— 16 —

Wine Grape Quality: When Is It Time to Pick?

Author

W. Gill Giese, Surry Community College

The quality of wine is predetermined at the point at which grapes are harvested. This monumental harvest decision is one of the most perplexing to new growers, and even seasoned growers agonize over the decision. Because the quality of wine is dependent on grape quality, both growers and winemakers should play a role in the harvest decision. Soluble solids concentration, titratable acidity, and pH contribute to grape composition and influence the harvest decision. However, these components do not constitute the sole basis of grape and wine potential quality; they often are used as quality indicators because they are easily measured. Other components, less easily quantified, play a significant role in defining grape and future wine quality, as well. Growers and winemakers should have some knowledge of how these components originate, interact, and relate to quantitative or subjective measures of grape and wine quality. Wine quality is dependent on the quality and composition of grapes best suited to make a desired style of wine. Wines of a certain style possess common characteristics that are recognized by consumers. For example, Chardonnay can be produced in a clean, crisp, somewhat austere style with cold, stainless steel fermentation. Chardonnay can also be produced in a more complex, barrel-fermented, on yeast lees (*sur lees*) style. The quality of either style of wine may range from low to high. High quality wines generally have visual, taste, or aroma characteristics that are perceived as above-average to a wide segment of wine consumers. "Very good" or "outstanding" wines are variations of the high quality theme and may be associated with numerical ratings (for example, a 50 to 100 point range) by trained panels, wine magazines, or individuals. Coombe and Iland, (2004) define wine grape (the raw product) quality as: "*...the suitability of a batch of grapes to produce a wine of the highest quality in a targeted style.*" In essence, we want to harvest grapes when they are "ripe" for their intended wine style. Often, "ripeness" and "maturity" are interchangeably used, although we prefer to use "ripe" to describe grapes that are suitable for harvest as defined by the intended wine style. Maturity may be defined as "physiological" (containing a viable seed, typically prior to veraison) or "commercial" (achievement of a desired state of use or suitability). Coome and Iland (2004) state that ripening, a stage of maturation, is "the alteration of physiologically mature fruit from an unfavorable to a favorable condition with respect to firmness, texture, color, flavor, and aroma." Grapes can ripen beyond the stage when they are considered optimum for the intended use. Tissue in overripe berries will break down, the grapes can shrivel or "raisin" due to dehydration or rot due to microbial invasion. The effects on yield can be substantial. Wines made from such grapes can be high in alcohol, unbalanced, and without elegance and aging potential.

Determination of berry ripeness will depend on objective, qualitative parameters (visual and sensory assessment) and quantitative parameters as well as subjective clues. Quantitative parameters include: titratable acidity, soluble solids concentration, pH, anthocyanin concentration and measures of volatile components (aromas) and other "secondary metabolites" (see glossary) that contribute to the sensory quality of wines, either immediately, or with wine aging. Visual indications may include: integrity of fruit, color intensity of skins, seed coat color, and degree of lignification of the cluster peduncle. Other indications of ripeness relate to berry skin integrity, seed coat texture, the degree of tannin "ripeness" when skins are chewed (harsh and astringent versus soft and supple), as well as juice taste and aroma. Practical/economic considerations often take precedence in harvest timing and may include: availability of labor, impending weather (advancing tropical storm), vineyard condition, or tank space in the winery. Attention must also be paid to late-season pesticide applications in order to avoid residues on grapes.

This chapter describes basic growth and physiological processes of the grape berry. It will focus on how to evaluate grape quality and how to apply that evaluation to the harvest decision. Grape berry development, structure and composition, and variability and ripening will be discussed. A brief description of berry assessment and sampling protocol for harvest determination is also provided.

GRAPEVINE PHENOLOGY

Grapevine phenology is the record of the annual developmental growth stages of the grapevine and how those growth stages relate to one or more aspects of environment, typically heat accumulation. The grapevine is a temperate zone, woody perennial that does not initiate growth until ambient temperatures reach approximately 50° F. Energy for this early growth comes from carbohydrates stored in the woody structure of the vine (trunk, cordons, and roots) the previous season. Following budburst, with sufficient heat, shoots can grow an inch or more per day. Shoot growth rate slows after anthesis (pollen shedding/flowering) and berry set, and progressively slows or, ideally, stops with the onset of fruit ripening, or veraison. Continued shoot growth concomitant with fruit ripening is common in a humid environment, but is not desirable from a wine quality standpoint (see chapter 6, page 124).

Grapevine development can be broadly partitioned between the vegetative growth cycle and the reproductive cycle. The vegetative cycle is completed within one growing season. Outwardly we can observe budbreak, rapid shoot extension and canopy development in early summer, followed by a cessation of shoot growth and, ultimately, senescence and shedding of leaves in the fall. During this season of vegetative development, the succulent, green shoots mature into woody canes, and older wood such as trunks and cordons develop another annual ring of xylem. The older, nonfunctional phloem and tissues external to the current year's phloem are destined to form the bark. Below ground, the roots also go through flushes of new growth and development, the first commencing at about the time of budburst, and a second flush of regrowth typically occurring post-harvest.

The reproductive cycle requires two seasons to complete (figure 16.1, page 284). In the first season, the reproductive structures that give rise to flower primordia are initiated and develop in the primary compound buds. Those flower clusters will be borne on the shoots of the following year. The current season's flowers, which were initiated last year, achieve their final form between budburst and bloom of the current year. During a period around bloom (flower opening and pollen shed) (figure 16.2, page 284), after temperatures reach 68° F, the subsequent year's flower cluster primordia are initiated to start the cycle again. The cycle from flower initiation to ripe fruit is 14 to 16 months in duration, during which the environment has substantial influence. The number and size of the flower clusters for the 2009 season were determined in the 2008 season. The number of berries that set, develop, and achieve the desired degree of maturity in 2008 is determined by the season and environment in 2008.

Identifiable developmental growth stages from dormancy through bud burst, bloom, veraison, harvest and leaf fall have been used to define grapevine phenology.

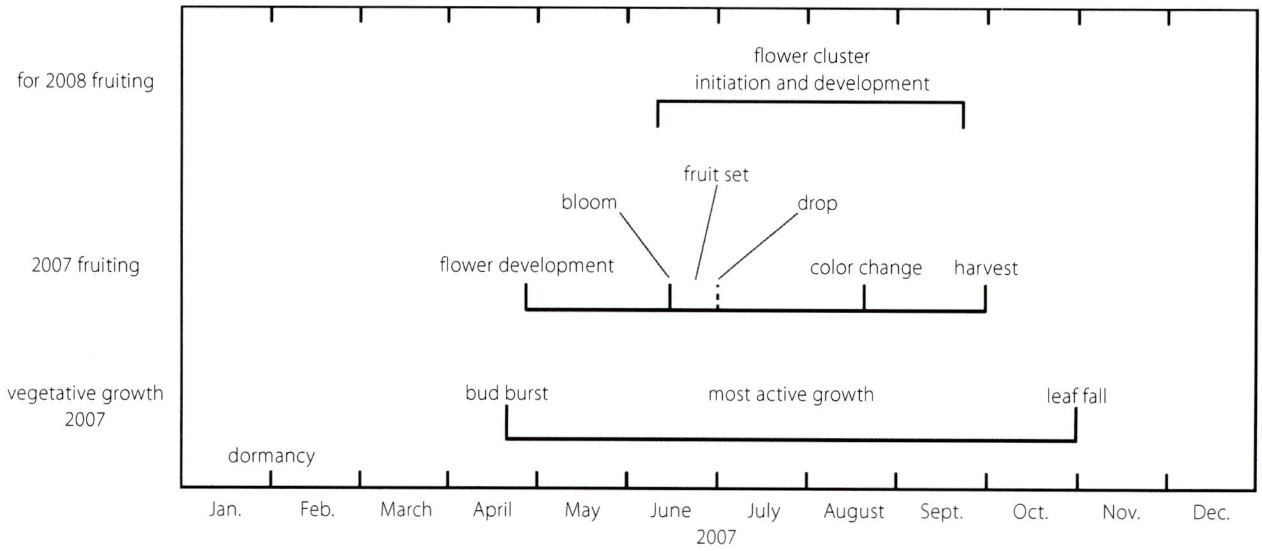

Figure 16.1 ▪ Generalized calendar of grapevine growth and development. *Adapted from Shaulis and Pratt (1965).*

One such system is the Eichorn and Lorenz system, named for the originators of the scheme, Eichorn and Lorenz, who published their system in 1977. Slightly modified (see Dry and Coombe, 2004), the Eichorn and Lorenz system assigns 47 so-called E-L stages to grapevines based on their phenological development. While grapevines complete each stage each season, the date that each stage is achieved varies by variety and environmental conditions of the season. There are 5 E-L stages of phenological growth commonly used by growers to chronicle grapevine development and to aid in vineyard management. The key stages are budburst, bloom, bunch closure, veraison, and harvest (figure 16.3).

Figure 16.2 ▪ Chardonnay flower at bloom. (19–23*)
Calyptra (cap) about to separate and expose anthers.
* = Phenological stages of the modified E-L system.

BERRY STRUCTURE AND DEVELOPMENT

The berry is connected to the rachis via the pedicel. The pedicel is the final vascular (xylem and phloem) connection from the vine into each berry. The berry flesh functions as storage tissue and contains most of the malate before veraison and sugars afterwards. Berry skin contains pigments and flavor compounds, and serves as both a protective covering and an active, metabolic site. From outside in, the skin comprises three layers: a waxy cuticle, a layer of epidermal cells, and finally a hypodermal layer. The dermal cells that bound the flesh of the berry are abundant in phenols and phytoalexins that aid in defense against some pathogens.

After fruit set, berry growth and development occurs in three stages represented by a double-sigmoid curve with two growth phases (figure 16.4, page 286). The first growth phase is "berry formation" in which cell division and tissue differentiation dominate growth processes. The second growth phase is "berry ripening" and is characterized by cell enlargement and water and carbohydrate import. A third, less defined phase, is referred to as the "lag phase." This lag phase is "physiologically artificial," representing a cessation of the first phase of growth. It may be quite transient. The embryos (seeds) begin to mature during the lag phase. When the embryos are mature, or become capable of germination, the "lag phase"—the first phase—is typically considered complete. This can be

a. Grape bud/shoot at stage of budburst (4*).

b. Grape flower cluster at 50% to 75% bloom (23*).

c. Beginning of bunch closure, berries touching if clusters are tight (32*).

d. Veraison (35*) color change, softening of berries, and accumulation of sugars.

e. Ripe (38*) 20 °Brix or greater for most varieties.

Figure 16.3 ▪ Readily recognized stages of growth commonly used by growers to chronicle grape vine development in a season. * = Phenological stages of the modified E-L system.

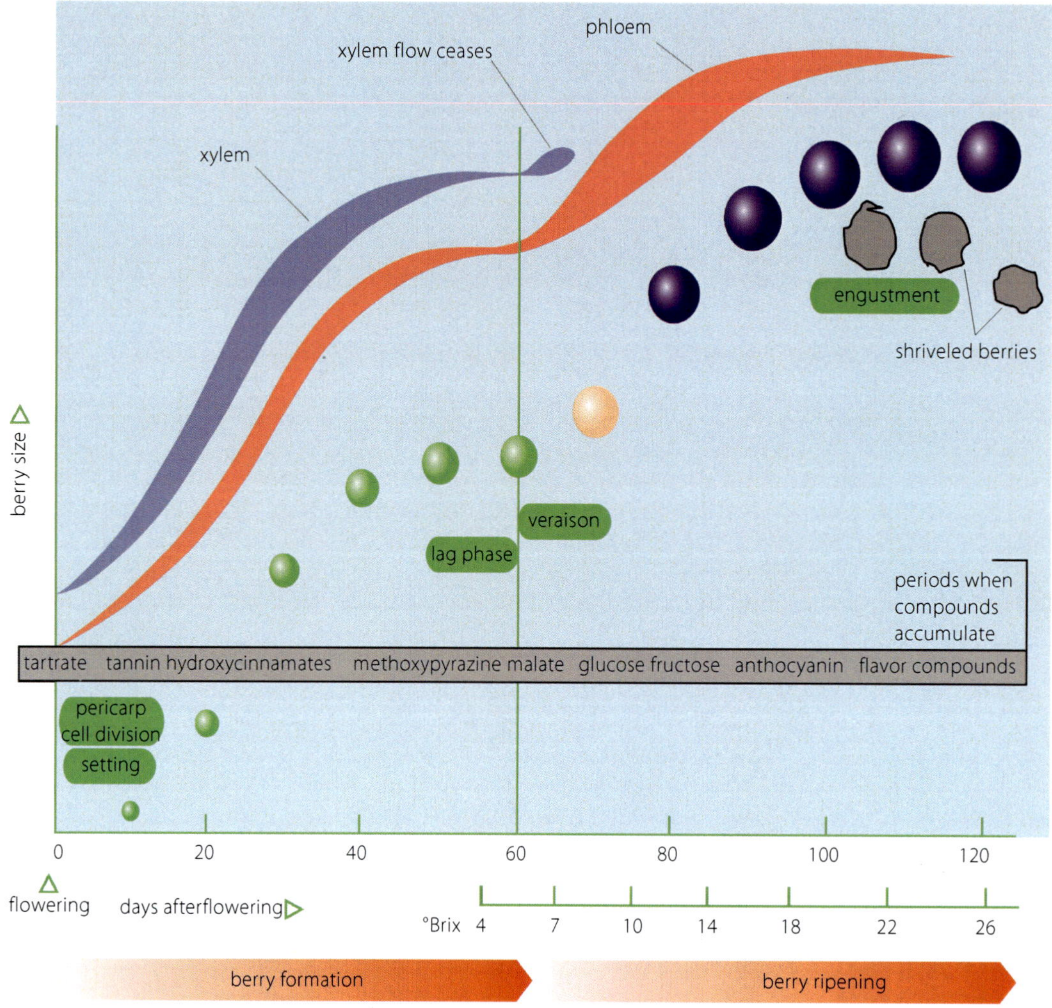

Figure 16.4 ▪ General representation of berry development and maturation including engustment and possible shriveling. Shriveling can occur due to extreme drought and berry dehydration or be caused by physiological disorders that alter water balance in the berries. *Source: Kennedy, J. (2002).* Practical Winery and Vineyard, *July/August, 14–23.*

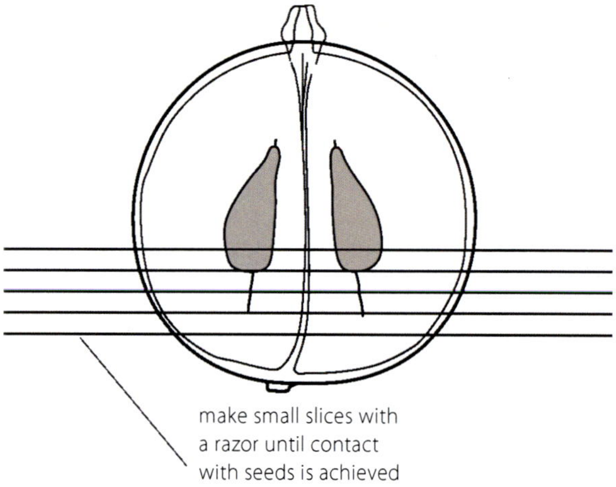

Figure 16.5 ▪ Technique used to determine onset of seed coat hardening.

gauged over a series of days or weeks by cutting through berries with a razor (figure 16.5). When the developing seeds can no longer be cut cleanly and the blade "pushes" the hardened seed through the berry flesh, lag phase is ended. This point has been used by some as a phenological benchmark for assessing mid-season cluster weights as part of crop estimation procedures (see chapter 7, page 135). However, it does not mean that berries are necessarily at 50% of their final weight.

The grape has the potential to set up to four seeds, and berry size is somewhat dependent upon seed number; small berries typically have fewer seeds and cells present at harvest than do large berries. In addition to seed number, final berry size is determined by variety (for example, Petit Verdot will usually have smaller berries than those of Merlot), by number and size of cells in the flesh and skin, temperature and water status (environmental), the rate of photosynthesis, leaf area, and the presence of competing "sinks."

BERRY RIPENING

Veraison marks the beginning of berry ripening and signals several physiological events:

1. Berry softening
2. Sugar concentration increases in the berry
3. Malate decrease (and acidity change)
4. Accelerated growth of the berry
5. Color change in skin
6. Increase in pH of cell sap or grape juice
7. Increase in respiratory quotient
8. Increase in concentration of proline and arginine (amino acids)

Berry softening and sugar accumulation begin simultaneously, followed in sequence by berry expansion and skin color development. Anthocyanin level and pH increase as ripening progresses. Seeds change from green to varied shades of brown, overall acid levels decrease and fresh berry weight may actually decline due to dehydration (figure 16.4). Variability in development (specifically berry size) between berries has been determined as early as fruit set and those differences can carry-through to veraison and harvest. Pronounced variability is ultimately detrimental to wine quality.

Ripening berries are a strong "sink" or organ of import, and carbohydrates are transported into them from photosynthesizing tissues and reserves via the phloem. Movement of carbohydrates (sugar) is driven by the gradient (difference in sugar concentration) between the "source" (leaves, where carbohydrate synthesis occurs) and the "sink" (fruit, where sugar accumulates). As berry sugar levels increase, the skin layer undergoes an increase in extensibility and berry enlargement occurs. Shortly after veraison, water flow via the xylem to the berries is diminished. The phloem becomes primarily responsible for further water and carbohydrate movement into the ripening berry (figure 16.4). Berry dry mass can increase as much as 400% between veraison and harvest. Engustment, "flavor development" (figure 16.4) occurs during the latter part of berry ripening. Engustment is detected as the varietal typicity of flavor and aroma in berries when they are crushed and tasted. This characteristic aroma and flavor is often discovered suddenly by growers and winemakers. Measurement of sugar and acid levels alone is an inadequate indicator of ripeness. Coombe and Iland (2004) and Bisson (2001) maintain that aroma and flavor develop independently of sugar accumulation, although there are rough, positive relationships.

BERRY COMPOSITION

Water

Water, whether derived from rainfall or irrigation, is the primary contributor to grape berry weight (70% to 80%). Other berry components determined by berry physiology and accumulation include sugar, acids, nitrogenous compounds, phenolics, aroma and flavor compounds and precursors, and vitamins.

Sugar

Glucose and fructose are the principal sugars in grape berries, constituting 95% to 99% of the grape berry sugar. Glucose begins to accumulate at veraison. The proportion of glucose to fructose is about equal in mature grapes, while fructose exceeds glucose content in overripe berries. Small amounts of sucrose—the transport form of sugar—are found in the skin, central bundle, and pedicel. Sugar is converted to alcohol and carbon dioxide during fermentation. The final alcohol concentration is therefore a function of sugar concentration (natural or additions to must). Sugar content is measured as soluble solids concentration (°Brix in the US). The progression of sugar accumulation in berries is one measure of ripening and can be easily tracked by collecting samples and testing soluble solids concentration on a regular basis starting at veraison. The period of ripening is the time from veraison to harvest, while the rate of ripening is the speed at which this period is completed. Research has provided strong evidence that grapes of a given variety within a particular region that ripen fastest (not necessarily earliest) are generally superior in quality to those that ripen slower. These observations argue against the practice or justification of an extended "hang time" (lingering ripening period) to increase grape or wine quality.

Acid/acidity

Tartaric and malic acid are the principal acids in grapes, although minor amounts of other organic acids can be found. Both tartaric and malic acids are synthesized in the berry starting after or during lag phase. Tartrate levels are generally constant after veraison although the concentration can be reduced as the berries import water, as often occurs with rains near harvest. Tartrate levels may increase slightly due to berry dehydration. Malate levels continue to decrease via respiration to a low of two to three grams per liter. Respiration of malate is increased by fruit heating. In cool climate regions, growers who aim to depress high acidity levels practice canopy management techniques that promote more direct exposure, and thus radiant heating, of fruit clusters. Excess grape acidity can lead to tart, acidic wines. Insufficient acidity will result in "flat" unbalanced wines with potentially elevated pH, and a short shelf life. Elevated pH can increase the potential for oxidation of wine, unwanted microbiological activity and poor color development/retention in red wines. Therefore, an appropriate acid level is desirable at harvest (although acid can be added to the must to alter must acidity and pH). The higher the pH of juice or wine, the less red color is expressed by anthocyanins.

The desired amounts of sugar and acid and corresponding pH of grapes at harvest for various wine styles are shown in table 16.1.

Table 16.1 ▪ Generally recommended ranges of soluble solids concentration, titratable acidity and pH in wine grapes at harvest for generic wine types

Type of wine	Soluble solids concentration (%)	Titratable acidity (g tartaric acid /L)	pH
Sparkling	17.0-20.0	7.0-9.0	2.8-3.2
White Table	19.0-23.0	7.0-8.0	3.0-3.3
Red Table	20.0-24.0	6.0-7.5	3.2-3.4
Sweet Table	22.0-25.0	6.5-8.0	3.2-3.4
Dessert	23.0-26.0	5.0-7.5	3.3-3.7

Adapted from Boulton, et al. (1996).

Nitrogenous Compounds

The presence of ammonia/ammonium and amino acids is required for yeast growth during fermentation. Arginine and proline are the most common of the 21 amino acids found in grapes. However, under winemaking conditions, proline is not utilized.

Phenolics

Phenolics are products produced in the berry from the amino acid phenylalanine, and most are stored in the form of glycosides (that is, bound to a sugar molecule, generally glucose). Phenolics are important for color, astringency, and aging of red wine. Anthocyanins, tannins, and flavanols are included in this group and are regarded as significant cardio-protective compounds.

Aroma and Flavor Compounds and Precursors

Grape quality is dependent on the production of aroma (nasal perception of air-borne volatiles) and flavor (perceptions in the mouth during tasting) compounds derived from secondary metabolites. Secondary metabolites include monoterpenes, norisoprenoids, and methoxypyrazines as well as many others. Conjugated forms, called "flavor precursors" lead to detectable flavors and aromas upon wine aging. No single compound is responsible for the full range of flavors of any one grape variety. However, some compounds are more important to the typicity of some wine styles than are others. Certain methoxypyrazines at moderate levels, for example, are important to the varietal intensity of Sauvignon blanc.

Vitamins

Vitamins, present in small amounts, generally increase with ripening and do not affect the harvest decision.

These are the principal grape constituents and much of what we do viticulturally is aimed at trying to optimize the levels of these compounds in grapes in order to achieve a desired wine style. We'll turn our attention to how the environment and our cultural management affect grape ripening.

ENVIRONMENT AND VINEYARD MANAGEMENT EFFECTS ON GRAPE RIPENING

This topic is briefly summarized here and is covered to some extent in chapters 2 and 6, in which site selection and canopy management are addressed. Temperature is the primary environmental determinant of grape ripening, although radiation (sunlight) can affect ripening in both direct and indirect ways. Direct sunlight can elevate berry temperature through radiant heating. It has been found that the positive effect of light on fruit composition is dependent on the degree to which it elevates berry temperature. Total soluble solids, berry pH, titratable acidity, berry growth, anthocyanins, and total phenols are affected by degree of light exposure. However, prolonged berry exposure to direct sunlight, with its concomitant radiant heating, can be detrimental to berry color and phenolics in warm regions. Diffuse or "dappled" sunlight exposure of clusters, visualized as partial cluster exposure for part of the day, is more desirable in regions that experience both high temperatures and high incident levels of sunlight. High temperatures, for example, greater than 85°F, during the ripening period are thought to increase the loss of volatile flavor and aroma compounds or their precursors. In addition to temperature and sunlight, water availability plays a key role in grape quality. As described in chapter 9, page 169, water drives growth and development processes, including fruit ripening. Severe drought stress can delay ripening or lead to poor quality fruit at harvest. Mild stress, depending upon when it occurs, can lead to smaller berries, a more open canopy, greater fruit exposure and, with the appropriate balance, increased fruit quality. Preveraison water deficits have a greater role in determining fruit size and composition compared to post-veraison water deficits. An influx of water near harvest can dilute flavor and aroma components. This may delay harvest as the fruit is allowed to dehydrate some of this moisture, assuming that climatic conditions permit.

Vineyard management to foster optimal grape quality requires a holistic approach. For example, it does little good to have perfect canopy management but lose the crop to powdery mildew because of a faulty disease management program. Optimal grape quality is rarely achieved when vines are unbalanced with respect to vegetative vigor and crop level, as noted in chapter 6, page 124. A "balanced" vine has sufficient vegetation to ripen a crop to the degree of desired maturity on a sustainable (year after year) basis.

DECIDING WHEN TO HARVEST: VARIABILITY EFFECTS ON GRAPE RIPENING

Regardless of how optimal grape maturity is defined, winemakers desire uniformity of optimally ripe fruit across a vineyard block at harvest. Non-uniformity invariably leads to reductions in average grape quality and can be introduced at various levels, as follows:

- **Block:** block layout not correlated to soil differences or irrigation zones
- **Vine:** non-uniform crop level, clogged irrigation emitters, disease, irregular pruning, variable vine age, wrong variety
- **Cluster:** lack of vine balance, excessive stress, non-uniform exposure, disease or disorders such as bunch stem necrosis
- **Berry:** uneven cluster exposure, poor weather at bloom, tight clusters, crowded canopies, trace element deficiencies, disease or insect infestation

There are three sources of variability in ripening berries:

1. between vines within the vineyard
2. between clusters within the vine
3. between berries within the cluster

Variability between vines within a vineyard can be largely attributed to soil type and condition in any particular vineyard. For example, small vines on eroded hills may exhibit low yields and advanced fruit maturity relative to vines growing in deeper soils within the same vineyard. For example, mapping with Global Positioning Systems (GPS) and analysis with Geographic Information System (GIS) illustrated that in a vineyard that produced an *average* yield of 3.13 tons per acre, actual yield ranged between 0.45 and 8.50 tons per acre. This variation in yield would be expected to reduce the average grape quality.

Variability between clusters within a vine can be substantial due to cluster position on the shoot (that is, proximal clusters which flowered first and retained their developmental advantage through to harvest). Clusters on opposite sides of the vine can differ considerably.

Variability between berries within a cluster is due to berry size differences (occasionally caused by disorders that lead to abnormal ripening (such as coulure and "shot berries") that can affect yield levels. Degree of fruit set can affect within-bunch variation by affecting berry size, rate of ripening, color, and composition. Small berries have a greater surface to volume ratio and thus contain more of the volatile compounds that enhance wine quality on a per weight basis. Asynchronous berry ripening also occurs in "normal" clusters due to uneven exposure to sunlight.

Premier wine grape quality has been associated with a rapid ripening period, although it may be preferable that that period coincide with a cooler period of the season in warm to hot grape growing regions. A brief ripening period minimizes the risk of environmental and biological insults (for example, untimely rain, wildlife depredation, disease development). However, under conditions of rapid development of flavor and aroma, the optimal time to harvest may be missed.

Sampling for Fruit Maturity Assessment

Pre-harvest grape samples are collected starting at or before veraison to track the ripening process. Samples may be collected every 7 to 10 days and continue over the 3-week period leading up to harvest. Sampling frequency may increase to every 2 or 3 days in the 10 days prior to harvest. Samples should be representative of the entire unit or block that will be harvested. How uniform is the vineyard? How uniform are vineyard treatments and vines? The definition of a blocking unit will depend on the characteristics of the vineyard. These include variety, vine age, soil type, incidence of disease, microclimate of the vine canopy, and season (rainy and cool versus warm/hot and dry). Individual berry sampling is statistically more robust than is cluster or whole-vine sampling, but is more susceptible to sampler bias. One may wish to sample clusters close to harvest, but it is difficult to get large enough samples and thorough crushing becomes difficult without specialized equipment. It may be more important to be aware of the progression or trend of berry ripening (provided by berry sampling) than to obtain a sample that emulates the conditions at the crusher (cluster sampling). Some growers and wineries utilize customized record-keeping for tracking the progress of ripening during the season. This approach provides a reference for future years, especially when each particular vineyard or block is followed through the winemaking process (figure 16.6).

Generally, a 200-berry sample is adequate for a particular sampling unit or block. The unit might be as small as several rows, or it might be as large as several acres. All vines within the sampling unit should be at the same stage of berry ripeness. The sample should be collected before 10:00 AM to ensure that the berries are relatively cool. Randomly or blindly select one berry per cluster sequentially from the top, middle, and bottom, as well as front and back of each cluster as you walk a row. Select berries from clusters distributed throughout the canopy or fruit zone. Pick berries from both sides of vertical canopies by reaching through the foliage. This ensures inclusion of berries that are more or less likely to have been exposed to the sun. This method requires walking diagonally across rows or "four-wheeling" (utilization of an all-terrain vehicle) up and down rows of large blocks. Avoid sampling row-end vines and outside rows. Place the berries in a plastic bag and crush thoroughly—ideally to the same extent as would be done by the press at the winery. The juice should then be decanted and the gross particulates allowed to settle several hours before taking a refractometer reading of the clear supernatant

Figure 16.7 ▪ Evaluation of grape juice soluble solids with temperature-compensated, hand-held refractometer.

BERRY SAMPLE ANALYSIS SCORES

Variety: ______________________________ Year: ____________ Block: ______________________

Date	Sampled by	Flavor	Aroma	°Brix	TA	pH	Tannins	Juice color	Brown seeds (%)	Comments
Harvest										

Figure 16.6 ▪ Fruit scoring and harvest record form. *Adapted from Chateau Morrisette, Floyd, VA.*

(figure 16.7). The same juice can be used to test pH and titratable acidity if you are equipped for those measurements. Otherwise the berry samples should be refrigerated and promptly shared with the winery that will purchase your grapes. Each sample should be properly labeled with information about the sampled unit/variety, date and time of sample, and name of sample collector.

MEASUREMENT OF GRAPE MATURITY AND QUALITY

Although they don't tell the whole story, sugar (as soluble solids concentration), pH, and titratable acidity are useful to describe grape maturity and to inform the harvest decision. Despite its limitations, measurement of sugar concentration (°Brix) is a common industry practice. Sugar concentration is important due to its impact on fruit quality (sweet taste) and its role in alcohol formation during fermentation (1.7% sugar = 1% alcohol). Actual alcohol yield is 51 to 60 percent and depends on starting sugar concentration, nutrient level of juice, and other winery-controlled factors. Sugar content is a good indicator of quality in cool climates; however, several studies have shown no or poor relationships between sugar levels and accumulation of grape berry flavorants.

Juice pH is important to the fitness of grape must and final wine. The pH affects free SO_2 levels, wine balance (especially the perception of "sourness" or tartness), aroma, as well as the microbiological and physiochemical stability of the must and wine. Wine, particularly white wine, with a pH greater than 3.5 to 3.6 may be flat or unbalanced, while wine with a pH near 4.0 will not age as well, is more prone to microbial growth, and may require acid additions which add to winery costs. Measurement of pH from a field sample can be misleading, as it does not account for crushing and fermentation effects. Titratable acidity is determined by measuring the amount of a strong base (sodium hydroxide) required to neutralize juice or wine to a given endpoint pH. The most common acids present at harvest are malic and tartaric. In the US, titratable acidity is expressed as either percentage or grams of tartaric acid per liter. Sugar measurement can be used with either pH or titratable acidity to yield an index or ratio that can help in the determination of the sugar/

acid balance. However, the sugar: acid ratio is a relative one that is based on the dynamic change of sugar and acid during grape ripening. The ratio is variable across different varieties and environments, and should be carried out over several consecutive years to be meaningful.

Sugar levels are used by many wineries to assess grape maturity, and growers should include refractometer readings as part of their grape maturity monitoring (figure 16.7). Growers can also learn to use several sensorial and visual clues to assess grape maturity. Tannin and acidity influence wine quality (and aging potential) and should be assessed as grapes mature. In red grape varieties, particularly, skin tannins become more extractable as skin integrity breaks down during ripening. The color of the grape skin will change as tannins progressively ripen. With some training and experience, most growers can taste differences in tannin maturity that accompany grape ripening. Tannins, especially "unripe" tannins, cause friction in the mouth when they bind with salivary proteins and saliva viscosity is reduced. An applied field test is accomplished by picking a few grapes, separating the berry skin from the pulp and then chewing the skins to release the tannins. At this point, the tongue is pressed and moved against the roof of the mouth, or palate. If the tongue sticks or binds against the palate the tannins are judged "green," "gripping," or less mature. If the tongue moves freely and "slides," the tannins are judged smooth or "silky" and more mature. Seed tannins, which are often undesirable, become less extractable during ripening. Although brown seed color has been mentioned in some sources as indicative of grape maturity, uniform and complete brown seed color is seldom achieved in eastern US vineyards. This is especially so in late-maturing varieties. The browning of seed coats can be monitored but should not be used independently of other maturity indices. Prior to tasting selected sample berries, apply slight pressure to the berry and judge its softness. If the berry is deformed by this slight squeeze and does not regain its original shape, it is judged to be near ripeness. As the berries are removed from the berry stem (pedicel), observe whether the pulp detaches cleanly from the pedicel. If so, that berry is judged to be ripe. If some skin and pulp are still attached to the pedicel, and the berry is somewhat harder to remove, then the berry is considered unripe.

The progression of flavor, aroma, and acidity development that characterize a particular grape variety can be evaluated as well. The intensity of varietal flavor may be as important as its specific character. In some white varieties, that progression may be characterized as grassiness → tree fruit → tropical fruit.

It should be apparent by this point in the discussion that no one single parameter should be used to judge crop ripeness. Rather, that decision should be based on a synthesis of both the analytic (sugar, pH, and titratable acidity) and the sensorial (visual, taste, and aroma) data. Practical considerations beyond the grape itself must also be factored into the harvest decision. Labor is necessary for hand-harvesting and should be available when grape maturity dictates, not when it is convenient. Selective or sequential harvest of the same block over time is sometimes used, but this adds expense. Thus, uniformity of vineyard blocks is important. Weather must be considered, especially in regions where late-season rains and/or storms or hurricanes are likely. Experience in one's vineyard will help in predicting whether a late-season rain will quickly pass and bring cooler temperatures, or linger and damage grapes by causing berry splitting and encouraging late-season rots. The decision to pick before or after a weather front passes must be weighed against further ripening potential, as judged by recording and tracking of sugar and pH, and the possible effects of the impending weather. Wildlife, birds, and fruit-feeding insects can pose a serious threat to grape yields and quality. Netting, fencing, exclusion devices, and deterrents, as well as effective and legal pesticides, should be considered to prevent these pests from driving the harvest decision (see chapter 14, page 272). Growers should also be aware of vineyard conditions that might influence the harvest time, such as soil characteristics, vine age, foliage quality and quantity, growth status of primary and lateral shoots, as well as seasonal effects such as temperature and precipitation patterns.

If you are growing grapes for sale to a winery, you must work closely with winery personnel both before and during the growing season to monitor the progress of the vineyard and to closely track ripening in advance of a mutually-agreed upon harvest decision (see chapter 15, page 278). Growers should communicate with wineries frequently as harvest approaches. To optimize quality, the grapes should be picked, delivered, and processed with minimal delay once the harvest decision is made. With simple contracts, it's in the grower's best interest to bring fruit to the minimum desired stage of quantifiable ripeness, while it's in the winery's best interest to obtain the maximum quality of fruit. Knowing where these two goals converge is easy when one is both grape grower and winemaker. However, finding that balance can be a significant challenge, especially where long-term relationships do not exist between the winery and the grower. An example of a winery score card, which can also be used by the grower, is shown in figure 16.6 (page 291).

CONCLUSION

Harvest timing is the setting for one of the most dynamic and critical decisions faced by grape growers and winemakers. Understanding grape berry development and physiology is vital for the understanding of grape maturity and wine quality. There is no single, definitive, objective, and convenient method of measuring grape quality. Instead, a range of analytic and sensorial parameters are weighed to determine harvest time. Each region, vineyard, and season will require its own unique assessment indices of grape maturity and composition. The harvest decision is greatly facilitated when winemaker and grape grower work in harmony and maintain open, objective lines of communication.

References

Bisson, L. 2001. In search of optimal grape maturity. *Practical Winery and Vineyard.* July/August pp. 32–43. San Rafael, CA.

Boulton, R.B., Singleton, V.L., Bisson, L.F. and R.E. Kunkee. 1996. *Principles and Practices of Winemaking.* Chapman and Hall, New York.

Coombe, B.G. and P.G. Iland. 2004. Grape berry development and winegrape quality. In: *Viticulture Volume 1 - Resources,* 2nd edition. P.R. Dry and B. G. Coombe (eds), pp 210–248. Winetitles, Adelaide.

Dry, P.R. and B.G. Coombe. 2004. Viticulture. Volume 1—Resources, 2nd edition. *Winetitles.* Adelaide, South Australia. 255 p.

Kennedy, J. 2002. Understanding grape berry development. *Practical Winery and Vineyard.* July/August, 14–23.

Shaulis, N.J. and C. Pratt. 1965. Grapes; their growth and development. *Farm Research Reprint No. 401,* NY State Agricultural Experiment Station, Geneva, NY.

Further Reading

Gladstones, J.S. 1992. *Viticulture and the Environment,* Winetitles, Adelaide, Australia.

Jackson, D.I., P.B. Lombard. 1993. Environmental and management practices affecting grape composition and wine quality—a review. *American Journal of Enology and Viticulture* 44: 409–430.

APPENDICES CONTENTS

APPENDIX 1

Target Soil and Petiole Values and Fertilizer Guidelines for Grapevines

Appendix 1 provides recommended response measures that are based on a range of soil and plant tissue analysis test results. The recommendations are presented for each essential element. The reader finds the target values that most closely match his or her own sample test results. The "AND" column may provide qualifying factors. The "THEN" column provides the recommended response. Footnotes provide additional information to tailor the grower's action.

NITROGEN

	Target Values				
	Soil[f]	Bloom Petiole	70–100 DAB*	AND	THEN
IF <		0.80 %	0.60 %	Inadequate vine size and/or other visual indicators of low N status (pronounced symptoms of N deficiency).	Apply heavy N application as indicated for rates, below.
IF <		1.00 %	0.60 %	Inadequate vine size and/or other visual indicators of low N status.	Apply moderate N application as indicated for rates, below.
IF <		1.20 %	0.80 %	Adequate vine size and/or other visual indicators of N adequacy.	Apply maintenance N.
IF =		1.2-2.2 %	0.80–1.20 %	Inadequate vine size.	Investigate other source of poor vine growth.
IF =		1.2-2.2 %	0.80-1.20 %	Adequate vine size and/or other visual indicators of N adequacy.	Apply maintenance N.
IF =		1.2-2.2 %	0.80-1.20 %	Excessive vine size.	Curtail "routine" or maintenance N applications; allow some weed competition for N.
IF >		2.2%	1.2%	Inadequate vine size.	Investigate other potential sources of poor vine growth. Otherwise no action.

Notes:

[f] Soil test results are generally not used to make N fertilization recommendations. *DAB = Days after Bloom

Sources:

Ammonium nitrate (32% N): acidic soil reaction (1 lb N = 1.8 lbs $CaCO_3$; It takes 1.8 lbs of lime to neutralize the acidic reaction of 1 lb of ammonium nitrate fertilizer), may be difficult to obtain due to explosive nature.

Urea (46% N): economical N source, acidic soil reaction (1 lb N = 1.8 lbs $CaCO_3$), subject to ammonia volatilization if not incorporated

Calcium Nitrate (15% N): more expensive, basic soil reaction.

Manure and other organic matter: variable N concentration, long-term, slow-release N.

Rates:

Rate depends on N need, desired vine size and soil conditions. Wine grapes rarely require more than 30-50 pounds actual N/acre/year. Use lower rate for grafted vines and higher rate for non-grafted. Heavily cropped V. labrusca may require up to 50-80 lbs N/acre as maintenance rate. Use experience to modify rates. Split applications or multiple applications via irrigation system (fertigation) may improve efficiency and allow a lower net use of N. Maintenance of elevated tissue levels of N can reduce the incidence of oxidant stipple and late-season bunch stem necrosis (BSN) on varieties susceptible to those disorders.

Heavy: 25 to 50 pounds actual N per acre between bud break and bloom, followed by an additional 25 pounds actual N between bloom and two weeks post-bloom. Apply 20 pounds of actual N per acre immediately after fruit harvest IF confident that rain or irrigation will infiltrate applied N at least 3 weeks prior to natural leaf senescence or frost.

Moderate: 20 to 30 pounds actual N per acre between bud break and bloom, followed by an additional 20 pounds actual N between bloom and two weeks post-bloom.

Light/maintenance: 20 to 30 pounds actual N per acre between bud break and bloom.

PHOSPHORUS					
	Target Values				
	Soil[f]	Bloom Petiole	70–100 DAB	AND	THEN
IF <	20 ppm	0.17 %	0.14 %	pH adequate and/or leaf reddening observed on red-fruited varieties.	Apply P_2O_5 fertilizer as indicated in notes.
IF <	20 ppm	0.17 %	0.14 %	Acid soil and/or leaf reddening observed on red-fruited varieties.	Adjust soil pH for long-term correction and supplement with P_2O_5 fertilizer as indicated in notes for short term correction.
IF =	20-50 ppm	0.17–0.30 %	0.14–0.30 %	No visual disorders.	No action necessary; repeat sampling in two years.
IF >	> 50 ppm	> 0.30%	> 0.30%		Monitor for micronutrient metal deficiencies (such as zinc), particularly at soil pH > 6.8 .

Notes:

[f] Multiply parts per million (PPM) by 2 to obtain equivalent pounds/acre.

Target soil values are typical for Mehlich-3 and Bray extraction methods. There are 4 different P extractions methods (Mehlich 1 & 3, Bray, and Morgan). Mehlich-3 and Bray are similar in value; Mehlich is somewhat lower for low pH soils. Morgan is much lower than all three. Virginia Tech currently uses Mehlich-1 but may change to 3, Cornell and Northeast universities use Morgan, Penn State and other mid-Atlantic universities use Mehlich-3. If interpreting your own soil tests, it is important to know the P extraction method used by your analytical lab.

Vineyard P disorders are often associated with low soil pH. In established vineyards, raise soil pH with limestone applications. Supplement with P fertilizer until desired soil pH and phosphorus availability is achieved.

Sources:

Monoammonium phosphate (MAP) (48% P_2O_5), also contains 11% N, acidic soil reaction (1 lb N = 3.5 lbs $CaCO_3$).

Diammonium phosphate (DAP) (46% P_2O_5), also contains 18% N, acidic soil reaction (1 lb N = 4.1 lbs $CaCO_3$).

Triple superphosphate (approximately 46% P_2O_5); use only if additional N is undesirable.

10-10-10, 10-20-10, or other blends, but only if N and K are also required. Materials may be custom-blended to meet particular needs.

Rates:

Pre-plant:

Soil test results Low or Very Low (< 10 ppm), apply 100 lbs P_2O_5/acre.

Soil test results Mod (< 15 ppm), apply 50 lbs P_2O_5/acre.

If decision is based solely on petiole test, start with an under-trellis, banded application of 50 lbs P_2O_5/acre as soon as convenient and resample petioles at or following bloom-time. Use higher P rate if petiole test used to confirm a low soil analysis, and/or if leaf reddening observed with red-fruited varieties.

POTASSIUM					
	Target Values				
	Soil	Bloom Petiole	70–100 DAB	AND	THEN
IF <	75 ppm	1.00 %	0.80 %	High soil Mg and K deficiency symptoms obvious.	Apply K fertilizer – Heavy rate.
IF <		1.50 %	1.20 %	Excessively dry or excessively wet.	Apply K fertilizer – maintenance rate or irrigate to increase soil K mobility.
IF =	75–100 ppm	1.50-2.50 %	1.20-2.00 %	Large crop.	Apply maintenance rate of K fertilizer.
IF =	75–100 ppm	1.5 -2.50 %	1.20- 2.00 %	Normal crop.	No action necessary; repeat sampling in two years.
IF >	100 ppm	2.50 %	2.00 %		Monitor for chronic Mg deficiency.

Notes:

Muriate of Potash (KCl) typically applied in the fall to allow K movement into the root zone and chloride leaching out of the root zone. Caution must be used on soil with a salinity problem (not common in the Northeast) or on shallow or poorly drained soils where the chloride cannot leach from the root zone.

Potassium is typically banded; however, broadcasting in vineyards with spreading root systems and no-till row-middle management is an option.

Factors to watch:

1. K-Mg competition, especially with changes in soil pH.

2. K demand, especially in high cropping systems.

3. K soil mobility, it decreases with decreasing soil moisture.

Sources:

Muriate of Potash (52% K, 62% K_2O), most common.

Sulfate of Potash (44% K, 53% K_2O), use if chloride toxicity is a potential problem.

Sulpomag (22% K_2O, 11% Mg), has both K and Mg, more expensive, considered organic.

Rates:

Preplant:

Desired soil K is 75–100 ppm. Soil should be amended with potash (K_2O) prior to planting to bring within this range, as follows:

Example:

(100 ppm K desired) - (50 ppm K from sample test) = 50 ppm K needed x 2.4 = 120 lbs K_2O/acre needed

(120 lbs K_2O/acre needed) x 1.67 (60% K_2O in KCl) = 200 lbs KCl/acre applied - or -

(120 lbs K_2O/acre needed) x 2 (50% K_2O in K_2SO_4) = 240 lbs K_2SO_4/acre applied

Existing vineyard (rates based on soil test, petiole test, visual observation, or combinations thereof)

Heavy: Apply up to 600 pounds/acre K_2O (see text for graduated response to deficiency symptoms)

Moderate: Apply up to 400 pounds/acre K_2O

Light/maintenance: Apply up to 150 pounds/acre K_2O

CALCIUM					
	Target Values				
	Soil*	Bloom Petiole	70–100 DAB	AND	THEN
IF <		1.00 %	1.00 %	Acid soil	Increase soil pH with limestone; Use calcitic limestone if Mg is high; use dolomitic limestone if Mg is also low.
IF <		1.00 %	1.00 %	Alkaline soil	Rare—investigate soil conditions. Calcium can be increased without increasing pH with gypsum.
IF =	500–2000 ppm	1.00–2.5 %	1.3–2.5 %		No action necessary; repeat sampling in two years.
IF >		3.00 %	3.00 %		Very rare in mid-Atlantic. Could reflect poor sampling or elevated pH. Re-evaluate soil pH and reduce with sulfur if > 7.5; monitor plant tissue levels of other nutrients to avoid potential deficiencies.

Notes:

* Target soil calcium values are not listed because they will change based on soil CEC. For example, soil Ca levels of < 500 ppm may be adequate in low CEC soils as long as soil pH is > 6.0.

Low calcium availability typically associated with low soil pH. Adjust with limestone.

Sources:

Limestone (variable % Ca).

Gypsum (calcium sulfate, 22% Ca), not used to adjust soil pH.

Rates:

Acid soil: Adjust soil pH to 6.5 (see lime recommendation table in text). Because of interactions with potassium, no more than 2 tons limestone/acre/year is recommended in established vineyards.

MAGNESIUM					
	Target Values				
	Soil	Bloom Petiole	70-100 DAB	AND	THEN
IF <	100 ppm	0.30 %	0.35 %	Acid soil	Adjust soil pH with dolomitic limestone and add MgO to soil as indicated in notes.
IF <	100 ppm	0.30 %	0.35 %	High K or wet season	High soil moisture (high K mobility) can cause transient Mg deficiency. Monitor and apply maintenance rate of Mg fertilizer if necessary.
IF <		0.30 %	0.35 %	Neutral or Alkaline soil	Deficiency rare in high pH soils. Adjust with Epsom salts if needed
IF =	150-250 ppm*	0.3-0.5 %	0.35-0.75 %		No action necessary; repeat sampling in two years.
IF >		0.50 %	0.75 %	Dry year	Low K availability in dry soil may inflate Mg readings — monitor.
IF >		0.50 %	0.75 %	Normal year — neutral soil	Monitor for K deficiency.

Notes:

*Influenced by soil CEC.

Low magnesium availability typically associated with low soil pH. Can be aggravated in acid soils with high K application. Adjust with dolomitic limestone in low pH vineyards. Use Epsom salts in neutral and high pH soils.

Excessive soil Mg (either natural or fertilizer applied) may cause K deficiency and vine size reduction. Monitor petiole K and Mg.

Sources:

Dolomitic limestone (variable % Mg), most common

Epsom salts (magnesium sulfate, 10% Mg)

Sulpomag (22% K_2O, 11% Mg), has both K and Mg, more expensive

Rates:

Acidic soil: Adjust soil pH to 6.5 (see lime recommendation table in text). Because of interactions with potassium, no more than 2 tons limestone/acre/year is recommended in established vineyards.

If pH acceptable: adjust Mg with $MgSO_4$

Example:

150 ppm desired - 50ppm on soil test = 100ppm desired x 2 = 200 pounds Mg/acre

200 pounds/acre x 10 (10% Mg in $MgSO_4$) = 2000 pounds/acre $MgSO_4$ (Epsom salts)

Foliar applications of Epsom salts (5-10 pounds/acre in 100 gallons water) can be used for short-term correction (see text).

BORON

	Target Values				
	Soil	Bloom Petiole	70–100 DAB	AND	THEN
IF <	0.3 ppm	20 ppm	20 ppm		Apply boron as recommended in notes.
IF =	0.3–2.0 ppm	25–50 ppm	25–50 ppm		No action necessary; repeat sampling in two years.
IF >	2.0 ppm	50 ppm	50 ppm		Monitor for Boron toxicity.

Sources:

Solubor (20% B), most common; can be applied to soil or to foliage
Borax (11% B)
Borate-46 (14% B)
Borate-65 (20% B)

Rates:

Soil application rates of 1 pound B/acre in medium to coarse textured soils to 2 pounds B/acre on heavy clay soils are recommended. Blending with other fertilizers (such as N) for broadcast application is suitable. Soluble B products can also be applied to the soil with an herbicide sprayer. Calculate sprayer rate based on actual area sprayed, as opposed to total vineyard acres sprayed. For example, assume we want 1 pound B per planted acre, or 5 pounds Solubor (20% B) per planted acre. If only spraying ⅓ of a planted acre (for example, a 36-inch herbicide band of 9 foot rows), use 1.7 pounds Solubor per planted acre.

Foliar applications of 0.2 pounds B/acre (1 lb. Solubor) are recommended and no more than 0.5 pound B/acre (2.5 lb. solubor) in one application. Spring foliar sprays are timed at 6–10 inch shoot growth and 14 days later. In California, fall (immediate post-harvest) foliar sprays have been more effective than spring foliar application in eliminating cluster and berry disorder.

To reduce the risk of foliar burn, do not apply boron sprays at less than 14-day intervals or tank-mixed with water-soluble packages, oil, or surfactants.

IRON

	Target Values				
	Soil	Bloom Petiole	70–100 DAB	AND	THEN
IF <	10 ppm	10 ppm	10 ppm	Alkaline soil and/or visual symptoms of Fe deficiency.	Lower soil pH to improve long-term iron availability and apply Fe foliar spray to correct current foliar deficiencies.
IF <	10 ppm	10 ppm	10 ppm	Waterlogged soil and/or visual symptoms of Fe deficiency.	Improve soil drainage to improve long-term iron availability and apply Fe foliar spray to correct current foliar deficiencies.
IF =	20 ppm*	30–100 ppm	30–100 ppm		No action necessary; repeat sampling in two years.
IF >	50 ppm			Acid soil	High soil Fe can limit P availability. Monitor and increase soil pH if needed.

Notes:

*Soil values between 11–19 ppm should be monitored for possible developing deficiency.

Iron deficiency is often associated with calcareous soils (high soil pH), low soil oxygen (waterlogging), and variety (*V. labrusca* more susceptible). Tissue standards for interpreting iron levels are not well defined. In the absence of visual deficiency symptoms, no corrective action is indicated.

Common deficiency treatments:

Lower soil pH by trenching in soil sulfur or using acidifying nitrogen fertilizers.
Improve soil drainage
Apply foliar iron sprays (only effective for existing foliage)
Apply iron chelates (expensive and short-lived)

MANGANESE

	Target Values			AND	THEN
	Soil	Bloom Petiole	70–100 DAB		
IF <	10 ppm	25 ppm	25 ppm	Manganese deficiency symptoms apparent.	Manganese deficiency is rare except on high pH soils. Apply Mn foliar spray if needed.
IF =	10 ppm	25–1000[f] ppm	25–1,000 ppm		No action necessary; repeat sampling in two years.
IF >	50 ppm			Acid soil	Manganese toxicity a potential problem at low soil pH. Adjust soil pH with limestone if needed.

Notes:

[f] Very high Mn levels may reflect contamination of tissue by Mn-containing fungicides.

High Mn in tissue can also be a result of wet soils.

Sources:

Manganese sulfate (32% Mn), Foliar spray.

Some fungicides (e.g., mancozeb) are also an effective source of Mn.

Rates:

Foliar spray: 2–3 pounds manganese sulfate/100 gal water.

COPPER

	Target Values			AND	THEN
	Soil	Bloom Petiole	70–100 DAB		
IF <	0.2 ppm	5 ppm	5 ppm		Deficiency rare, but possible on sandy soils or soils with very high organic matter content. Apply copper as indicated in notes.
IF =	0.5 ppm	5–15 ppm	5–15 ppm		No action necessary; repeat sampling in two years.
IF >		20 ppm	20 ppm		Potential toxicity reported when copper sprays repeatedly used, leading to copper accumulation in low soil pH vineyards. Symptoms similar to lime-induced chlorosis (iron deficiency).

Notes:

Apply foliar copper as Bordeaux mixture or other copper fungicide if grape variety is tolerant of foliar copper application.

If pre-plant situation, broadcast elemental copper at up to 5.0 pounds/acre.

Very high Cu levels may reflect contamination of tissue by Cu-containing fungicides.

ZINC

	Target Values				
	Soil	Bloom Petiole	70–100 DAB	AND	THEN
IF <	2 ppm	25 ppm	25 ppm		Apply Zinc to soil (pre-plant) or foliage as indicated in notes.
IF =	2 ppm	25 ppm	25 ppm		No action necessary; repeat sampling in two years.
IF >					(We lack experience with Zn toxicity.)

Notes:

If pre-plant situation: broadcast up to 5.0 pounds/acre of elemental zinc.

Zinc sulfate should be applied with equal amounts of hydrated spray lime (1–4 pounds/100 gal) at the 3- to 5-inch shoot stage. Repeat in 14 days as needed.

Zn-chelate may be used at manufacturer's recommended rates.

MOLYBDENUM

	Target Values				
	Soil	Bloom Petiole	70–100 DAB	AND	THEN
IF <	*	0.5 ppm	0.5 ppm	Vines chronically exhibit symptoms of millerandage ("hens and chicks" uneven berry size).	Apply molybdenum to foliage as indicated in notes.
IF >	*	0.5 ppm	0.5 ppm	No visual symptoms of poor growth or poor fruit set.	(We lack experience with Mo toxicity.)

Notes:

We lack evidence of molybdenum deficiency symptoms in eastern North America, although it has been reported in Australia, particularly with Merlot, where poor vine growth and poor fruit set are peculiar features of the deficiency. The 0.5 ppm action threshold is therefore a provisional threshold for eastern North America.

Apply sodium molybdate to canopy at 0.50 pound per acre in at least 100 gallons of water per acre. Apply at the 3- to 5-inch shoot stage and repeat in 14 days, or just prior to bloom.

* We lack specific knowledge of appropriate soil concentrations of molybdenum.

APPENDIX 2
Determining Crop Coefficient, K_c

A crop coefficient (K_c) is used to adjust the potential evapotranspiration rate (ET_o) to derive a vineyard-specific ET rate (equation 9.1, page 178). To estimate the amount of water required for peak irrigation, the crop coefficient (K_c) should be estimated during July and August at its peak value. As stated in the text, the K_c is a function of both the canopy development of the vine and the presence or absence of weeds and row middle cover crops. Larry Williams, with the University of California, demonstrated that grapevine K_c values for a range of different training systems correlate well to the vineyard floor area shaded by the canopy at solar noon (roughly 12:30 to 1:30 pm in midsummer) with a linear trend and a slope of 0.017. Thus, the percent shaded area (PSA) multiplied by 0.017 gives a reasonable approximation of grapevine K_c. The PSA can be estimated at solar noon by one of two methods: 1) measure the average width of shaded area (figure A2.1) beneath the vine row; or 2) estimate the canopy shaded area from a 4-foot x 4-foot board with 6-inch gridlines (figure A2.2) placed in the shade beneath the vine row. Row and vine spacing must be known to determine the PSA. An example calculation by method 1 follows:

A = Row width = 9 feet
B = Vine spacing within row = 6 feet
C = Area per vine = A x B = 9 feet x 6 feet = 54 square feet
D = Average width of measured shaded area between two vines = 32.5 inches or about 2.7 feet (Use many measurements to get a good average width)
E = Shaded area per vine = B x D = 6 feet x 2.7 feet = 16.2 square feet
PSA = E / C = 16.2 square feet / 54 square feet = 0.30 or 30%
K_c = PSA x 0.017 = 0.51 (no ground cover)

The presence of row middle cover crops will increase the K_c by approximately 20%. Therefore, the Kc can be multiplied by 1.2 to adjust for the presence of cover crops:

K_c = PSA x 0.017 x 1.2 = 0.61 (row middle cover crop present)

It must be understood that crop water use is a function of canopy size and sunlight exposure so it can vary from vineyard to vineyard. The canopy will change as the vines grow over a season, so calculations should be done periodically if water use is going to be estimated for differing vineyard conditions. The contribution of weeds under the trellis and row middle cover crops will also affect the calculation of K_c and, consequently, the estimation of ET (equation 9.1) and the estimation of vineyard irrigation needs (equation 9.2, page 179). We will assume that most of the irrigation water pumped into the vineyard with drip irrigation will benefit the grapevines and that relatively little of it will be used by row middle cover crops that occupy 40 to 50% of the vineyard floor area.

Ed Hellman, Texas Agri Life Extension Service

Figure A2.1 ▪ Measuring the average width of canopy shade with a tape or yardstick.

Ed Hellman, Texas Agri Life Extension Service

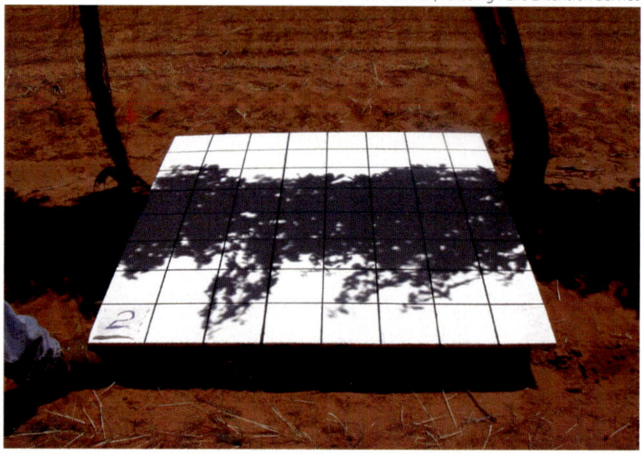

Figure A2.2 ▪ Estimating canopy shade with use of a 4-foot x 4-foot board with 6-inch gridlines.

APPENDIX 3
Irrigation Math and Physics

This appendix provides an opportunity to address some irrigation calculations and to discuss numerical values and conversion units. Example problems are solved to illustrate computations.

Area

An acre of land is made up of 43,560 square feet. This is an area of a square 209 feet x 209 feet. If a vineyard has a row width of 9 feet, then an acre of land would allow 43,560 / 9 = 4,840 feet of row length. The length of drip tape or irrigation tubing required per acre for a vineyard with 9-foot rows is therefore 4,840 feet.

Volume

An acre-inch of water is equal to 27,154 gallons. This is the volume of water one inch deep on one acre of land. One cubic foot of water is equal to 7.48 gallons of water and weighs 62.4 pounds. Drip irrigation is about 90% efficient; therefore, 27,154/0.90 = 30,170 gallons would be needed per acre of land to supply an acre-inch of water.

Pressure

One cubic foot of water weighs 62.4 pounds. If this is a cube with 12 inches on a side, then 62.4 pounds rest on 1 square foot of ground. This is a pressure of 62.4 pounds per square foot or (with 12 x 12 = 144 square inches of surface) 62.4/144 or 0.433 pounds per square inch (psi). A 1-foot column of water resting on a 1-square inch area weighs 0.433 pounds and, thus, exerts 0.433 psi. If a 1-foot column of water equals 0.433 psi, a 100-foot column of water equals 43.3 psi. Pressure is created by gravity's effect on a vertical column of water or by a mechanical pump. Thus, a water tower creates static pressure equal to the height of a water column and a pump creates pressure by mechanical force.

A water column 1-foot high divided by 0.433 psi equals 2.31 feet of water per psi.

2.31 feet of water head = 1 psi
2.31 feet of elevation = 1 psi

Elevation change in a pipe line causes a pressure change as follows: 1 foot / 0.433 psi = 2.31 feet / psi. Thus, for each 2.31 feet of elevation change the water pressure in a pipeline changes 1 psi. The pressure in a pipeline going up a hill decreases 1 psi for each 2.31 feet of ascent. Conversely, the pressure increases 1 psi for each 2.31 feet of vertical descent.

Friction Loss

Friction loss is a pressure loss or an energy loss that occurs when water flows through the confined space of a pipeline. Although a pipe feels smooth inside, when water flows through it there is some roughness that causes turbulence and drag of the water against the pipe walls. The faster the water flows, the greater the turbulence and energy loss. Energy is lost forcing the water through the pipe. This loss is called "friction loss." For pipes, it is expressed as "psi per 100 ft" of length or "feet of water".

Friction Loss Charts are available for different types and sizes of polyethylene (PE) pipe (see table A3.1, page 306). As a rule, the water velocity in pipes should be held to a maximum of 5 feet per second (fps) to avoid "water hammering", a condition of great force if a valve is suddenly closed on a fast moving mass of water. In table A3.1, one can see the friction loss in each of four pipe sizes for increasing flow rates. To select a PE pipe size for carrying 30 gallons per minute (gpm), go down the left-hand column to 30 gpm. To the right, there are three pipe sizes with friction loss values shown. First, note that the 1-inch pipe is not available for that flow rate. The 1.25-inch pipe has a flow rate below the underlined flow rate; the footnote indicates the water velocity is too high. The 1.5-inch pipe has a friction loss of 2.59 psi per 100 feet of pipe. The 2-inch pipe has a friction loss of 0.77 psi per 100 feet of pipe. If an application requires a relatively short length of pipe then the 1.5-inch will work. Note, the 1.5-inch pipe is near the 5 fps velocity limit and the pressure loss (friction) is high. There is little room for increasing the flow in the future if this is the main line and, for a low pressure system, the pressure loss will be relatively large if the pipeline is very long. For distances greater than 300 feet, the 2-inch diameter pipe will preserve water pressure, and would be a better choice. The system pressure was not given in this example; however, as a rule of thumb, the pressure loss in a main should not exceed 1 psi per 100 feet of length. This is an economic guideline for sizing all pipes except the laterals.

Table A3.1 ▪ Friction Loss Characteristics for Polyethylene (PE) SDR-Pressure Rated Pipe

	Nominal pipe size, in inches				
	0.75	1	1.25	1.5	2.0
Flow, gpm	Friction loss, psi per 100 ft of pipe[a]				
3	0.95	0.29	0.08	0.04	0.01
5	2.44	0.76	0.20	0.09	0.03
10	8.82	2.73	0.72	0.34	0.10
14		5.08	1.34	0.63	0.19
20		9.85	2.59	1.22	0.36
24			3.63	1.72	0.51
30			5.49	2.59	0.77
35			7.31	3.45	1.02
40				4.42	1.31
45					1.63
50					1.98
55					2.36
60					2.78

[a] Water velocity should not exceed 5 feet per second (fps). Underlined friction loss value for each pipe size indicates the maximum flow allowed to keep water velocity below 5 fps, except for 0.75 inch pipe where flow should not exceed 8 gallons per minute (gpm).

In general, the faster the water moves in a pipe, the greater the frictional loss of pressure. A look at table 9.8 will illustrate this when one follows the column downward for any size pipe. Non-uniform pressure in submains or laterals will result in decreased uniformity in the application of water on the crop. The pressure loss in submains should be small so that the lateral pressures are nearly uniform.

As indicated above, the allowed pressure loss in a mainline pipe is equal to or less than 1 psi per 100 feet. There is an economic factor involved in selecting a larger pipe size or accepting more friction loss. Drip emitters operate at 8 to 12 psi, usually about 10 psi, so submains or manifolds serving lateral lines should have little friction loss to permit lateral lines to operate near their proper pressure. Elevation changes can be a large factor on rolling terrain so pressure regulators are used to deliver the proper water pressure into lateral lines placed on the contour (same elevation). A friction loss chart is used to select pipe based on length and the pressure loss that will occur for any given water flow rate.

System designers must take into account all these factors in designing the system.

Pump Sizing

A pump is selected primarily on the basis of two factors: (a) the flow rate of water, gpm, and (b) the total pressure head. The total pressure head is the sum of suction lift, pump lift to field, friction loss, and nozzle or orifice operating pressure. This pressure head is often given in terms of feet of water head or in psi. This point is made because many growers have asked "What horsepower pump do I need?" A dealer will look at additional factors but flow rate and pressure head will be the key deciding factors.

The pump's power unit is sized by taking the pump pressure head divided by the efficiency of the power unit and transmission to get the horsepower of the engine or motor needed to drive the pump.

With drip irrigation, a smaller pump means that single phase 110- or 220-volt pumps of less than 7.5 horsepower can be used in many systems. This saves on installing power to the site.

APPENDIX 4
Identification Keys for Insects

Keys are presented for the identification of the main grape feeding adult and larval insects and their signs and symptoms. This is an "artificial key," meaning that actual systematic relationships and their taxonomic characters are not used, so that keys may be used by readers with little entomological training. The following terms are used to simplify the keys:

- **pronotum** — the plate on a beetle's body immediately behind the head.
- **proleg** — the fleshy legs on the abdomen of a caterpillar, in addition to the 6 true, jointed legs.
- **cornicle** — one of a pair of tubes projecting from the posterior end of an aphid.

To use the key to insect and mite pests, begin at the first couplet. If the insect in question has wings, proceed to the second couplet, as directed in the right-hand column. If the insect does not have wings, proceed to the twenty-third couplet. Continue on to succeeding couplets, as directed, until the identity of the insect is reached.

A word of caution is necessary in the use of the keys: Many insects that are not included in this guide may feed on grape. Descriptions in the text should be used to confirm identifications. Occasionally the key to damage resorts to insect characteristics; this is necessary because of the similarity between some types of damage.

Key to Insect and Mite Pests of Grape

Couplet Number		Proceed to Couplet
1a	Wings present (may be covered with hardened wing covers)	2
1b	Wings absent	23
2a	Beetle	3
2b	Butterfly or moth	13
2c	Other	21
3a	Head prolonged into a snout	4
3b	Head not so prolonged	6
4a	A weevil about 3 mm long, body elongate-oval in dorsal view	5
4b	A weevil about 3 mm long, brown, almost round in dorsal view	Grape curculio (page 246)
5a	Weevil a reddish-brown color	Grape cane gallmaker (page 256)
5b	Weevil black in color	Grape cane girdler (page 256)
6a	Clubbed antennae	7
6b	Antennae not clubbed	10
7a	Green beetle with copper-colored wings	Japanese beetle (page 247)
7b	Otherwise	8
8a	Tan beetle with eight black spots	Spotted pelidnota (page 254)
8b	Otherwise	9
9a	Yellow-brown beetle, 8–10 mm long, with spiny legs	Rose chafer (page 246)
9b	Otherwise	10
10a	Cylindrical beetle, about 2 mm long, with head hidden in dorsal view	Ambrosia beetle (page 256)
10b	Cylindrical beetle, 6–9 mm long	Apple twig borer (page 257)
11a	Metallic blue-black beetle, very active, jumps when disturbed	Grape flea beetle (page 244)
11b	Brown beetle	12
12a	Pronotum longer than wide	Grape rootworm (page 260)
12b	Pronotum wider than long	Grape colaspis (page 256)

Couplet Number		Proceed to Couplet
13a	Forewings brown	14
13b	Wings variously patterned	15
14a	Hindwings clear; moth wasp-like in appearance	Grape root borer (page 258)
14b	Hindwings brown	Grapeleaf skeletonizer (page 255)
15a	White or yellow spots on wings	16
15b	Wings various shades of brown or grey	17
16a	Two white spots on forewing, 1 spot on hind wing	Grape leaffolder (page 255)
16b	Two yellow spots on forewing, 2 white spots on hind wing	Eightspotted forester (page 254)
17a	Each forewing and hind wing divided into narrow plumes	Grape plume moth (page 253)
17b	Wings not so divided	18
18a	A rust-colored stripe on forewing, forming a "V" when at rest	Redbanded leafroller (page 247)
18b	Otherwise	19
19a	Mottled brown with a blue-grey patch on forewing, 1 cm long	Grape berry moth (page 242)
19b	Mottled grey, brown, or green, larger in size	20
20a	Hind wings about ⅔ length of forewings, of normal width. Mottled grey or brown	Climbing cutworms (page 245)
20b	Hind wings about ½ length of forewings, narrowed. Mottled green or brown. Heavy bodied	Sphinx moths (page 254)
21a	Wings clear	22
21b	Wings patterned with pale yellow, black, and red	45
22a	Body small, black, and globose	Grapevine aphid (page 255)
22b	Body about 3.5 cm long; black with red-orange markings	Periodical cicada (page 258)
23a	Beetle or moth larva (grub or caterpillar); long, slender body with six jointed legs near front	24
23b	Otherwise	35
24a	Abdomen with prolegs (fleshy, unjointed legs)	25
24b	Abdomen lacking prolegs	34
25a	Found in fruit or root	26
25b	Found on foliage	27
26a	Found in fruit	Grape berry moth (page 242)
26b	Found in root	Grape root borer (page 258)
27a	With fine transverse black and white stripes, orange at each end	Eightspotted forester (page 254)
27b	Otherwise	28
28a	Body light green with dense white hairs; generally between two tied leaves	Grape plume moth (page 253)
28b	Otherwise	29
29a	Caterpillar entirely green (including head), usually tied in leaves or webbed in cluster	Redbanded leafroller (page 247)
29b	Otherwise	30
30a	Grey-green caterpillar with black markings near head, generally tied or rolled in a leaf	Grape leaffolder (page 255)
30b	Otherwise	31
31a	Yellow larvae, each segment with black tubercles, generally feeds in a group on a leaf	Grapeleaf skeletonizer (page 255)
31b	Otherwise	32
32a	With a horn-like "tail" protruding from abdomen	Sphinx moth (page 254)
32b	Otherwise	33

Couplet Number		Proceed to Couplet
33a	Abdomen with fleshy prolegs limited to last two segments	Grapevine looper (page 254)
33b	Abdomen with prolegs on additional segments; generally mottled grey or brown; hides in ground cover by day, feeds at night	Climbing cutworm (page 245)
34a	Body with dark tubercles on segments, feeds on foliage	Grape flea beetle (page 244)
34b	Feeds on roots	Grape rootworm (page 260)
35a	Long antennae, each with a white band	Banded grape bug (page 245)
35b	Otherwise	36
36a	Tiny (less than 1 mm; magnification usually required), oval in shape	37
36b	Otherwise	40
37a	Bright yellow, six legs	38
37b	Red or straw-colored, eight legs	39
38a	Near galls on foliage	Grape phylloxera (page 250)
38b	On trunk, canes; in June only	Grape scale (page 257)
39a	Brick red with small white tubercles	European red mite (page 249)
39b	Pale yellow with 2 large dark spots	Two-spotted spider mite (page 250)
40a	Elongated triangular in shape, widest at front	44
40b	Otherwise	41
41a	Relatively globose in shape, dark brown or black, with a pair of cornicles (horn-like projections from hind end)	Grapevine aphid (page 255)
41b	Otherwise	42
42a	Oval in shape; white, powdery in appearance; with waxy filaments from body margins	Grape mealybug (page 253)
42b	Otherwise	43
43a	Softshell scale, about half as tall as long, brown	European fruit lecanium (page 257)
43b	Hardshell scale, flattened, light grey	Grape scale (page 257)
44a	Light green with 6 white spots on front edge of pronotum	Potato leafhopper (page 252)
44b	Otherwise	Grape leafhopper (page 251)
45a	A beetle about 6 mm long, light to dark pink, orange or red, with black spots, and a black M-shaped mark against a white pronotum	Multicolored Asian lady beetle (page 247)
45b	Elongate triangular in shape, about 3 mm long	Grape leafhopper (page 251)

Key to Damage by Grape Pests

1a	Foliage affected	2
1b	Fruit or young clusters affected	14
1c	Swelling primary buds affected	21
1d	Shoots, canes, or trunk affected	22
1e	Roots affected	28
2a	Foliage rolled or folded, tied with silk	3
2b	Foliage cut or fed upon	5
2c	Foliage discolored	11
2d	Foliage with honeydew deposits	12
2e	Foliage with galls	13
3a	Leaf rolled into tube-like structure	Grape leaffolder (page 255)
3b	Leaf folded upon itself or two leaves tied together early in season	Grape plume moth (page 253)
3c	As (3b), but may be seen also in later season	4

Couplet Number		Proceed to Couplet
4a	Green caterpillar with green head	Redbanded leafroller (page 247)
4b	Grey green caterpillar with dark markings near head	Grape leaffolder (page 255)
5a	Leaf with a semicircular flap cut and folded	Grape berry moth (page 242)
5b	Foliage with feeding damage	6
6a	Leaves skeletonized (tissue between veins eaten)	7
6b	Leaves with more general feeding	8
6c	Leaves with chain-like feeding marks	10
6d	Leaves with small, elongate oval marks	Grape colaspis (page 256)
7a	Green and copper colored beetles from late June through August (abundant)	Japanese beetle (page 247)
7b	Groups of caterpillars feeding on a leaf	Grapeleaf skeletonizer (page 255)
7c	Brown beetle with black spots (rare)	Spotted pelidnota (page 254)
7d	Looper caterpillar feeding mainly near leaf edges	Grapevine looper (page 254)
8a	Green and copper-colored beetles, late June through August (abundant)	Japanese beetle (page 247)
8b	Spring, early summer	9
8c	Hornworm larvae, most abundant in August	Sphinx moth (page 254)
9a	Caterpillar with many fine black and white transverse stripes, with orange at each end.	Eightspotted forester (page 254)
9b	Usually with tied leaves, hairy, light green caterpillar	Grape plume moth (page 253)
10a	Brown beetle larva	Grape flea beetle (page 244)
10b	Adult beetle	Grape rootworm (page 260)
11a	Leaves with coarse, light-colored stippling, especially near larger veins	Grape leafhopper (page 251)
11b	Leaves with very fine stippling, or bronzing in more severe cases	European red mite (page 249)
		Two-spotted spider mite (page 250)
11c	Leaf edges curled, yellowed	Potato leafhopper (page 252)
12a	Growing tips or clusters with dark aphids	Grapevine aphid (page 255)
12b	Leaves or shoots with brown softshell scale	European fruit lecanium (page 257)
13a	Galls on leaf undersides, often many per leaf; galls bumpy; a less prominent swelling on upper surface	Grape phylloxera (page 250)
13b	Gall blisterlike on upper surface of leaf, with a mass of white, hair-like growth on underside	Grape erineum mite (page 255)
13c	Galls blister-like or fleshy; often red, on petioles or leaves, various forms	Probably gall midges (page 246)
14a	Immature fruit attacked	15
14b	Ripe or almost ripe fruit attacked	19
15a	Fruit with a small puncture on side	16
15b	Fruit with exterior feeding	17
15c	Fruit with sticky honeydew	18
15d	Prebloom rachis with damaged or lost fruitlets	Banded grape bug (page 245)
16a	Red spot often surrounding injury, cluster may be webbed; caterpillar (with prolegs) feeding inside berry	Grape berry moth (page 242)
16b	No webbing or red spot present, a legless C-shaped grub feeding inside berry	Grape curculio (page 246)
17a	Webbing often present in clusters, green caterpillar feeding in cluster on outside of berries	Redbanded leafroller (page 247)
17b	No webbing, damage occurs about bloom time, especially near light sandy soil	Rose chafer (page 246)
18a	Dark colored aphids feeding in cluster	Grapevine aphid (page 255)
18b	White insect with powdery appearance, with filaments projecting from body margins	Grape mealybug (page 253)

Couplet Number		Proceed to Couplet
19a	Caterpillar, with prolegs, in berry	Grape berry moth (page 242)
19b	Exterior damage	20
20a	Webbing present, green caterpillar in cluster	Redbanded leafroller (page 247)
20b	Large gouges into fruit	Probably yellowjackets (page 245), possibly multicolored Asian lady beetle (page 247)
21a	Bud with discrete hole eaten into side, with bud content gouged out. May be most common along wooded vineyard margins	Grape flea beetle (page 244)
21b	Bud eaten away in a more general fashion, presenting a ragged appearance. May be more common in vineyards with a weedy ground cover	Climbing cutworm (page 245)
	Note: Grapevine looper may cause either type of damage if active early enough, although this is normally not the case.	
22a	Punctures or swellings on shoots, canes or trunk	23
22b	Punctures or swellings absent	26
23a	Punctures into trunk or possibly cane	24
23b	Swelling or gall in addition to puncture on shoots or canes	25
23c	A series of oviposition slits several centimeters long, on canes about half centimeter diameter	Periodical cicada (page 258)
24a	Tunnel opening about 1 mm in diameter, with copious sap flow in mid to late April	Ambrosia beetles (page 256)
24b	Tunnel opening about 4 mm in diameter, often at junction of trunk and branch	Grape cane borer (page 257)
24c	Shoot or cane girdled with two rings of small punctures, shoot may break at one or both girdles	Grape cane girdler (page 256)
25a	A longitudinal series of small punctures on cane that develops into a reddish swelling about 5 cm long	Grape cane gallmaker (page 256)
25b	Galls appearing otherwise, usually on succulent growing shoots, often red, various forms	Probably gall midges (page 246)
26a	Light grey scales on bark of trunk, no honeydew; vine may be low in vigor	Grape scale (page 257)
26b	Otherwise	27
27a	Honeydew accumulation, dark aphids present	Grapevine aphid (page 255)
27b	Honeydew accumulation, brown softshell scale present	European fruit lecanium (page 257)
28a	Galls on roots	Grape phylloxera (page 250)
28b	Galls absent	29
29a	Larger roots fed upon near crown by large white caterpillar (with abdominal prolegs), smaller roots fed upon farther from crown by smaller larvae	Grape root borer (page 258)
29b	Small roots fed upon by beetle larvae (no prolegs); adults may be seen feeding on leaves	Grape rootworm (page 260)

APPENDIX 5
Resources on Grape Pest Management

The web links given were current at the time of publication. URL addresses are subject to change.

New York and Pennsylvania Pest Management for Grapes HTTP://IPMGUIDELINES.ORG/GRAPES/

North Carolina State University, College of Agriculture and Life Sciences, *2008 North Carolina Agricultural Chemicals Manual*, HTTP://IPM.NCSU.EDU/AGCHEM/AGCHEM.HTML (chapter 7, "Insect and Disease Control of Fruits"; table 7-6. Winegrape Spray Program)

Ohio State University Extension; Agriculture and Natural Resources web site (main section, "Crops"; subsection, "Grapes"); HTTP://NEWFARM.OSU.EDU/CROPS/GRAPES.HTML

Ontario Ministry of Agriculture, Food and Rural Affairs, Grapes in Ontario, HTTP://WWW.OMAFRA.GOV.ON.CA/ENGLISH/CROPS/HORT/GRAPE.HTML

Penn State, College of Agricultural Sciences, Publications on Fruits, HTTP://PUBS.CAS.PSU.EDU/PUBSUBJECT.ASP?VARSUBJECT=FRUITS (see *2006 New York and Pennsylvania Pest Management Recommendations for Grapes*, Publication AGRS-063)

Rutgers Cooperative Extension; New Jersey Agricultural Experiment Station (NJAES); *2008 Commercial Grape Pest Control Recommendations for New Jersey* (Cooperative Extension Bulletin E283); Daniel Ward (editor), Bradley Majek, Peter Oudemans, Peter Shearer; HTTP://NJAES.RUTGERS.EDU/PUBS/PUBLICATION.ASP?PID=E283

University of Tennessee Extension, *Insect and Plant Disease Control Manual*, HTTP://EPPSERVER.AG.UTK.EDU/REDBOOK/REDBOOK.HTM (section, "Fruits and Nuts," HTTP://EPPSERVER.AG.UTK.EDU/REDBOOK/PDF/FRUITSNUTS.PDF)

Virginia Cooperative Extension, *2008 Pest Management Guides*, HTTP://WWW.EXT.VT.EDU/PUBS/PMG/

University of Connecticut Cooperative Extension System, Integrated Pest Management (IPM) Program, HTTP://WWW.HORT.UCONN.EDU/IPM/

University of Florida, IFAS (Institute of Food and Agricultural Sciences) Extension, "Insect Management in Grapes," by Susan E. Webb, HTTP://EDIS.IFAS.UFL.EDU/IG071

University of Georgia, College of Agricultural and Environmental Sciences, *2008 Georgia Pest Management Handbook*, HTTP://WWW.ENT.UGA.EDU/PMH/

University of Maryland, College of Agriculture and Natural Resources, Maryland Grapes and Fruit Page, HTTP://WWW.GRAPESANDFRUIT.UMD.EDU

University of Massachusetts, UMass Fruit Advisor, *New England Small Fruit Pest Management Guide 2003–04*, HTTP://WWW.UMASS.EDU/FRUITADVISOR/NESFPMG/INDEX.HTM

APPENDIX 6
Pest Management Schedule

Dormant or delayed dormant

- Monitor spurs for mealybugs (treat now if >20%)
- Monitor cordons for eggs of European red mite

Bud swell

- Monitor buds for injury by grape flea beetle (especially at vineyard edges near woods) and climbing cutworms
- Observe trunks and cordons for tunneling beetles

Shoots 4–8 inches (10–20 cm)

- Place traps for grape berry moth; consider placing dispensers for grape berry moth mating disruption
- Monitor foliage for spider mites
- Monitor rachis for banded grape bug
- Place yellow sticky traps for sharpshooters in Pierce's disease (PD) areas; monitor foliage for sharpshooters in PD areas

Bloom

- Monitor clusters for rose chafer, grape berry moth; monitor traps for grape berry moth; alternative time for pheromone dispenser placement for grape berry moth mating disruption

Early summer

- Monitor foliage for adult grape rootworms, larval grape flea beetles, other defoliators, spider mites
- Monitor pheromone traps and clusters for grape berry moth
- Monitor yellow sticky traps for sharpshooters in PD areas; monitor foliage for sharpshooters

Mid Summer

- Monitor foliage for Japanese beetle, green June beetle, spider mites
- Monitor clusters for grape berry moth; continue monitoring traps for grape berry moth
- Place pheromone traps for grape root borer; begin observing soil surface for grape root borer exuviae
- Monitor yellow sticky traps for sharpshooters in PD areas; monitor foliage for sharpshooters

Late Summer

- Monitor clusters for grape berry moth; continue monitoring pheromone traps for grape berry moth; consult labels for preharvest intervals at this stage
- Continue monitoring pheromone traps for grape root borer and examining soil surface for exuviae
- Monitor foliage for spider mites, Japanese beetle, green June beetle
- Monitor yellow sticky traps for sharpshooters in PD areas; monitor foliage for sharpshooters

Preharvest

- Monitor clusters for grape berry moth, multicolored Asian lady beetle, yellowjackets

Harvest

- Note injury or presence of grape berry moth, yellowjackets, multicolored Asian lady beetle, grape mealybugs

Postharvest

- Collect petiole samples from suspect vines in PD areas and submit to testing laboratory to confirm presence of *Xylella fastidiosa* bacterium.
- Observe foliage, cordons for European red mite eggs, grape mealybugs

APPENDIX 7
Wine Grape Contract

_____________Vintage

This agreement made on __________________________ by and between (Vineyard Name and Location), hereinafter referred to as Grower, and ____, hereinafter referred to as Winery, who agrees to buy wine grapes at the prices and subject to the conditions of sale as set forth below.

Whereas, Grower has 100 acres of wine grapes located at the address ____, it is agreed as follows:

Grower agrees to sell, and Winery agrees to buy, the following grape varieties and quantities determined by tonnage or on an acreage basis:

1. ____
2. ____
3. ____
4. ____

Winery agrees to pay Grower the following prices:

By Acreage

Variety and clone: ____

Price per acre: ____

Notes: ____

Total acres: ____

Maximum price per ton: ______________________________________

Field: ____

Row numbers: ____

Variety and clone: ____

Price per acre: ____

Notes: ____

Total acres: ____

Maximum price per ton: ______________________________________

Field: ____

Row numbers: ____

By Tonnage

Variety and clone: ____

Tons: _________________________________

Price: _________________________________

Quality parameters: _______________________

Variety and clone: ______________________________

Tons: ______________________________

Price: ______________________________

Quality parameters: ______________________________

Additional Grapes: ______________________________

____

* It is the intention of Winery to hold the crop level to about ________ton(s) per acre. Therefore, Winery has the right to state specific thinning instructions, and Grower will execute those instructions to the best of Grower's ability. To assist in these thinning determinations, Grower will estimate yields by use of lag phase cluster weights. If crop falls below 1.8 tons per acre without thinning instructions from Winery and thinning by Grower, special pricing considerations will apply.

The above prices are based upon 5% or less MOG, rot and mildew. If there is more than 5% MOG, Winery has the right to renegotiate the price of grapes. Minimum sugar content is 19 degrees Brix. Below 19 degrees Brix, Winery has the right to renegotiate the price of grapes. *Note:* Grower and Winery understand that sugar content is only one of many criteria used in determining fruit ripeness. Grower will cooperate with Winery to harvest grapes at optimal ripeness using all maturity indices. Final prices may change contingent upon post-harvest assessments of wine quality and evaluation of the vintage and Grower performance.

OTHER CONDITIONS OF SALE

Transportation: Grower agrees to transport grapes to Winery if Winery is located within ______ miles of Grower. Out of state customers must supply their own shipping containers and transportation. Grower will pay Winery up to $50 per ton if Winery handles pickup and delivery. All Grower grape containers must be returned clean and undamaged.

Containers: Unless otherwise stated, Grower will deliver fruit in four-foot by four-foot by two-foot standard grape bins. Off-loading of bins at Winery's facility is the responsibility of Winery. If there is an unreasonable delay, Grower will charge Winery a waiting fee of $50 per hour. A charge of $200 per ton will be added to the price of grapes if Winery asks Grower to use 30-pound grape lug boxes. Any special grape containers will be provided by Winery to Grower with the approval of Grower.

Determination of Weight: Unless otherwise agreed, Grower will provide Winery with gross and tare vehicle weights from a licensed state scale. If delivery is scheduled for after-hours or on Sunday, Grower will make closest possible estimate of tare weight based on experience.

Determination of Juice Composition: The measurement of grape sugars or other parameters of ripeness that may relate to this contract shall be the responsibility of Grower, unless otherwise agreed upon in writing by Grower and Winery. Sugar readings shall be performed by temperature-compensated refractometer.

Tonnage Reduction Notice: In the event Grower cannot deliver the tonnage or acreage of grapes requested by Winery, Grower shall notify Winery at the earliest knowledge of such conditions and shall not be subject to penalty.

Vineyard Practices: Grower agrees to follow the highest standard of viticultural practices to produce quality wine grapes for Winery. Grower agrees to keep Winery informed of all use of pesticides and other practices that might affect fruit quality. If Winery requests additional cultural practices (for example, shoot thinning or leaf pulling) or additional pesticide applications beyond the normal practice of Grower, Winery will be expected to compensate Grower for additional farming expenses incurred by Grower.

Harvest: Grower shall inform Winery of all sampling data at harvest. Grower shall allow winery access to grapes for sampling and tasting prior to harvest. Grower and Winery shall cooperate to determine harvest date. Grower shall provide Winery with a minimum of 24 hours notice prior to picking. Grower retains final right to determine date at which grapes will be harvested and delivered to Winery.

Title: Title to grapes shall pass from Grower to Winery upon departure of grapes from Grower's property. At such time Winery assumes all responsibility for possession and payment for wine grapes to Grower.

Force Majeure: In the event that Grower or Winery is prevented or unable to perform its obligations herein stated because of natural disasters, acts of God, government restrictions, wars, insurrections or any other cause beyond the reasonable control of the affected party, that party shall be excused from performing its obligations as stated herein.

Life: The life of this contract shall extend until the final payment for grapes is made from Winery to Grower.

Integration: Grower and Winery agree that this written agreement supersedes all prior oral understandings, contains the entire agreement between the parties, and forms the sole basis for governance of this agreement. All oral understandings not expressed in this agreement are void. Amendments to this agreement shall be made in writing and signed by both Winery and Grower representatives.

Correspondence: Correspondence between Grower and Winery shall be addressed to the following:

Grower: ______________________________

Address: ______________________________

____

City, State, Zip: ______________________________

This contract shall be governed by and interpreted in accordance with the laws of the State of Pennsylvania and shall be binding upon the parties hereto and their successors and assignees.

Payment: Payments from Winery to Grower shall be made promptly according to the schedule and terms set forth herein. Winery agrees to keep Grower informed and agrees to provide explanation of any delay in payments. Winery agrees to pay a late payment charge of 1.5% per month on all past due amounts. In the event of any default, the undersigned Winery agrees to pay all attorney fees and the costs of collection to the extent permitted by law. A 5% discount shall apply to all payments in full within 30 days of last grape delivery under contract.

Payment Schedule

__________ % of Net Price within 30 days of delivery of grapes to Winery

__________ % of Balance on ______________________________

__________ % of Balance on ______________________________

Attorney fees

In the event that a suit or action is undertaken in an effort to enforce any of the provisions of this contract, the losing party agrees to pay such sums as the trial court may adjudge reasonable such as attorney fees and court costs. If an appeal is made of any judgment or decree of such trial court, the losing party further promises to pay such sum as the appellate court shall adjudge reasonable.

Other conditions: ______________________________

____

____

____

Accepted for Grower by: ______________________________

Title: ______________________________

Date: ______________________________

Accepted for Winery by: ______________________________

Title: ______________________________

Date: ______________________________

APPENDIX 8

Conversion Table

	To Convert This:	*To This:*	*Multiply by This:*
Weight/Volume	pounds	kilograms	0.454
	ounces	grams	28.35
	quarts	liters	0.946
	gallons	liters	3.78
	gallons (water)	pounds	8.34
	tons (English)	pounds	2,000
	tons (English)	tonne (metric ton)	0.907
Area	acres	hectares	0.405
	acres	square feet	43,560
	acres	square yards	4,840
	square miles	acres	640
Length	inches	centimeters	2.54
	inches	millimeters	25.4
	feet	meters	0.304
	yards	meters	0.914
	miles	kilometers	1.6
Pressure	pounds per square inch (psi)	bar	0.069
	pounds per square inch (psi)	kilopascal (kPa)	6.895
Rate	pounds per acre	kilograms per hectare	1.12
	tons per acre	metric tons per hectare	2.24
	miles per hour (mph)	feet per second	1.47
	miles per hour (mph)	kilometers per hour	1.6
	miles per hour (mph)	meters per second	0.447
Temperature	degrees Fahrenheit (°F)	degrees Celsius (°C)	(°F - 32) / 1.8
	degrees Celsius (°C)	degrees Fahrenheit (°F)	(1.8 x °C) + 32
Water measurement	acre-inches	cubic meter	102.8
	cubic feet per second	cubic meter per hour	101.9
	US gallons per minute	cubic meter per hour	0.227
Concentrations	percent (%)	gram per kilogram	10
	parts per million (ppm)	milligram per kilogram	1

Glossary

abiotic – associated with or caused by non-biological factors such as frost, desiccation, or hail.

acclimation – phase during late summer when shoots stop growing and become brown and woody, and tissues acquire increased cold hardiness.

acervuli – small fungal structures that consist primarily of a mass of slimy, rain-dispersed spores.

acreage contract – an agreement in which the winery purchases the entire grape production from a vineyard's specified acreage for a flat price per acre.

advective freeze – temperatures below 32°F, caused by the turbulent passage of large frontal systems of cold air, during which little stratification of air temperature occurs with changes in elevation; *see* **radiational freeze.**

airblast sprayer – a traditional design sprayer that uses a large, axial fan which directs spray into the canopy and up into the air.

allelopathy – the suppression or destruction of one plant species from toxic substances produced by other, nearby plants.

alleyways – intentional breaks in long vineyard rows to facilitate the lateral movement of equipment and workers.

alluvial soil – soil originating from alluvium – gravel, sand, silt and clay deposited by or in transit by watercourses.

ambient – normally referring to conditions of the natural environment. Ambient temperature is the air temperature within the environment occupied by plants and animals.

anoxia – deficiency of oxygen.

anthocyanins – phenolic compounds responsible for the red and blue colors in grape berries and in vegetative portions of grapevines.

apical dominance – ability of shoots near the distal end of the cane to produce hormones that retard growth of more basal shoots.

arm – variable length, usually horizontal, woody perennial (two-year-old or older) portion of the vine. Arms usually arise from the "head" or other renewal region of the vine.

ascospore – a sexually produced spore of the Ascomycete group of fungi.

aspect – compass direction toward which the slope faces.

balanced pruning – pruning system that determines the number of nodes to retain based on weight of one-year-old canes removed at dormant pruning.

basal – in the direction of the roots or base of vine; contrast with distal.

bearing units – canes or spurs.

bilateral cordon – a training system consisting of two horizontal extensions of the trunk(s) upon which bearing units (canes or spurs) are borne.

Brix – a measure of a fluid's total soluble solids concentration, usually expressed in degrees or as a percentage of the weight of sugar in the solution (for example, 20 pounds of sugar in 100 pounds of grape juice is expressed as 20° Brix or 20% soluble solids concentration.

bud – normally refers to the compound structure from which shoots arise in the spring. Buds are borne at *nodes* of the cane which are separated from each other by *internodes*. "Base" buds (sometimes called "non-count" buds) are those dormant buds that are borne at points other than nodes, often at the *base* of canes and spurs.

bud fruitfulness – ability of the bud to produce fruit and measurable as clusters per shoot or weight of fruit per shoot (*see* **node fruitfulness**).

bud necrosis – death of grapevine buds due to reasons other than winter injury, pathogens or pests.

bunch – bunch and cluster are used synonymously and refer to the fruiting unit of grapes comprising berries and a cluster stem (rachis).

bunch stem necrosis – complex physiological disorder that affects the cluster rachis and prevents normal berry development of the affected portion of cluster. When it occurs near veraison, the disorder may also be called "water berry."

callus – undifferentiated mass of plant tissue often caused by wounding, as in grafting, or by crown gall development.

cane – a woody, mature shoot after defoliation.

canopy – the shoots of the vine including stems, leaves and fruit.

canopy division – training systems that divide a grapevine canopy into at least two distinct sub-canopies—either vertically (for example, Scott Henry) or horizontally (for example, Geneva Double Curtain).

capacity – a term loosely synonymous with vine size as measured by dormant cane pruning weights.

cation – generally a positively charged ion.

cation exchange capacity – the sum total of exchangeable cations that a soil can adsorb.

chelate – a chemical compound comprising a metal ion coordinately bonded to an organic molecule.

chlorosis – abnormal yellowing of foliage.

cliestothecia (syn. chasmothecia) – overwintering structures of some fungi that contain sexually-produced spores (ascospores).

clone – one or more vines that originated from an individual vine, which was in some way unique from other vines of the same variety.

colluvial soil – soil that develops from colluvium – deposit of soil and rock at the base of slopes as a result of erosion and gravity.

combing – often referred to as "shoot positioning", combing is the act of creating two or more discrete canopies within a divided canopy system such as a Geneva Double Curtain or Scott Henry. Combing is also used to describe the positioning of shoots downward on bilateral cordon systems such as the Hudson River umbrella (high bilateral cordon).

conidium (conidia = plural) an asexual fungus spore.

cordon – long, horizontal extension or two-year-old or older wood; long arm.

coulure – excessive shedding of flowers, leading to poor or no fruitset.

cover crop – plant species grown between and/or in the vineyard rows to reduce soil erosion or for other benefits; *see* **ground cover**.

crop level – nodes per vine or clusters per vine/ shoot retained.

cropload – the ratio of crop weight to cane pruning weight for a given year.

crop yield (or crop size) – yield per unit of land area, for example, yield per acre.

cultivar – a named, cultivated variety.

diapause – a period of arrested growth or development in which metabolic activity is diminished, as in certain insects.

distal – end of the stem towards the growing tip; *see* **basal**.

divided – canopy training system – training system whereby the canopy is partitioned into two (or more) discrete sub-canopies, such as with Geneva Double Curtain or Scott-Henry.

dormancy – time between leaf-fall in autumn and bud-break in the spring, apparent as the absence of visible growth.

dormant pruning – annual removal of wood during the vine's dormancy.

double pruning – one pruning cut in late winter or early spring followed by a second pruning cut after the threat of frost is past but before appreciable shoot growth has occurred. Practiced where spring frosts are common.

endophyte – a plant, fungus, or microorganism that lives within a plant either parasitically or symbiotically, without causing obvious symptoms.

evapotranspiration – the combined transpiration, or loss of water through stomata, and evaporation of water from the soil surface.

field capacity – the amount of water held by soil after gravitational drainage of excess water.

Force Majeure – literally "greater force". A contract clause included to excuse a party from contract obligations if some unforeseen event, such as a devastating natural disaster, beyond the control of that party, prevents the party from performing its obligations under the normal terms of the contract.

girdling – the constriction or cessation of vascular transport in a plant part by intentional or unintentional (for example, damage to trunk) removal of certain vascular tissues, by restriction materials, or by tissue overgrowth such as crown gall.

graft union – point at which the rootstock is joined to the scion.

ground cover – material (non-living) or living plant species maintained in the row middles or under the trellis for suppression of weeds, mitigation of soil erosion, and possibly intentional competition with vines.

head – upper portion of vine consisting of the top of the trunk(s) and junction of the arms and/or origin of canes with head-trained vine.

headland – area at the end of the rows used for vehicle turning.

hedging – pruning during the growing season, usually removing only shoot tops and retaining only the nodes and leaves needed for adequate fruit and wood maturation.

hilling – protecting the graft union and a portion of the trunk with mounded soil in the fall.

humus – generally stable soil residue following partial organic matter decomposition.

hypha – (plural – hyphae) a single branch of mycelium.

ideotype – a complex model system describing an ideal scenario. In the grapevine context, a canopy ideotype describes an ideal grapevine canopy in terms of leaf layers, leaf color, leaf size, cluster exposure, and several other variables.

inclinometer – instrument that measures a plane's degree of inclination from the horizontal (as in the slope of a parcel of land).

inoculum – the unit or part of a pathogen that infects its host.

internode – the portion of the stem between nodes.

isotherm – a contour line (as on a map) that connects points of equal (air) temperature.

kicker cane – extra cane or canes retained during dormant pruning for subsequent removal during the growing season, intended to divert vigor from the retained, bearing shoots of the vine.

lag phase cluster weight – the mid-season, average weight of a grape cluster at the point where berry development temporarily slows as seed coats begin to lignify.

lees – wine sediment, notably dead yeast, that accumulates during and after fermentation.

lenticels – surface pores on berries and certain other tissues that may be bordered by darkened, raised, corky cells. Riesling and Vidal berries, for example, have distinct lenticels.

liana – a climbing vine.

lignify – the process that begins late in the summer, whereby shoots or other vine organs accumulate a polymeric compound called lignin, which allows the stems to become resistant to cold and water loss. Lignin is incorporated into a complex tissue known as *periderm* on the surface of grapevine canes.

macroclimate – climate of a large geographical region, such as a continent.

meristem – zone or tissue of active and indefinite cell division.

mesoclimate – climatic conditions within 10 feet of the ground and peculiar to a local site.

microclimate – environment within a specific small area, such as a grapevine canopy.

millerandage – reproductive disorder in which grape clusters bear significantly different sized berries within the same cluster; the so-called "hens and chicks" berry appearance.

mistblower – a canopy sprayer which creates a fine spray which is directed into the canopy.

MOG – Material Other than Grapes. An acronym used to describe contaminants such as leaves that are associated with harvested grapes, particularly machine-harvested grapes.

monoterpene – aroma compounds that are largely responsible for the typical aromas of muscat, Riesling, and other aromatic white grape varieties.

mycelium – vegetative portion of fungi, may be observed under magnification as a mass of branching, thread-like filaments or *hyphae*.

mychorrhizae – symbiotic association of certain fungal mycelia and plant roots wherein the fungus derives a source of carbon from the host plant and the plant derives an increased access to mineral nutrients.

must – crushed grapes or grape juice prior to fermentation.

necrosis – death.

nematodes – small, often microscopic worms, with non-segmented bodies. The viticulturally important nematodes feed on vine roots and some can transmit viruses to the vine.

node – conspicuous joints of shoots and canes. Count nodes have clearly defined internodes in both directions on the cane.

node fruitfulness – the ability of the node to produce fruit and measurable as clusters per node or weight of fruit per node and affected by percent budburst (shoots per node) (*see* **bud fruitfulness**).

nodosities – small, hook-like swellings caused by root feeding of phylloxera.

oospore – a sexually produced spore of the Oomycete group of fungus-like water molds (for example, downy mildew).

parthenogenetically – reproduction by development of an unfertilized egg or gamete, common among some insects and lower plants.

pathogen – a biological entity that can cause disease.

pedicel – stem of an individual berry or flower.

peduncle – the portion of the rachis from the shoot stem to the first branch of the cluster.

periderm – bark.

phenology – the study and record of recurring biological phenomena such as bud break, bloom, and veraison on the basis of elapsed time, accumulated heat units, or other climatic conditions.

pheromone – a substance produced by one individual, as with certain insects, that elicits a specific reaction by others of the same species, such as sex attractants or alarm substances.

phenols – (adjective = phenolic); secondary plant metabolites largely responsible for astringent and bitter tastes in grapes and wines, but also important in stabilization of wine color.

phloem – vascular tissue of plants that transports products of photosynthesis (sugars) and other solutes throughout the plant.

photosynthesis – the complex biochemical process whereby plants convert light energy to chemical energy using carbon dioxide and water as source materials and oxygen and sucrose as products.

phylloxera – tiny, aphid-like insects of the order Homoptera. By feeding on roots, grape phylloxera (*Phylloxera vitifoliae*) can weaken or kill own-rooted *Vitis vinifera* and some hybrid grapevines, which is a principal reason for grafting these vines to pest-resistant rootstocks.

phytoalexin – chemical produced by a plant (grapevine) that inhibits or reduces the growth of an attacking parasite, such as a fungal pathogen.

phytotoxicity – injury or death of plant or plant part, as from natural or man-made compounds.

pith – central part of a shoot or cane.

plant available water (PAW) – the amount of water easily available to plants, typically that volume held between field capacity and the wilting point.

point quadrats – transects of the canopy with a thin probe, emulating incident sunlight, performed to quantify canopy leaf layers, fruit exposure, and canopy gaps.

pollinator – vine planted to supply pollen.

postemergence herbicide – weed control chemical that is applied after weeds germinate and are visible; see **preemergence herbicide**.

preemergence herbicide – weed control chemical that is applied prior to weed seed germination; *see* **postemergence herbicide**.

primordia – (singular = primordium) growing point origin of a developing organ, such as a flower bud or shoot tip.

procumbent – with respect to shoot growth habit, the tendency for shoots to droop rather than grow upright. *Vitis labruscana* varieties tend to have procumbent shoot growth.

pycnidia – fruiting structures of some fungi; pycnidia produce and release spores that are dispersed over short distances, usually by splashing rain.

radiational freeze – sub-freezing temperatures that develop during calm, clear sky conditions when the ground cools by radiating long-wave radiation to space, thereby cooling the air immediately adjacent to the earth; *see* **advective freeze**.

rake wire – a movable trellis wire that is used to facilitate shoot positioning as in canopy division.

Ravaz Index – after the French viticulturist, L. Ravaz, the ratio of crop size (crop yield per vine) to vine size (weight of cane prunings per vine); *see* **cropload**.

refractometer – an optical instrument used to measure the solute (such as sugar) concentration of liquids such as grape juice.

renewal region (of canopy) – part of the canopy where buds for next year's crop develop (usually the fruiting region).

rhizosphere – the physical, chemical and biological soil environment immediately adjacent to, and affected by, plant roots.

rootstock – variety used as rootstock of grafted vine.

rotovate – to cultivate soil using a rotary plow or tiller.

scion – above-graft part of a grafted vine, including leaf- and fruit-bearing parts.

secondary metabolites – organic compounds that are not directly involved in normal growth, development or reproduction of the vine, but rather provide defense, attractant (flavor, color and aroma), or fit other ecological roles important to the fitness of the plant species.

self-fruitful – able to set fruit with pollen of the same variety.

shoot – succulent growth arising from a bud, including stem, leaves, and fruit.

shoot positioning – manipulation of shoot orientation to optimize canopy sunlight interception.

signs – in pathology, the pathogen body or reproductive structures observed on the host plant; *see* **symptoms**.

sporangia – specialized, asexual spores or sacs of spores of certain disease-causing fungi; singular = sporangium.

spur – cane that has been pruned to 1 to 4 nodes.

stomata – leaf pores that allow gas exchange between leaves and the environment.

subtending – in a botanical sense, the relative position of tissues or organs that lie or originate below or beside other tissues or organs.

summer pruning – hedging or removing vegetation during the growing season; *see* **hedging**.

symptoms – the observed responses or reactions of the host plant to disease; *see* **signs**.

tendril – twining organs of shoots, located opposite leaves at nodes, that can coil around objects and provide shoot support.

training – a physical manipulation of a plant's form.

transects – *see* **point quadrats**.

transpiration – the loss of water vapor from the plant, primarily through stomata.

trunk – vertical support structure that connects the root system with the fruit-bearing wood of the vine.

vascular cambium – tissue of canes and older wood that generates new xylem and phloem cells annually.

veraison – the period or stage at which fruit begins a third stage of ripening characterized by softening, color change, and perceptible increases in sugar and decreases in acidity.

vigor – rate of shoot growth.

xylem – vascular tissue of stems that primarily serves to transport water and minerals and forms wood with secondary development.

Nelson J. Shaulis Fund for the Advancement of Viticulture supports *Wine Grape Production Guide for Eastern North America*

The Shaulis Fund is pleased to support the production of the *Wine Grape Guide for Eastern North America* because we believe this publication will lead to enhanced knowledge and understanding of viticulture for all grape producers in the region.

In 1978 in order to give lasting recognition to Dr. Shaulis and his research, a group of grape industry leaders established the Nelson J. Shaulis Fund for the Advancement of Research. Over the years, proceeds from the fund have been used to enhance and encourage viticulture education. Specifically, income has been used to

1. Sponsor and encourage graduate study in viticulture, and
2. Sponsor seminars and lecture programs that will enhance the knowledge of viticulture by industry producers, academia, and processor and winery representatives.

Nelson J. Shaulis was a professor of viticulture at Cornell's New York State Agriculture Experiment Station. At the time of this death in January 2000, his peers and associates recognized him as *the* leading viticulturist of his time. During his 34-year career at the New York State Agricultural Experiment Station, he was instrumental in the development of many of the techniques and technologies that are still being utilized by the grape industry both here in the United States and in other grape growing regions all over the world.

His pioneering research led to the development and implementation of:

- The Geneva Double Curtain training system, which enables higher production of higher quality grapes
- The use of soil and petiole sampling to determine vineyard nutrient needs so that applications of correct amounts are applied only when necessary in order to maximize production
- The use of herbicides to replace tillage, which has resulted in reduced production costs and increased vine size, and
- The development of the mechanical grape harvester, which revolutionized grape production throughout the world.

Dr. Shaulis was instrumental in training associates, producers, and processor field representatives resulting in increased production and improved quality. He was a mentor to many of today's leading viticulture scientists, and his viticulture principles continue to be the basis for much of today's viticulture research.

Dr. Nelson J. Shaulis

Dr. Shaulis, in a presentation at the NYS Horticulture Society in 1956, made the following statements, which are as true today as they were 52 years ago. Speaking at that conference, he stated, "The vineyardist of tomorrow will still have to follow the rule of many years ago, when it was said that the best fertilizer was the owner's footprint in the vineyard." He went on to say that, "In the most highly mechanized crop production, as in sugarcane and pineapple, there is an increasing emphasis on regular inspection and sampling of the crop. And this is done by the most competent men in the enterprise, not done when they are spraying or cultivating, but strictly on an inspection trip."

Because of what he stood for and believed in, Dr. Shaulis would be proud to support the publication of *Wine Grape Ptoduction Guide for Eastern North America.*

— Thomas G. Davenport,
Director of viticulture for National Grape Cooperative
and member of the Shaulis Fund Committee

The Atlantic Seaboard Wine Association (formerly Vinifera Wine Growers Association), focuses on promoting the East Coast as a quality wine producing region

The ASWA is pleased to be a sponsor and fully supports the educational and technical objectives of the *Wine Grape Production Guide for Eastern North America*. As one of the earliest established wine trade associations on the East Coast (1973), the VWGA, now the ASWA, has, for the past 35 years, promoted quality wine production, consumer education and the marketability of East Coast wines.

Anyone interested in the promotion of sustainable viticultural and enological growth practices, as well as the economic viability of the wine industry of the East Coast of the United States, is welcome to become a supporting member of the ASWA. For a review of ASWA's wine industry support programs and membership information, click onto www.vwga.org. For further information, e-mail Fairfax@earthlink.net or thewinexchange@aol.com, phone 703-922-7049, ASWA, PO Box 10045, Alexandria, VA 22310.

ASWA Objectives:

- Promote general public wine education and appreciation;
- Support state, national and international wine educational forums and research;
- Provide business, health, technical, cultural, historical and international wine information through the periodic e-mail dissemination of pertinent new and evolving subject material;
- Promote consumer public wine enjoyment and responsible moderate consumption as part of a healthy life style;
- Support state and national production of quality wine and the marketability of American wines;
- Vigorously support both state and federal legislation that is favorable to the growth and economic viability of the US wine industry;
- Strengthen cooperation with other wine organizations in addressing wine issues of common concern;
- Sponsor an annual Atlantic Seaboard Wine Competition and public wine events and festivals.
- Provide special focus on the promotion of the East Coast as a quality wine producing region.

Viticulture Consortium

The United States Department of Agriculture—Cooperative State Research, Education, and Extension Service (USDA-CSREES), formed a Viticulture Consortium (VC) in 1996 to promote research and extension education programs that support the growth and competitive future of the US viticulture industry. Cornell University and the University of California (UC) Division of Agriculture and Natural Resources (ANR) cooperate in the management of this competitive grants program. UC administers the Viticulture Consortium West (VCW) and Cornell University administers the Viticulture Consortium East (VCE). The Request for Proposals for both VCE and VCW are given wide distribution to grape research and extension personnel, plus Directors of Research and Extension at Universities in the respective regions. All proposals are peer-reviewed by qualified scientists having expertise in the subject area. Peer reviews, along with the proposal and the progress report, are given to Steering Committees who make the funding recommendations. The Steering Committees are appointed by the lead Universities and are made up of grape growers, winery owners, grape processors and academic personnel.

The Viticulture Consortium is funded by the USDA-CSREES and the funding is further supplemented by other Federal (Hatch) and state sources and by several state and national viticulture industry groups. As such, the coordination of high priority research and extension projects is being pursued to extend the ability of both the VC and industry to address critical needs. Coordination of joint priority setting and matching industry funding continues as a basic philosophy of the Viticulture Consortium, the industry, and CSREES.

The Viticulture Consortium grant permits funding of applied, mission-specific research relevant to grape growing which would not normally be attractive for other competitive grant programs.

SPONSORS

Platinum Sponsors

Gold Sponsors

BAYER
Bayer CropScience
ELITE®
FLINT®
SCALA
Protection. Weather or not.
Adament
BAYTHROID® XL
Sevin®
Brand Carbaryl Insecticide
PROVADO®
LEVERAGE®
Rely®
HERBICIDE
Dan DiGiacomandrea "Danny D" Senior Technical Sales Representative
585-330-3263 Fax 585-425-8774
dan.digiacomandrea@bayercropscience.com
Website http://www.bayercropscienceus.com/

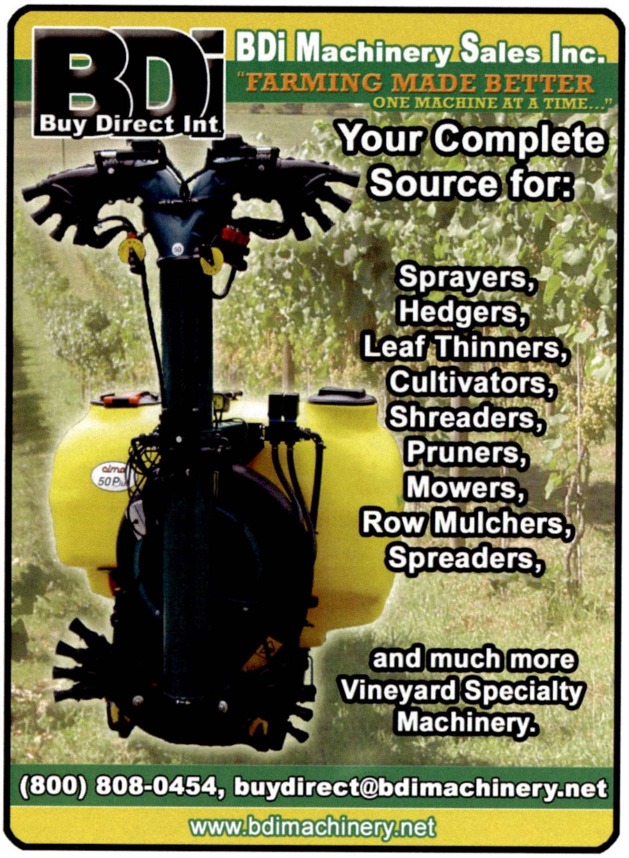
BDi
Buy Direct Int
BDi Machinery Sales Inc.
"FARMING MADE BETTER
ONE MACHINE AT A TIME..."
Your Complete Source for:
Sprayers,
Hedgers,
Leaf Thinners,
Cultivators,
Shreaders,
Pruners,
Mowers,
Row Mulchers,
Spreaders,
and much more Vineyard Specialty Machinery.
(800) 808-0454, buydirect@bdimachinery.net
www.bdimachinery.net

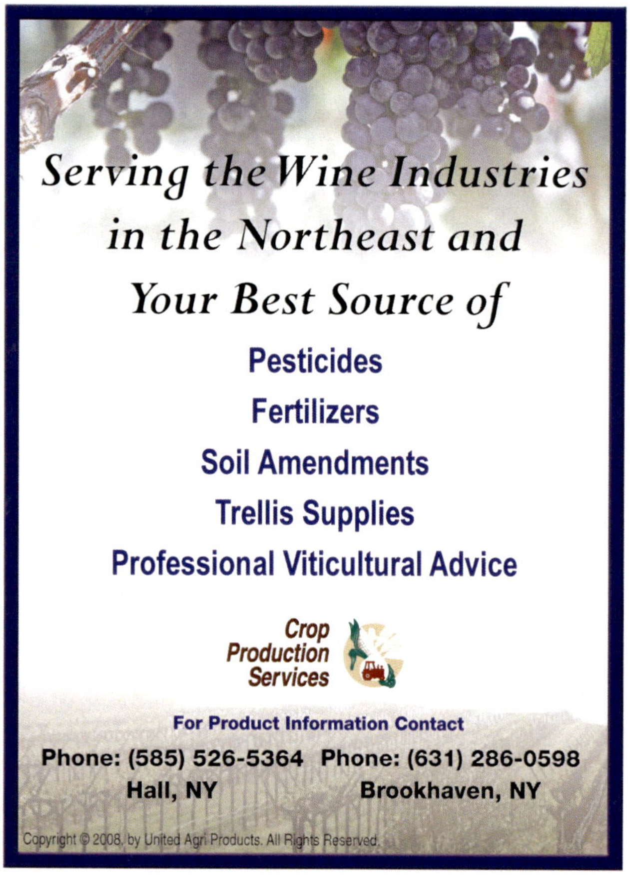
Serving the Wine Industries in the Northeast and Your Best Source of
Pesticides
Fertilizers
Soil Amendments
Trellis Supplies
Professional Viticultural Advice
Crop Production Services
For Product Information Contact
Phone: (585) 526-5364 Phone: (631) 286-0598
Hall, NY Brookhaven, NY
Copyright © 2008 by United Agri Products. All Rights Reserved.

Corners, Nooks and Crannies.
Dow AgroSciences
Quintec®
Fungicide
Quintec® fungicide has a unique mode of action that moves around the grape bunches, getting to hard-to-reach areas and offering long-lasting bunch protection against powdery mildew.
www.QuintecFungicide.com
®Trademark of Dow AgroSciences LLC Always read and follow label directions.
L38-136-005 (10/08) BR 010-32627 DAAGQUIN8013

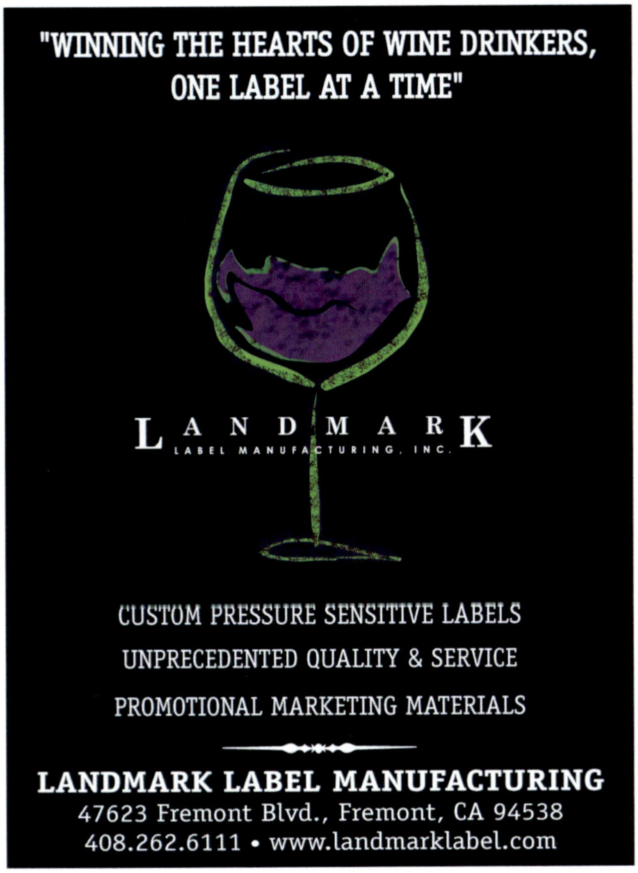
"WINNING THE HEARTS OF WINE DRINKERS, ONE LABEL AT A TIME"
LANDMARK
LABEL MANUFACTURING, INC.
CUSTOM PRESSURE SENSITIVE LABELS
UNPRECEDENTED QUALITY & SERVICE
PROMOTIONAL MARKETING MATERIALS
LANDMARK LABEL MANUFACTURING
47623 Fremont Blvd., Fremont, CA 94538
408.262.6111 • www.landmarklabel.com

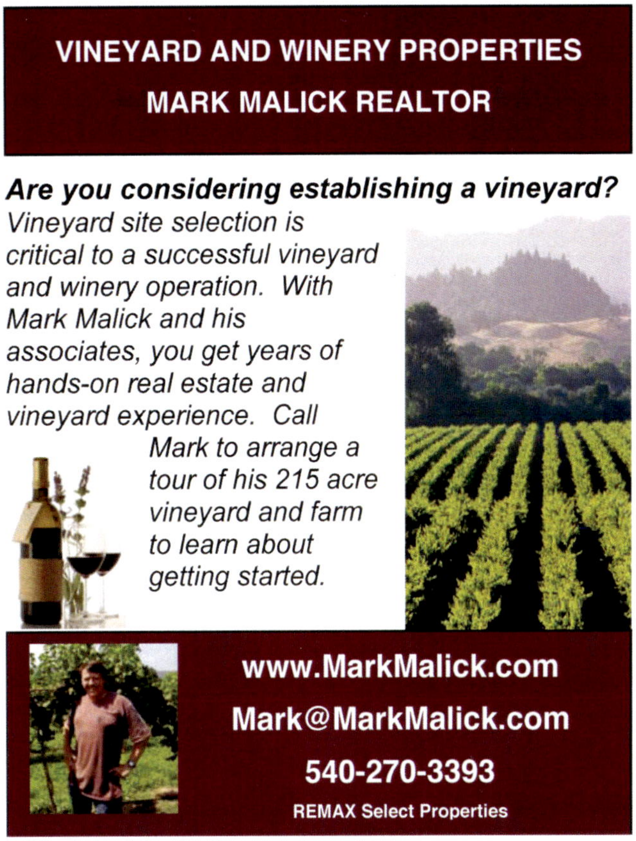
VINEYARD AND WINERY PROPERTIES
MARK MALICK REALTOR
Are you considering establishing a vineyard?
Vineyard site selection is critical to a successful vineyard and winery operation. With Mark Malick and his associates, you get years of hands-on real estate and vineyard experience. Call Mark to arrange a tour of his 215 acre vineyard and farm to learn about getting started.
www.MarkMalick.com
Mark@MarkMalick.com
540-270-3393
REMAX Select Properties

Custom Tanks & Equipment
• High quality, stainless steel equipment
• Custom-designed and manufactured
• State-of-the-art designs, meticulously crafted
• Prompt, "easy to work with" service for clients large and small, worldwide
• Innovative ideas and solutions that meet your needs
• Reliable, on-time delivery
• Competitive pricing
Contact us at today for more information and an estimate for your next job.
THE BELLI CORPORATION
WESTEC
TANK & EQUIPMENT CO.
Custom tanks, catwalks and equipment for the wine and beverage industries
Tel: (707) 431-9342
Fax: (707) 431-8809
www.westectank.com
info@westectank.com

Vineyard & Winery Management Magazine
Serving the North American wine industry for over 30 years
Vineyard & Winery Management Magazine
• Bimonthly Trade Magazine
• Wineries Unlimited
• Tasting Room Profitability Conference & Trade Show
• Annual Wine Industry Index
• Wineries of North America on CD
• Grand Harvest Awards Wine Competition
• West Coast Wine Competition
• International Eastern Wine Competition
(800) 535-5670
vwm-online.com

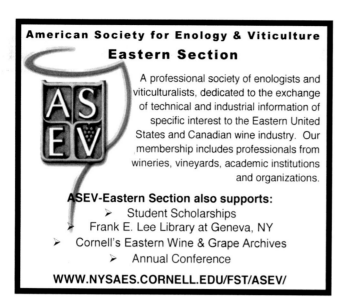

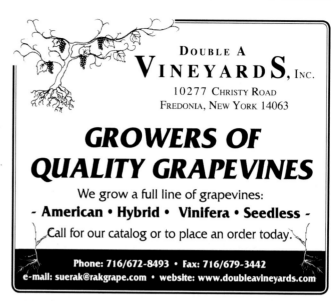

PEER REVIEWERS

The following persons reviewed one of more chapters in the original manuscript. Their suggestions resulted in significant improvements in the manuscript.

Alan R. Biggs
West Virginia University

Bruce Bordelon
Purdue University

Jake Bowman
University of Delaware

Frank L. Caruso
University of Massachusetts

Lailiang Cheng
Cornell University

Paul D. Curtis
Cornell University

Imed Dami
OARDC/Ohio State University

Joan Davenport
Washington State University

Dragana Dimitrijevic
formerly with Cornell University (Geneva)

Mike Ellis
OARDC/Ohio State University

Jodi Creasap Gee
Cornell Cooperative Extension

Emilio Gil
Universitat Politecnica de Catalunya (Spain)

Daniel Gilrein
Cornell Cooperative Extension of Suffolk County

Edward Hellman
Texas A&M and Texas Tech University

Rufus Isaacs
Michigan State University

Albert R. Jarrett
Penn State University

Richard Kiyomoto
University of Connecticut

S. Kaan Kurtural
formerly with University of Kentucky

John Lea-Cox
University of Maryland

Greg Loeb
Cornell University (NYSAES, Geneva)

Tim Martinson
Cornell Cooperative Extension

Stephen D. Menke
Colorado State University

Ian A. Merwin
Cornell University

Eric Miller
Chaddsford Winery, Pennsylvania

Jim Parkhurst
Virginia Tech

Gary C. Pavlis
Rutgers Cooperative Extension

Paul E. Read
University of Nebraska

Bruce Reisch
Cornell University (NYSAES, Geneva)

Andrew Reynolds
Brock University (Canada)

Annemiek C. Schilder
Michigan State University

Sonia Schloemann
University of Massachusetts Extension

Bill Sciarappa
Rutgers Cooperative Extension

Andy Senesac
Cornell Cooperative Extension of Suffolk County

Sara E. Spayd
North Carolina State University Extension

Ed Stover
USDA, ARS, PWA

Hans Walter-Peterson
Cornell Cooperative Extension

Timothy Weigle
Cornell Cooperative Extension

Gerald White
Cornell University

Alice Wise
Cornell Cooperative Extension of Suffolk County

Thomas J. Zabadal
Michigan State University

Index

Page numbers in **bold** type include photos, illustrations or tabular representation of the index terms. Glossary (*see* page 319) terms are not indexed but should be consulted for specific definitions.

H

I

J

K

L

M

N

O

P

U

V

W

X, Y, Z